Acupuncture Pulse Diagnosis AND THE *Constitutional Conditional Paradigm*

Acupuncture Pulse Diagnosis AND THE *Constitutional Conditional Paradigm*

MORE DONKEY BUSINESS:
ENIGMAS *in the* CLASSICAL CHINESE TEXTS
and THEIR ELUCIDATION *by* PULSE DIAGNOSIS

Peter Eckman, M.D., Ph.D., M.AC. (UK)

"Of what malady will he die?"

This etching with aquatint by Francisco Goya was distributed in Madrid in 1799. In 1792 at the height of his career, Goya contracted cholera with subsequent sequelae of paralysis, partial blindness, vertigo, tinnitus and eventual deafness. That he could still produce works like this is astounding.

What was Goya thinking when he made this etching? The physician depicted as a donkey in a suit is ostensibly how Goya viewed the medical community of his time. In the background we see the hooded figures waiting for this gentleman's death, while the physician can do nothing but take the man's pulse. While the etching has been interpreted (no doubt correctly on one level) as satirizing the medical profession as "jackasses", this donkey strikes me as a rather serious but kindly figure, both focused and humane, allowing for another level of interpretation. His efforts at pulse diagnosis may imply a degree of coarseness given the hooves (not having access to the elaborate Chinese treatises on the pulse), however what better method did physicians of his era possess? If this be merely satire, I would expect to see grim scenes of bleeding, purging or any of the other gut-wrenching interventions common in that era.[1] Perhaps rather than, or in addition to satire, Goya was unconsciously offering prophecy. I find it curious that today's medical profession has little to provide those suffering from these same conditions of paralysis, blindness, vertigo, tinnitus and deafness. Interestingly, these are maladies that have been treated with varying degrees of success by acupuncture. And the information gleaned from taking the pulse has turned out to be an invaluable guide to diagnosis and treatment in this author's personal experience.

Credit: National Library of Medicine - History of Medicine at https://flickr.com/photos/47756470@N03/4379551038

1 See many of the other Goya etchings in the "Los Caprichos" series from which this etching comes.

CONTENTS

Overview

Acupuncture Pulse Diagnosis and the Constitutional Conditional Paradigm; More Donkey Business: Enigmas in the Classical Chinese Texts and their Elucidation by Pulse Diagnosis is the third part of a trilogy that started with *The Compleat Acupuncturist* (2014), which was followed by *Grasping the Donkey's Tail* (2017). All three parts together comprise an evolving exposition of the author's style of acupuncture, called Constitutional Conditional Acupuncture (CCA). CCA was developed based on the author's 50 years of full time practice as an acupuncturist, together with years of study under several Master practitioners of international repute, accompanied by continual study of the Chinese medical classics, their commentaries, and the interpetations (occasionally disputed by this author) of several English translators. I wish to emphasize the word 'evolving' because it is my belief that developing an understanding of the many uses of acupuncture is a never finished accomplishment, thus CCA is always a work in progress. I think of acupuncture as a branch of Asian medical science, and like other sciences one should never claim to have the final word. Corrections are always possible as new discoveries emerge, but this author has found the present guidelines remarkably effective for improving the health of practically any kind of presentation, running the gamut from "possession" to cancer. The thread that holds this approach together is its grounding in the practical art of pulse diagnosis.

To briefly summarise the discoveries that were revealed by this focus on pulse examinations and other corroborating techniques, here is a partial list that spans all three treatises:

1. Every individual has one of 20 Constitutions that are unique in their compostion of Organ/Meridian strengths and weaknesses, encoded in yīn yáng and wǔ xíng (Five Element) terms, as revealed by several corroborative pulse findings. Two polarities (yīn yáng) times five phases (wǔ xíng) times 2 tendencies (xū shí) produce 20 unique Constitutional types.
2. Separately, each individual at any given moment also exhibits their present Condition via the pulses, and their health can be improved by

strategies aimed at transforming the Conditional pulse state to their Constitutional pulse state.

3. The traditional Chinese pulse exams at the wrist positions known as cùn, guān and chǐ are reflective of wave signals that can only provide reliable information if their locations are specified precisely. These specifications for the pulses of both the zàng and fǔ Organs and their corresponding Meridians are now identified with nuanced variations from how they were originally described in the classics.
4. The correspondences of the healthy pulse depths at cùn, guān and chǐ with the Element of their associated Organ, as specified in the *Nánjīng* (*Classic of Difficulies*), has been confirmed, and treatment strategies to return abnormal pulse depths to their correct levels are described.
5. An alternate description of the Meridian associations at cùn, guān and chǐ was presented in the *Màijīng* (*Pulse Classic*), but is rarely employed by contempotary practitioners. Its true significance and clinical utility is described, and has been incorporated in CCA practice.
6. The most clearly revealed method of pulse diagnosis in the *Nèijīng*, comparing the pulses of the carotid and radial arteries (Rényíng Cùnkǒu <RYCK> diagnosis), has largely been abandoned by contemporary practitioners. I believe this has resulted, in part, from misunderstandings and mistranslations of the classical texts. Alternative etymological interpretations of the crucial terms zào (躁) and jìng (靜) are presented, as they are key pulse qualities for elucidating the RYCK diagnosis, as well as for every other aspect of the Conditional diagnosis.
7. The Tridosha system of Ayurvedic (Indian) medicine also includes both radial pulse diagnosis and a theory of Five Elements, but neither of these are the same as those used in Chinese medicine. However, the author has discovered a concordance between the Indian and Chinese systems that allows translation of information from one system to the other, greatly expanding the information that pulses can provide in guiding the practice of acupuncture.
8. The Korean system of Sasang Constitutional Medicine (SCM) is shown to be based on passages in the Chinese classical texts. The treatises that make up this trilogy explain the pulse diagnosis methods devised by several Korean Master practitioners and additionally include this author's discovery of how to use kinesiology testing of four foods to

corroborate the SCM classification, thus simplifying Constitutional diagnosis. This kinesiology method was used by the author on every patient treated during three workshops conducted in China, with very successful results.

9. The circulation rhythms (50 cycles) of yíng (營) and wèi (衛) qì were repeatedly discussed in the classics, but are rarely applied in the contemporary practice of acupuncture. I believe that Dr. Igor Simonov's interpretation of the original meanings of these passages form the basis for a more effective use of biorhythmic acupuncture strategies than those in current vogue. As Simonov has not published his ideas as yet, I have incorporated them in my presentation of CCA, with Simonov's permission. Biorhythmic treatment has proven very effective as a form of preventative acupuncture in my experience, but depends on an accurate Constitutional diagnosis for the most reliable results.
10. The Extraordinary Meridians (Vessels) are usually considered as individual components of the Meridian system of the human organism. My studies have revealed pulse phenomena that show them to actually be highly inter-related. These findings make diagnosis of Extraordinary Meridian malfunctions much easier to diagnose and treat with acupuncture.
11. The modern Japanese practitioner Akabane Kobe devised an eponymous diagnostic test that has been widely incorporated in many styles of contemporary acupuncture practice, due to its efficacy in guiding treatment. I have discovered a pulse signature that corresponds to abnormalities revealed by Akabane's heating test method, thus offering confirmation of the heating test's results, and establishing a more directly perceptual diagnostic method, in keeping with the traditional four examinations of classical Chinese medicine.
12. JR Worsley taught a method for both diagnosing and treating a category of illness traditionally labeled as "possession" in Chinese medicine. I have redefined the meaning of this diagnosis following my discovery of its consistant abnormality on pulse examination. Worsley's teachings in this regard were transmitted to him by Hsiu Yangchai, his Chinese teacher, and claims that Worsley invented this methodology can now be refuted.
13. The role of pathogenic factors in Constitutional styles of acupuncture has largely been ignored to date. A chapter on this topic in the

present treatise explores this subject in great detail, and proposes a way to incorporate this consideration in CCA, based on passages in the *Nèijīng*.

14. Cancer is one of the most feared diagnoses by people around the world. I have discovered a pulse phenomenon that I call the "metastatic signal" which correlates with the presence of any type of neoplastic illness. A chapter in this treatise explains how to use pulse diagnosis to detect cancers, even in their incipient stage, and additionally how to structure an apparently highly effective acupuncture protocol for actually reversing the neoplastic process itself, with the potential to possibly lead to a cure. Much more research is needed to confirm the reliability of this important discovery.

Prophecies

"...only he who knows how to maintain and strengthen the connatural[2] pulses in the human body—or to bring them back and restore them, preternatural and unchanged, to their original condition[3]...only he will be able to maintain life, look into one's health, and cure diseases. For they understand one's condition based on pulses that have been restored to their proper, natural condition[4], likewise that the twelve paths[5] returned to their primeval state; and likewise, life is borne by those inborn qualities in spirits and blood[6], when they are properly moderated; and the twelve sources[7]...distribute it throughout the body."

Michal Boym, in *The Medical Key To The Doctrine Of The Chinese On Pulses* (*Clavis Medica ad Chinarum doctrinam de pulsibus,* 1686)[8].

When I came across that neighboring corner
They started to speak and talk oddly, as in old ways,
They delighted in odd passages in old books
And doubtful passages gave up new meanings.

Táo Yuānmíng 陶淵明 (c.365-427), a Buddhist/Naturalist Chinese poet.[9]

2 i.e. inborn.

3 i.e. the Constitutional state.

4 i.e. the state of the Conditional pulses must be restored to that of the Constitutional pulses.

5 Meridians, Vessels, or Channels. In Chinese the zhèng jīng 正經.

6 Spirits and Blood are here translations of qì 氣 and xuè 血.

7 The zàng fǔ 臟腑.

8 English translation by Shawn Daniels, pages15-16. Soul Care Publishing, 2018, Vancouver.

9 Translated by Richard Bertschinger and posted on Facebook, 5/29/20.

Dedication

Ting Hor [10]

Dedicated to Ting Hor (賀霆), former professor in the anthropology department of the Yunnan University of Traditional Chinese Medicine, and father of the beautiful children riding the donkey in the photograph below, under his watchful eyes.

I first met Professor Hor in 2014 at the annual symposium of the British Acupuncture Council where I was an invited speaker. His specialty, or I should say his passion, is gathering the widespread materials produced as a result of the migration of Chinese medicine to countries of the Western world. To this end he has created a wonderful museum in Kunming of original artifacts, photographs and out of print publications. But his efforts are not directed towards merely satisfying a personal curiosity. Rather he recognizes that there are potential "Treasure Houses" in these foreign countries, some of which originated in China, but are no longer to be found there.

10 Source: Anne Blanchard-Laize.

Thus he longs to bring home a part of the heritage that China has forgotten, but the West has maintained, all for the sake of creating the best version of Chinese medicine, which has now become a world medicine.

As our relationship grew, I was first a guest Skype participant in the annual Forum on the Transmission of Chinese Medicine to the West in 2014 in Kunming. I was subsequently invited to make an in person presentation to the Forum in 2017, and following my offer to conduct a pre-Forum clinical workshop, my teaching career in China was launched in Kunming, guided by Professor Hor's encouragement and organizational skills. This workshop on Constitutional Conditional Acupuncture was an experience I would never have even dreamed of in the past. 2018 felt like another giant step forward. Professor Hor inspired the creation of an international organization focused on his field of interest, CRAOMC[11], with its inaugural meeting in Paris, where I was asked to serve as a vice-president., but with it came an invitation to expand my teaching in China from a three day workshop to a five day workshop, this time in Beijing. The organizational challenges made this a touch and go situation for quite some time, but Professor Hor pulled it all together, and in the end it was a remarkably successful class, and the response has been most gratifying. In no small part it has led me to the writing of this present treatise. As I write this, I have completed another 5 day workshop in Beijing in October 2019, followed by a presentation of my work at both the Beijing University of Chinese Medicine and the Guangdong Provincial Hospital of Chinese Medicine, all due to Professor Hor's tireless efforts.

Dear Ting: May your children someday figure out how their father knew I needed a picture of them on a donkey, years before that humble creature became my muse.

---

11 Centre de Recherche sur les Apports occidentaux à la Médecine chinoise (CRAOMC).

The Hor Family in Normandy [12]

12 Source: Kai Wu.

Foreword

I first met Dr. Peter Eckman through the Internet medium of Chinese medical forums, particularly engaging in the group Scholars of Chinese Medicine. It was here I took notice of his participation in discussions on medical ideas that are not typically discussed in Chinese medical schools, such as the discussion on the exact finger positioning when taking the radial or the carotid pulse, and their aberrations. Now, there is a standardized method that is taught for the radial pulse, but not a nuanced discussion on its significance and the interpretation that comes about from such positioning. Inquiry on it is rare; many just take what is taught and don't question it. Of course, it is a valid medical perspective, and that is why it is systematically taught in such a standardized way, but once one begins exposing oneself to various lineage-based methodologies it all begins to become confusing. Dr. Peter Eckman's work represents a medical inquiry into matters such as this, to pierce closer into the reality of clinical certainty, away from the muddled confusion that comes with diagnostics. One of the gifts and curses of Chinese medicine is its vast array of diagnostic theories and models; this gives clinicians a flexible perspective, but it also fills Chinese medicine physicians with an almost endless profusion of methodologies that are hard to distinguish from each other, and thus master individually. We are left to our own devices of when to use each one. And though there are certain frameworks to follow for each one, the essence of each one is difficult to distill and therefore apply.

It is refreshing to see Dr. Peter Eckman not only introduce various diagnostic models that are not typically known in the West, particularly Korean models, but also elucidate his experience with each one. Like any good diagnostician, Peter approaches all methods he has learned with a certain sense of scientific skepticism, but one that is respectful of the process required to effectively understand these diagnostic methods. Often Dr. Eckman, when he could not reproduce the diagnostic accuracy of his teachers, instead of questioning the validity of the methods, he directed scrutiny onto his own application and understanding of them. This I believe has led to the refinement of the diagnostic skills that Dr. Eckman has synthesized. This approach has given rise to insights of his own into the methods, rather than to the foregoing of the methods all together because they weren't working for him. This synthesis of methods has become the basis of the constitutional model that is presented. This constitutional model is unique because to arrive at a diagnosis, an amalgamation of various pulse approaches is used to double-check each of the other method's validity. As Dr. Eckman notes, when one pulse method is directly contraindicatory to another's, then the physician has erred somewhere down the line and must go back and examine again. Now, it must be mentioned again, these are approaches not typically taught in TCM schools, as they are lineage-based approaches or systems that arose from a different theoretical offspring, such as the Korean constitutional models and their unique terminological interpretation of taiyang, shaoyang, taiyin, and shaoyin.

Chinese medicine has bloomed different flowers depending on the place it has taken root. Historically, Chinese medicine doctors have assimilated medical concepts and botanical medicine from other systems and integrated them into its own doctrine. We see this in the introduction of various foreign medicinals in Materia Medicas, and conceptually in the work of Sun Simiao, the Medicine King. Dr. Eckman has followed in the footsteps of this revered figure and integrated Ayurvedic lineage-based approaches and diagnostic concepts into his system of approach. Now what is even more important is keeping the identity of each model separate, so as not to conflate the meanings of the findings for each one. It is my great pleasure to see how Dr. Peter Eckman analytically manages to avoid this conceptual trap where theories are mixed together, and then lose their individual meaning. This would ultimately leave the diagnostician muddled, because there is no clarity.

Undoubtedly, Dr. Eckman has spent his time seeking clarity and making sure whatever he didn't understand continually becomes elucidated by the refinement of the diagnostic process. Examples range from the nuances of the finger positioning on the pulse, to the diagnostic truth of Ren Ying Cun Kou, to a dynamic Nanjing based 5 phase model, and most importantly to the theoretical and philosophical constructs underlying each one. This is paramount to understand, because without this foundational rationale, we will always be chasing a shadow and never understanding why we chase it in the first place. Dr. Eckman thus illuminates for us the source of this constitutional model and the constituents that form it. This gives the reader insight into the essence of what constitutes a Chinese medicine physician. The hallmark of a Chinese medicine physician is unique in that they cultivate looking at phenomena and the cosmos from a multifaceted perspective. This culminates in a fluidity that is shared by all great Chinese medicine thinkers and physicians. We see Dr. Eckman continually challenging his understanding through intellectual and pragmatic rigor to arrive at the revelation that will best capture the image of a patient's innate nature and unlock their healing potential.

We see this finally play out in the thought-provoking case studies Dr. Eckman shares. It is here where the rubber meets the road, and he illustrates for us his harmonic diagnostic dance. But what makes these case studies particularly fascinating, besides the patients themselves, is the setting that they take place in. Not too many Chinese medicine doctors in the West are invited to the heartland of China to treat recalcitrant cases that other Chinese medicine physicians there are struggling to treat. This alone should speak volumes on the methods Dr. Eckman has refined and presented for us to take in. He thus invites us on a physician's journey to use and take these methods to enhance our own diagnostic process and clinical results. And in the same spirit that he pondered his teachers' conclusions and cultivated his own understanding, Peter ultimately invites the reader to do their own inquiry so they can attain the skill rooted in flexibility that he has. But his penetrating insight into matters that are simply not discussed much in the West should not be ignored, as they are of great utility to any aspiring clinician seeking clarity. I myself learned several vantage points and lineage-based approaches I had not considered and will take them to heart.

In Dao, Ivan Zavala (also known as Ivan Cloud),
Founder of Classical Chinese Medicine (Facebook group)

Acknowledgments

Throughout this book I will be referring to the work of Kuon Dowon, a brilliant Korean acupuncturist who is approaching his 100th birthday in 2021. As I am writing this book, he has recently retired after a lifetime of research and practice. He was kind enough to allow me to observe at his clinic in the 1980s, and my development of Constitutional Conditional Acupuncture (CCA) would never have gotten started without his generosity. Another Korean acupuncturist, Yoo Tae Woo, the originator of Korean Hand Acupuncture, was also willing to mentor me, for which I am very grateful. However, I would never have met either of these seminal teachers and practitioners were it not for the apprenticeship and personal introductions to these Masters of Korean acupuncture offered to me by Chae Woo Lew, whose untimely passing I mourn.

On a brighter note, new relationships have been added to those of the Masters who guided my development. I must thank Ting Hor, to whom this book is dedicated, for opening up my teaching world to include China, birthplace of acupuncture. Professor Zhou Xiaofei, of the Beijing University of Chinese Medicine, has been most helpful in coordinating my workshops in China's capital, and in pursuing a publisher for my books in Chinese. A big thank you to Ivan Cloud Zavala, who graciously provided the Foreword. Ivan embodies the best in the younger generation of scholars and practitioners of Chinese medicine, and I am honored to have his viewpoint on my work. I would also like to thank Sylvie Martin, a French colleague and friend, who introduced me to the work of the late Jacques Pialoux. Readers will probably not be familiar with Pialoux's work, but due to Sylvie's efforts it has been introduced into China for approximately the last decade. Pialoux, a leading member of Acupuncturists Without Borders, was a repository of the French acupuncture tradition starting with Soulié de Morant. His immediate teacher was Jean Borsarello who in turn was a generous correspondent of mine, and with whom I collaborated on his last publication, shortly before his passing. I have integrated some of Pialoux's teachings into CCA and reformulated them, initially spurred on by the request from Sylvie Martin's Chinese students that I help them understand how Eight Extraordinary Meridian pulse diagnosis is carried out. Several of these Chinese students saw me demonstrate this part of pulse diagnosis in Beijing, and were impressed with the clinical results I achieved in class, but also with their ability to duplicate this in their own acupuncture practices. So, I scheduled a 5 day workshop in Shanghai for September of 2020, but it had to be postponed due to the Covid-19 pandemic. I ardently hope to resume teaching in China whenever it becomes possible.

Once again I am indebted to my old friend Neal White, who contributed his signature drawings, but who also alerted me to Goya's Donkey etching. At first I didn't believe that there could be such a famous image of a donkey taking a patient's pulse! While on the subject of old friends, I wish to thank Stuart Kutchins, founding President of the NCCA, for our career long friendship. He has shared all of my Korean adventures from the very beginning, but more importantly he has been, and continues to be, my soul brother. I could never ask for a truer friend. As for new friends, I would like to acknowledge Igor Simonov, whose idiosyncratic interpretations of classical Chinese texts always challenge me to consider alternatives to commonly held dogmas. His influence shows up at several places in this treatise. I have used quotations from many other colleagues whose work has inspired me, but I would like to specifically thank Ted Kaptchuk, David S. White and Will Ceurvels who each offered me personal assistance. Although their approaches to acupuncture differ vastly from my own, they embody what is best in our community of scholar practitioners: mutual respect, a friendly attitude, and a willingness to share their ideas with their peers. And of course I am very grateful to my apprentice friends, Olga Fedina and Lucas Garcerá Ramírez, who have hosted my classes in Spain, but more importantly, who are committed to maintaining the lineage of CCA after I am gone. And, of course, I am thankful for the many patients who have put their trust in me for almost half a century. It is perplexing when patients beg me never to retire. I ardently hope that there will be more than a couple of practitioners of CCA to carry on this work after me, and perhaps this treatise will further that purpose.

Finally I wish to thank my wife Marina, who has understood the importance of my work, and my lengthy teaching trips to China. Since meeting her, no demon has dared to disrupt the equanimity of my life, in what are turning out to be very challenging times indeed. I have surely been blessed by fate.

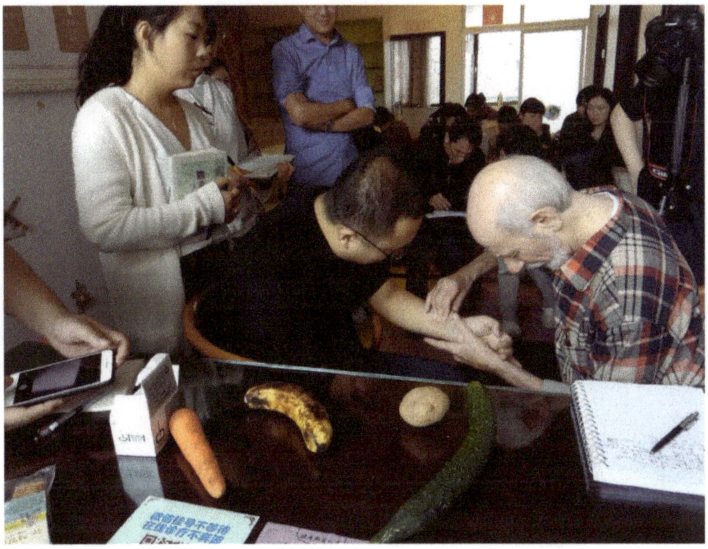

Kunming Workshop, 2017 [13]

13 Source: Olga Fedina

Exhortation

What is an exhortation and what is it doing here? "The gift of exhortation is often called the gift of encouragement. The Greek word for this gift is parakaleo (παρακαλώ). It means to beseech, exhort, call upon, to encourage and to strengthen." [14]

There can be no question about the historical importance of pulse diagnosis in Chinese medicine.[15] I first introduced the pulse diagnosis based system Constitutional Conditional Acupuncture (CCA) in 2014 in my text *The Compleat Acupuncturist*, and have since taught seminars and workshops on this method around the world. Frankly, it is not an easy method to master, and I am still on that path myself. However, the depth, power and scope of this approach is well worth the effort, as the reader will see in the case histories, especially those that deal with cancer. Just this week, one of my students in Beijing wrote me to say that he is not yet very successful in his use of pulse diagnosis. On the other hand, another one of my students in Valencia wrote me that he has already been amazed by the successful results of his use of my cancer diagnosis and treatment recommendations. Not everyone has equal skills in pulse palpation to start with, but I believe that given enough practice and correct instruction, anyone can learn to apply this methodology to help their patients. As I've stated in my prior works, the ideas expressed in this text are not graven in stone, but merely communicate my present understanding of the material discussed. Some of these findings may need revision as future experiences lead to different conclusions. That is the way of science. There can be no final word that is beyond future revision as new discoveries are made. But sciences, including the classical approach to acupuncture, are not abandoned because they might later be revised. Rather I conceive of such endeavors as the stepping-stones upon which our knowledge of the natural world is built.

---

14 spiritualgiftstest.com/spiritual-gift-exhortation

15 "Pulse diagnosis became the supreme diagnostic tool for elite medicine throughout the Chinese empire, and it retains primary importance for practitioners of TCM today." Michael Stanley-Baker and Vivienne Lo in the *Oxford Handbook of the History of Medicine*, Mark Jackson ed. Oxford University Press, 2011.

In *The Analects*, Confucius is quoted thusly regarding the *Book of Changes*, "If some years were added to my life, I would give fifty to the study of the *Yi*, and then I might come to be without great faults."[16] In a similar vein, I can humbly state, "If some years were added to my life, I would give fifty to the study of the pulse, and then I might come to diagnose without great mistakes." I do not make this statement lightly. As I am writing this draft, every few days I find myself learning something new and amazing in the pulses of my patients. These discoveries are leading me into the ever deeper unity underlying the various aspects of pulse diagnosis, and how CCA has the potential to reflect the realities of my patients' lives at all levels of their being. Like their lives, this path of study is a long and winding road, but also a wonderful journey.

I am reminded of a quotation from Wáng Shūhé in the *Màijīng*, 'The patterns of the pulse are finely intermeshed and their forms difficult to differentiate, and although they possessed the most subtle of faculties, Biǎn Què and Yī Hé nevertheless had to think twice sometimes."[17] What did Wáng mean by "think twice"? Although we cannot know for sure, I'd suggest that he meant they had to change their diagnoses when the results of treatment were not what they expected. I have had to do this many times, as have all of my masterful teachers of every style of acupuncture. A favorite book of mine is *30 Years of Kanpo* by Otsuka Keisetsu[18], the Japanese lineage Master of Kanpo herbal medicine, in which he cites case after case where he had to redo his diagnosis based on clinical results. So I say to the readers and all of my students, "do not give up when things turn out differently than you expected. Always ask yourself why, and then go back and start over, examining the pulses, until you find the answer." In such an endeavor you will be in the company of the greatest physicians of Chinese medicine throughout its long and glorious history.

16 James Legge translation of the Analects, Chapter 51

17 Cited by Miranda Brown in Med Hist. 2012 Jul; 56(3): 366–389.

18 Oriental Healing Arts Institute; First Edition (1984)

Preface

Let me preface this preface by noting that the practice of Constitutional Conditional Acupuncture (CCA) can never be fully captured in either a written document or a set of formal guidelines. Every person who presents for treatment is a unique individual, and the practitioner must pay close attention to everything going on in the treatment room, including but not limited to all the pulse signals detected. Synthesizing this input into a diagnosis and treatment plan is a creative act that rides the border between knowledge and intuition. The knowledge base in CCA keeps evolving, while the intuitive aspect cannot be put into words. What I am presenting in this treatise is a formulation of CCA derived from experiences over years of uncounted observations and treatments, with their ensuing results. Particularly in reading the case histories, there are likely going to be some perplexed feelings generated due to the different time periods during which these encounters happened, eventuating in different descriptions of the pulse findings. I have tried to reduce such inevitable discordances to a minimum by sharing only the most illustrative aspects of each case. Familiarity with all the chapters, themselves written during different time periods, should help the reader develop a sense of how to apply CCA in a flexible, as opposed to a rigid, manner.

I have previously remarked on the responsibility I feel for innovators to pass on their discoveries for the benefit of future generations, and not keep them as family secrets, nor wait until every aspect of their work is absolutely perfect. My reason for bringing this up is that at 75 years old, as I start collating this treatise, I feel the pressure of time more acutely than I did even 10 years ago. I note that one of my most important mentors, the aforementioned Kuon Dowon, only retired from acupuncture practice at the age of 96 in 2017, never having managed to write a textbook on his widely respected style of practice (Korean Constitutional Acupuncture or KCA, later renamed Eight Constitutions Medicine or ECM), but many brilliant practitioners did not even reach my current age. That explains the timing of this treatise.

Since the publication of my second book on Constitutional Conditional Acupuncture, *Grasping the Donkey's Tail*, my professional life has entered a new phase. While I had already begun teaching my CCA approach to colleagues and students in the United States and Europe, I suddenly was offered the opportunity to do so in the People's Republic of China, the cradle of acupuncture itself. I have now taught three workshops in China, and I think that I learned as much as the attendees did from these workshops. The first was in Kunming in 2017 and the next two were in Beijing in 2018 and 2019. These workshops were arranged by my Chinese hosts, so the format they proposed was new to me: examine, diagnose and treat patients every morning, then give both didactic presentations and hands on practical exercises with individual feedback to all participants every afternoon and evening. In addition all the patients treated were invited to return on the subsequent days to report on their responses to treatment, and close to 100 percent of them complied. Previously my teaching had been mostly comprised of lectures, with the addition of practical exercises, but I had never experienced being in the teaching role of attending physician for multiple patients of every stripe, day after day (3 days in Kunming, 5 days in Beijing each time), all expecting my treatments to prove their rapid efficacy. Since my primary diagnostic method consists of pulse examination while my primary treatment method is the application of acupuncture needles, you could say I was in a position of putting my hands where my mouth was, or more colloquially, "put up or shut up". Thankfully CCA treatment seems to have helped in every single case, an outcome beyond my most optimistic expectations. As a result of the first two workshops, I received an invitation to conduct a third 5 day workshop in Beijing, followed by a presentation of my work in Guangzhou at the Guangdong Provincial Hospital of Chinese Medicine in November of 2019, an annual event that had previously been restricted to Chinese practitioners, but was opened to foreigners then for the first time. Needless to say, it was an unforgettable experience. If I had to summarize what I have learned from teaching in China, it would be something like this: CCA is an effective form of treatment for virtually any patient, and as its methodology is firmly grounded in the classical literature, it validates the position that all students wishing to learn the art and science of acupuncture should study the classics regardless of how obscure they may at first seem. Their buried wisdom is even now in the process of being rediscovered.

I have two distinct purposes in writing this book. The first is to introduce the new discoveries (and "tweaks" to the old ones) that I have made, since the publication of *Grasping the Donkey's Tail*, in the clinical practice of pulse guided CCA. But behind this lies a deeper purpose. What I have formulated as CCA is a process of diagnosis and treatment by acupuncture based on, but not limited to, classical texts. Almost daily, in my clinical practice, I am faced with something (usually in the pulse) that I do not yet understand. I want to emphasize that such experiences are the norm, and provide us with the possibility of making new discoveries. So when faced with the unknown, or even apparent self-contradictions, I begin to ponder on possible explanations that I can test for both reproducibility and clinical efficacy. The results of my application of this ongoing process form a substantial part of this treatise. Some might call this process a kind of meditation, and the answers that emerge a kind of divine revelation. I believe that some of my teachers have viewed this process in precisely that way[19]. I have no way of knowing how the process of discovery actually works, but I can attest to the fact that it does indeed work. I choose to interpret such matters as the experiential basis for an Eastern scientific methodology. The discoveries I have made are small steps on a potentially endless path, and I truly believe that what I have done can be repeatedly done by anyone seriously committed to study, observation and contemplation, together with responsible clinical research. There is certainly no shortage of people who need the best of our loving care. The more we learn, the more we can provide this invaluable human service. Finally, it is my observation that the daily practice of any endeavor can become boring if it is carried out with a totally standardized methodology. There will always be new frontiers to investigate, so as to enlarge our understanding of natural law as it applies to health and illness. With that in mind, every day of practice becomes an adventure into the vast mystery of life.

19 My inspiring mentor Kuon Dowon has explicitly stated in private conversation that divine revelation was the source of his discoveries.

Introduction

I must assume that the reader has perused both *The Compleat Acupuncturist* and *Grasping the Donkey's Tail*, as I cannot present all of that material here. These books were of somewhat different natures: the former being an account of how Constitutional Conditional Acupuncture (CCA) was developed, and an introduction to its clinical practice, while the latter was an exploration of topics in the classical Chinese medical literature about which I had some novel and often unorthodox ideas. In a nutshell, CCA is based on the thesis that human beings differ from each other from the moment of conception, having their own unique Constitution. Maintaining this inborn state leads to health, while deviating from it leads to illness. Acupuncture treatment in CCA is thus aimed at transforming whatever present Condition a person has developed, back towards their original Constitutional state. Pulse diagnosis is the main examination method for determining both the Constitution and the Condition. This present text is more or less an amalgamation of those two previous kinds of presentations, as it clearly focuses on my personal reinterpretation of classical textual material, but also explains how these new interpretations lead to changes in clinical diagnostic and therapeutic methodology.[20] To illustrate the latter aspect, I have once again provided a chapter of case histories.

The subtitle is of course tongue in cheek. For some, *Monkey Business* (the operative pun along with *Donkey's Tail*), has a rather pejorative connotation, but for me it is quite the opposite (a position that should find a receptive audience among both Daoists and aficionados of yīn yáng[21] theory). I grew up with the Marx Brothers' movies, among which one can find "Monkey Business" (Fig. 1). These movies are hilarious in many ways, not the least

20 I have also found it necessary in writing this treatise to investigate historical lines of teachings, or lineages, in order to explain where some of my ideas come from, and how they relate to mainstream acupuncture practices, as well as to passages from the classics. This "third hat" I wear (historian) was developed in composing my first book, *In the Footsteps of the Yellow Emperor*.

21 Wherever possible, I have included the Pinyin Romanization of Chinese terms, accompanied by their tonal marks. Ignoring the tonal marks may render the terms more familiar to English readers.

of which is their constant challenge to conventional strictures. Though different in style, I see echoes of Zhuāng zǐ in their antics. What is yáng will always become yīn when pushed to its extreme and vice-versa. And the Marx Brothers were nothing, if not extreme. The lesson for me here is that being fully alive means not being attached to any doctrine as ultimate truth (expressed in Chinese medicine as having an "empty heart"). One must study the classics and one must also challenge them. In so doing I have discovered both long lost teachings and precious tools for helping to treat the suffering of us humans, subject as we are to the "slings and arrows of outrageous fortune". It is my fervent hope that my students will surpass me in expanding the applications of CCA, and that they too will, in this endeavor, feel some degree of enlightenment, as I have.

Figure 1: "Monkey Business"

The four Marx Brothers stowing away on an ocean vessel by hiding in fish barrels, in this promotional still for Monkey Business. *Left to right: Harpo, Zeppo, Chico, Groucho. 1931* [22]

22 Source: https://en.wikipedia.org/wiki/Monkey_Business_(1931_film). (Category: Publicity photograph with no terms, which qualifies as fair use under Copyright law of the United States.)

Perhaps not coincidentally, I was born in the year of the Monkey (Fig. 2A). In addition, although there are no Donkeys in the Chinese zodiac, let's imagine that a donkey is a very unusual type of horse (Fig. 2B). Why not? The Chinese have been discussing whether or not a white horse is, or is not, a horse since c.300 BCE when Gōngsūn Lóng (公孫龍 of the School of Names) wrote the White Horse Dialogue (白馬非馬 *Bái Mǎ Fēi Mǎ*).[23] Not only white horses, but also grey ones (donkeys?) have made their appearance in this ongoing paradoxical discussion that included commentaries by Liè zǐ, Zhuāng zǐ, Mèng zǐ, Hán Fēi zǐ and Mò zǐ. Zodiacal Monkeys and Horses are both celebrated for their intelligence, but Horses are said to be stubborn (like donkeys), while Monkeys are proverbially mischevious. So dear reader, do not take any of my claimed discoveries as Gospel, but test them out for yourself. Perserverence furthers. (元 亨 利 貞 yuán hēng lì zhēn).

Figure 2A: Funerary Earthenware Zodiac Sculpture: Monkey

23 Igor Simonov helped me understand why this paradox plays such an important role in Chinese philosophy, and indirectly in CCA. To say a white horse is a horse is to ignore what makes it unique and different from other beings called horses. Every being in creation is like this, unique (having its own Original Nature or Constitution), even as it shares characteristics with other beings. CCA does not espouse a way to treat "human beings," but rather a way to treat specific individuals.

Figure 2B: Funerary Earthenware Zodiac Sculpture: Horse [24]

Figure 2C: Zodiac Netsuke of Monkey on Horse [25]

24 2A and 2B are from a collection of 12 zodiac sculptures recovered from Yunnan province, and dated as artifacts of the Yuan dynasty 1279-1368. These sculptures are now in the Los Angeles County Museum of Art. Source: Wikimedia commons.

25 This 19th century Japanese ivory netsuke is from the collection of the Metropolitan Museum of Art. Source: Wikimedia Commons.

CHAPTER 1

Constitution and Condition Revisited

Figure 3: The Four Temperaments [26]

The classics of Chinese medicine are usually stated as originating with the *Huángdì Nèijīng* and the *Shānghán Lùn*. Where in these foundational texts can we find any discussions of the distinction between Constitution and Condition? In fact, where is Constitution even mentioned?[27] Of course, the answer to these questions depends on what one accepts as the Chinese term for Constitution. Possibilities include xiān tiān zhī qì (先 天 之氣), xiān tiān

---

26 Open source image attested as such and published by Mateo Sol at LonerWolf.com. From left to right the temperaments depicted are melancholic, phlegmatic, choleric and sanguine.

27 *Língshū* chapter 64 is often cited as a presentation of constitutional theory, but it does not have any term that may be translated as constitution. Literally, its title is, "Yīn Yáng 25 People" (陰陽二十五人). *Língshū* chapter 72 reprises this subject, but again avoids using a term denoting constitution: "(There are) major yīn people, minor yīn people, major yáng people, minor yáng people, yīn yáng harmoniously balanced people." (太陰之人，少陰之人，太陽之人，少陽之人，陰陽和平之人). "(The) five (times) five, 25 people are not the same as the people with the five attitudes (just described)" (五五二十五人之政，而陰陽之人不與焉)

zhī jīng (先天之精), yuán qì (元氣) and yuán qì (原氣).[28] Surprisingly, in my digital search I found none of these terms even once in the above texts, a rather unexpected finding! It is not until the (putatively) later *Nánjīng* that the latter two terms are introduced into the medical corpus.[29] Let me start with what I see as a prelude to its concept of Constitution in *Nánjīng* 8, which introduces the idea that human life is rooted in the "moving qì between the Kidneys (腎間動氣)". This "moving qì" is called yuán (原), meaning origin or source. What is it the source of? The answer is clearly stated as the 12 jīng mài (Principal or Regular Meridians), the 5 zàng and 6 fŭ (Organs), and the Triple Heater (function). Additionally it is said to be the "Gate of inhalation and exhalation" and the "Spirit guarding against Evil".[30] In this "Difficult Issue" one can find a solid basis for the fundamental concepts of CCA. The implications include the idea of human beings having a Constitution stored in mìng mén (Gate of Life), which is the "root" of the Organs, the Meridians and their pulses (mài). It is common to think of the moving qì at mìng mén as reflecting the Essence (jīng), but this passage clearly states that it is also reflective of the Spirit (shén) which guards us against Evil (xié qì). Where did this idea about "moving qì," which is not found in the *Nèijīng*, come from? Liao P'ing (廖坪, 1851-1914), a commentator on the *Nánjīng* (who also vehemently disparaged it!), tells us that this idea originated with the alchemists (i.e. Daoists), which accounts for the significant differences in focus between the *Nèijīng* and the *Nánjīng*. I believe he is correct in this assertion, as the moving qì between the Kidneys parallels the alchemical notion of the dān tián. Liao P'ing said Difficulty 8 emphasizes the role of the Kidneys, establishing designations for the Gate of Life based on the alchemist teachings.

28 It has been suggested that xìng (性) might be the closest Chinese term to Worsley's concept of the "Causative Factor" (CF), which is not identical with the term Constitution, as I use it. In my research I found this character 12 times in the *Sù Wèn*, with the clearest example being in chapter 67. This chapter deals with the relationship of the 6 climatic factors with the 5 Elements, and the nature (性) of each Element is generally expressed in terms of one of the climatic factors. In any case, this term is not used there to classify people into different groups, whether one chooses 5 or 20 groupings, so in my opinion it is not specific enough for designating human Constitutional typology. It adds nothing to using the 5 Elements as a basic classificatory concept.

29 The concept of yuán qì (元氣) was believed to have been first proposed by Dŏng Zhòng shū in the second century BCE, but was not in common usage until the Eastern Hàn. Dŏng Zhòng shū (董仲舒, Wade–Giles: Tung Chung-shu, 179–104 BC) was a Hàn Dynasty Chinese scholar. He is traditionally associated with the promotion of Confucianism as the official ideology of the Chinese imperial state. He favored heaven worship over the tradition of the five elements. This citation about 元氣 is mentioned in Unschuld's *Nánjīng* translation, p.29. Simonov (personal communication) has speculated that the extant copies of the *Nánjīng* are most likely earlier in composition than the extant copies of the *Nèijīng*.

30 From *Nánjīng* 8: 諸十二經脈者，皆係於生氣之原。所謂生氣之原者，謂十二經之根本也，謂腎間動氣也。此五藏六府之本，十二經脈之根，呼吸之門，三焦之原。一名守邪之神。

These teachings were applied in *Nánjīng* 8 in a medical doctrine, possibly for the first time. Previously, the *Nèijīng* had emphasized the role of the Stomach (i.e. the sea of grains and liquids responsible for the production of postnatal qì and blood)[31], whereas the *Nánjīng* changed the focus to the role of the Kidneys (responsible for our prenatal original qì). I find this to be a useful image to keep in mind when trying to understand both the underlying principles of CCA, and the different roles played by our primordial qì and our acquired qì. What is primordial never changes (our Constitution), while what is acquired is constantly in flux and can lead to either health or illness (our Condition). Having laid this groundwork, the *Nánjīng* proceeds to elaborate on this new concept of Original qì, specifically mentioned as a medical technical term for the first time. CCA is based on the new material introduced by the *Nánjīng*, however without discarding the prior teachings of the *Nèijīng*.

> *Nánjīng* 14 says, "Just as a person has feet, a tree has roots. Even though branches and leaves may wither, the roots can live by themselves. Similarly the vessels have their root, and that is where man has his primordial influences (yuán qì 元氣)".
>
> *Nánjīng* 25 tells us, "The Heart Master and the Triple Heater represent outside and inside (biǎo lǐ). Both have a name, but no form." The relevance of this passage will become evident as the presentation proceeds.
>
> Nánjīng 36 says, "The two Kidneys are not both Kidneys. The one on the left is the Kidney, the one on the right is the Gate of Life (命門 mìng mén). The Gate of Life is where the Spirit Essence (神精 shén jīng) lodges. It is the place to which the original influences (原氣 yuán qì) are tied."
>
> *Nánjīng* 38 adds, "The Triple Heater provides an additional source of original influences (原氣 yuán qì); it governs all the influences (qì)

31 " ... there is a lot of material in the *Nèijīng* which would reinforce a soil (Earth Element) centered view of the qì dynamic." Posted on the Facebook group 'Scholars of Chinese Medicine' by Par Scott, and accessed 3/23/19.

> circulating in the body.[32] It has a name but no form."
>
> Finally, *Nánjīng* 66 ties these passages together thusly, "The shū (stream) points of the five zàng (臟) are the locations where the influences that are sent out by the Triple Heater stop. The influences moving below the navel and between the Kidneys constitute man's life. They are the source (本 běn) of the 12 Meridians (經脈 jīng mài). Hence they are called original influences (原氣 yuán qì). The Triple Heater is the special envoy that transmits the original influences (原氣 yuán qì)."

These citations from the *Nánjīng* allow us to begin to understand the concept of Constitution in Chinese medicine, and the role it plays in human energetic physiology (氣機 qì jī). Yuán qì (元氣) can be interpreted thusly: yuán 元 is the image of a standing man with two lines on top emphasizing what is above, thus implying Heaven (天 tiān) and its input, Spirit (神 shén[33]). These make the non-material part of our Constitution. The homonym Yuán qì (原氣) has a somewhat different denotation: yuán 原 is the image of a hill with a spring gushing out[34] emphasizing what is from below, thus implying Earth (地 dì) and Essence (精 jīng). These make the material part of our Constitution. Classically, a new individual is conceived when the Essences of the two parents (jīng) unite, and at that very moment the Spirit appears (shén), as if attracted by the pull of these unique Essences, like the opposite poles of a magnet attracting each other. From this perspective we can see that our Constitution is partly an aspect of our physical nature and partly an aspect of our non-physical nature. Any theory of Constitution that separates

32 If the Triple Heater is an additional source of Original influences, what is it in addition to? I imagine that the Gate of Life, carrying the Original influences, bifurcates into the Circulation function (also called Pericardium, Heart Governor and other names) and the Triple Heater function. These are the pair that is responsible for distributing our original 神精 shén jīng to the Organs and Meridians throughout the body.

33 Wieger discusses the origins of this character thusly: L. 3D explains the radical on the left as the influx from heaven. In its non-contracted form it depicts two horizontal lines on top indicating what is above (i.e. heaven) and three vertical lines below indicating the sun, moon and stars, which are used to calculate heaven's astrological effects on man. All characters using this radical (113) relate to transcendental matters. L. 150C explains the phonetic on the right as an ancient form depicting two lowered hands extending a rope (thus pulling down the influences from above).

34 Wieger L. 59C explains that the radical on the left (27) is a primitive form depicting a cliff (i.e. a high place facing heaven), while the phonetic on the right (L. 125F) is also a primitive form depicting the spring bubbling up from the earth, which then expands and spreads to flow on the surface. This imagery nicely captures the notion of our Constitutional qì being distributed from its origin internally to its distribution externally, resulting in the principal meridians (jing mài) flowing like rivers on the body's surface.

these two aspects would necessarily be reductionist, and miss part of what is so important about our individual Constitution.[35]

What then is the sense in which the Constitution was understood in the *Nánjīng*? Being present from conception, it is an innate aspect of our nature, not an acquired one.[36] The Heart (Imperial Fire) stores the Spirit (shén), while the Kidney stores the Essence (jīng). But how are their influences (qì) able to reach and order the 12 Officials (臟腑 zàng fǔ)? Starting from their location as the "moving influences between the Kidneys" (aka the Gate of Life), there are two special Officials whose job it is to facilitate this distribution: the Triple Heater and the Envelope of the Heart or Heart Master (sān jiāo and xīn bāo luò or xīn zhǔ). These are the two Ministerial Fire Officials that were described as having functions but not forms. One of their tasks is to transport the Original Qì (Constitutional information) to the ten formed Organs (Officials). *Nánjīng* 66 says this explicitly for the Triple Heater, but I think we are justified in seeing a parallel function for the Envelope of the Heart, which is tasked with being the envoy of the Heart. These two Ministerial Fire Officials are linked, not only by the fact that their Meridians share a biǎo/lǐ (exterior/interior) relationship, but also by their common function as envoys of the Original Qì.[37] Just as the Triple Heater has its Back Shū Point adjacent to that of the Kidney (UB 22 and 23), the Heart Master has its Back Shū Point adjacent to that of the Heart (UB 14 and 15). These two Ministerial Fire Officials can be viewed as the conveyors of Spirit (shén) to the Heart to be stored there, and of Essence (jīng) to the Kidneys to be stored there. This interpretation even informs us about the depth of pathological disruption as reflected in the sequence of Meridians attacked in cold invasion, a subject that has proponents still contending with each other about which level is the

35 My mentor, Kuon Dowon, believed that one's Constitution was necessarily identical to that of either of their parents. I have often observed identical Constitutions in close relatives, but this is not always the case. Kuon's belief would support the idea that Constitution was solely determined by one's inherited essence (jīng). In my practice I have been treating a husband, wife and son with successful outcomes. However, they all have different Constitutions. Their CF Elements are Metal, Wood and Fire respectively, their Sasang affiliations are Taeyangin, Taeumin and Soeumin respectively, and their Meridian affiliations are Yangming, Jueyin and Shaoyang respectively. I can only assume that their Spirit (shén) might be the dominant factor in determining Constitution, in this instance at least.

36 The Constitution is clearly tied to Heaven and Earth through their postnatal representatives, Fire and Water. In the *Book of Changes* (*Yìjīng* 易經), the Fú Xī order of trigrams has Heaven on top and Earth on bottom (prenatal). The Wén Wáng order has Fire on top and Water on bottom (postnatal).

37 *Língshū* 8 specifies the Heart Master's function as envoy, while *Nánjīng* 66 calls the Triple Heater the special envoy of the Original Qi.

deepest. Shǎoyīn includes the Heart and Kidney systems, which are deepest on a structural level, but Ministerial Fire including the Juéyīn Heart Master represents attack on an even deeper, non-material level, and that may explain why it is presented as such in both the *Nèijīng* and *Shānghán Lùn*.

Historically, there has been a school of thought that the *Nánjīng* was not really an attempt to clarify some of the challenging passages in the *Nèijīng*, but rather was an almost heretical departure from the Yellow Emperor's classic. Liao P'ing, mentioned above, was clearly of that opinion.[38] I have a very different point of view about that question. Although, as I have said, the idea of people having a Constitution that is separate from their Condition was not stated explicitly in the *Nèijīng*, there are some chapters that I believe are foundations on which the *Nánjīng* was able to establish a paradigm that includes Constitutional thinking. Here I am referring, for example, to those chapters devoted to the Rényíng Cùnkǒu (RYCK) method of pulse diagnosis and acupuncture treatment. While I will discuss this topic in a later chapter, I will point out here that this method specifically instructs the practitioner to choose treatment points from the distal portion of the implicated Meridians. It has been claimed that one departure the *Nánjīng* made from the *Nèijīng* was to focus treatment on these distal points rather than on the whole bodily trajectory of the Meridians to be treated. These chapters show that the seed of such an idea is clearly present in the *Nèijīng*. In using the RYCK method, it is always one of the Five Elements that is identified for treatment, using both its Yīn and Yáng coupled Meridians, and selecting distal Points for needling. My experience with extensive use of the RYCK approach is that the Element identified is that of the individual's Constitution, regardless of what Condition may be present. Thus the *Nèijīng*

38 This denigration of the *Nánjīng* continues today, with a number of widely respected authors challenging its authenticity. For example, Jean-Sylvain Prot posted this online 6/4/21, "The earliest commentaries (on the *Nèijīng*), like that of the *Nánjīng*, already have presented misunderstandings of the *Nèijīng*'s fundamental theories." Edward Neal more explicitly has written, "Simply stated, if a clinical practice can trace its theories and practices back to the central ideas found within the *Nèijīng*, it is a practice that exists within the tradition of Chinese medicine. Conversely, if it cannot do this, it does not exist within this tradition." Regarding the assignment of Elemental qualities associated with the běn shū (本輸) points, Neal wrote, "Specifically, these ideas appear to be built on mistaken descriptions within the *Nánjīng*." Journal of Chinese Medicine # 125, 2021, pp. 17-28. Now, if Neal is correct, the entirety of the schools of Five Element acupuncture, including those based in England, Japan and Korea, are simply wrong in aligning themselves with traditional Chinese medicine, and of course CCA must be even worse, and yet these styles of practice are all clinically effective, with uncountable testimonials. As I see it, such strictures as Neal and others propose prevent traditional acupuncture from ever developing new ideas based on experience and research, even though the authors of the *Nèijīng* came to their conclusions based on just such methods, and didn't even always agree with each other!

was presenting a Constitutional style of diagnosis and treatment without being explicitly aware of it. It took the *Nánjīng* to begin unraveling the implications of buried information, such as Constitutional orientation, in the *Nèijīng*, and develop a framework for discussing such ideas.

When I have been teaching in China, my translators have been troubled by the lack of a suitable Chinese equivalent to the terms Constitution and Condition. One version of yuán qì could be selected for Constitution, but that would omit the other version of yuán qì. After much discussion, I have decided to stay with the terminology I used in *The Compleat Acupuncturist*, tǐ zhì (體質). Not the least of my reasons for this choice is that it was the term used by Kuon Dowon, originator of Korean Constitutional Acupuncture (KCA), whose work was the inspiration for my long-standing interest in the concept of Constitution. My choice of a corresponding term for Condition has led me to zhuàng tài (狀態). These terms are mentioned merely as a guide for Chinese speaking readers and potential translators of my works. Otherwise there is no real need for non-Chinese practitioners and lay readers to bother with them.

To briefly summarize the main proposition of my previous publications, I might say that we are all born with a unique individual Constitution. There are however many different ways that we can classify Constitutions: They can be Yīn or Yáng as transmitted by Chinese medicine, Vata, Pitta or Kapha as transmitted by Indian medicine, or Soyangin, Taeyangin, Soeumin or Taeumin[39] as transmitted by Korean medicine, for example. This short list is only the beginning of the increasingly complex ways we can categorize people's Constitutions. I like to think of them as "lenses" through which we look at a specific aspect of anyone's Constitution. While this list of potential lenses is I believe infinite (no two individuals have identical Constitutions in every respect), the importance for acupuncture practitioners of Constitutional theory lies in its clinical utility. In that regard, I have found that dividing people

39 I translate these Korean terms as follows: Young Yáng, Old Yáng, Young Yīn and Old Yīn respectively. This English terminology helps to avoid confusion with the 6 Great Meridians of Yīn and Yáng (3 Yīn, 3 Yáng) whose terminology is partially overlapping with it in Chinese. Both the 4 part Korean "Sasang" and the Chinese 6 part "Liù Jīng" ideas are outgrowths of the basic teachings found in the *Yìjīng* (*Classic of Changes*), however they serve as an excellent illustration of the difference between using a 4 part versus a 6 part "lens".

into 20 Constitutional types is the highest level of classification that I have been able to apply with reproducible clinical results. The division of people into 20 types includes all the smaller groupings that I have alluded to, so that this approach is not in conflict with any of the fundamental ideas in Chinese, Indian or Korean medicine. As examples I can ascertain from the 20 type classification whether an individual is of yīn or yáng type, whether their Elemental resonance is with Wood, Fire, Earth, Metal or Water, and which of the 6 Great Meridians (Tàiyáng, Shăoyáng, Yángmíng, Tàiyīn, Shăoyīn or Juéyīn) is pivotal in their make-up. Thus 20 Constitutional type theory is a very informative lens for understanding the individuals we are called upon to treat. As a shorthand, I often refer to excess or deficiency of one of the 10 formed Organs/Officials (zàng fŭ) as a Constitutional type, but within each type the strengths and weaknesses of the other 9 Organs/Officials are each designated by Five Element (wŭ xíng) theory, so there is an additional relational aspect to Constitutional diagnosis. The same is true for Conditional diagnosis, which I will discuss further on. As a form of shorthand, I might call a Condition Spleen deficiency in someone with a Liver Excess Constitution. This is not a complete Conditional diagnosis, because the Condition includes both Spleen deficiency and Liver excess, each of which will show in the Conditional examination of the pulses.

Most styles of acupuncture practice are based on the patient's present Condition rather than their Constitution, so my approach (Constitutional Conditional Acupuncture, or CCA) is atypical in that respect. Five Element acupuncture (FEA), as formulated and transmitted by J.R. Worsley is also based on determining the individual's Constitution[40], however what I believe it lacks is any explanation of the laws governing the relationship between one's Constitution and Condition. My teacher, Kuon Dowon, was the person whose ideas started me on the path of investigating the connection between Constitution and Condition, and those ideas have been invaluable in the development of my present understanding of this relationship. Subsequent chapters will expound on this critical part of diagnosis, illustrating how even very complex diseases, such as cancer, can be more effectively treated by uncovering the connection

40 Worsley coined the term "causative factor" or CF to indicate the Element that was central to the treatment of each person's state of health, however he never referred to the CF as the Constitution. Many of his students have come to the conclusion that CF and Constitution are merely different names for the same aspect of diagnosis.

between Constitution and Condition. This connection is a phenomenon that follows natural law, as we should expect from any system of diagnosis and treatment that strives to merit the designation of being scientific.

One corollary of adopting CCA diagnosis is the idea that ultimately all acupuncture treatment should be based on the Spirit (神 shén), as this aspect of our existence is a fundamental and unchanging part of our original or true nature (Constitution)[41]. Remaining in touch with or returning to our true nature is thus the basis for maintaining health and curing disease. The following citations, supporting the fundamental role of Spirit-Essence/Constitution in acupuncture, are given in the order of their appearance in the *Nèijīng*, and commented upon in subsequent Chinese medical texts, to demonstrate that this idea is not merely a modern, New Age, interpretation of classical thought, and certainly not a radical refutation of *Nèijīng* teachings introduced by the *Nánjīng*, but rather was from the beginning an essential understanding underpinning the rationale for the various modalities of traditional Chinese medicine[42]:

> *Sù Wèn* 13, Discourse on Moving the Essence and Changing the Qì: "when (the people) in antiquity treated a disease, they simply moved the Essence and changed the qì. They were able to invoke the origin and (any disease) came to an end."

Wú Kūn's (吴昆) later commentary on this passage clarifies that Spirit is not to be separated from Essence: He wrote, "They altered the Essence Spirit and they transformed the qì in the depots (zàng)." 'To invoke the origin (祝由 zhù yóu)' has a number of implications. Invocation is certainly a reference to the Spiritual component of reality, while 'the origin' can reasonably be interpreted as referring to the

41 In the classics, the terms for Spirit, Essence and Essence-Spirit are often used interchangeably. This makes perfect sense in CCA, as both Spirit and Essence are combined in the formation of our Constitution.

42 I am indebted to Ted Kaptchuk, who gathered these quotations and used them as part of his presentation to the annual symposium of the Pacific College of Oriental Medicine in October of 2018. The English translations I cite are briefly excerpted passages from the published works of Paul Unschuld unless specified otherwise.

person's Constitution, consisting of their unique combination of shén and jīng.[43] Gāo Jìwǔ (高紀武) put it thusly, "when the physicians in ancient times treated a disease, they merely changed the thoughts and Spirits of the patient, (thereby) cutting the root of the disease to cure the disease." In CCA this root is reflective of the individual's Constitution. It is certainly possible to treat the Condition directly, but if such treatment does not harmonize and restore control to the Constitution, then the root of the disease has not been cut off. I should add here that treating the Constitution in CCA is not merely achieved by inserting acupuncture needles. The relationship between practitioner and patient allows the practitioner to help guide the patient in discovering which ways they have allowed themselves to deviate from their original nature (Constitution), and thus be in a position to "change their thoughts and Spirits" so as to move in the direction of health. In this regard, please see case history number 43, where the relationship I worked to establish with this patient appears to me to have been crucial to the rapid resolution of an aggressive cancer within a week of beginning acupuncture treatment.

> *Sù Wèn* 14, Discourse on Decoctions and Wines: "This applies to the way of needles and (pointed) stones. When the Essence Spirit fails to enter… a disease cannot be healed."

Huá Shòu (滑寿) commented, "it is clear why the nature of a disease in its initial phase appears as a weakening of the Essence Spirit which is unable to actively support the treatment…. The Essence Spirit weakens, the Spirit qì disperses and cannot provide any appropriate response to the treatment." By equating Essence Spirit with Constitution, a powerful case is made for why treating the Constitution in addition to the Condition is firmly rooted in the *Nèijīng*. What

43 What might we understand to be the meaning of "invoke the origin" (zhù yóu 祝由)? Kroll glosses yóu (由) as "deriving from" or "following from". He also cites its use as a time-word meaning "beginning, starting from, all along". Does it not make sense to equate this origin with the Constitution? The Constitution begins at conception, our growth and development follow from it, and it is the one thing that remains with us, unchanged all along the duration of our lives. So, from this *Sù Wèn* passage, we can hypothesize that the "Constitution" might indeed have played a role in the most ancient ways of healing in Chinese medicine. Invocation (zhù 祝) is reminiscent, for me, of the way Worsley taught "emotion testing" as a path toward identifying the CF (Constitution), and also as a way of initiating the healing process. There are many anecdotes in Chinese, and other works relating traditional Asian medical case histories, where emotional reactions induced by the practitioner (via invocation?), were used to initiate a cure. In fact, this approach is still practiced in China today, as discussed by Heiner Fruehauf in varied communications about his experiences with his teachers in China. The *Sù Wèn* text here asserts that in ancient times, the application of such invocation of the origin (Constitution) led to any disease coming to an end. In this light, CCA, by using needles as one way to invoke the Constitution, can be thought of as a modern fruit derived from this thousands of years old Classical text.

does it mean to say that the Essence Spirit weakens? The nature of a person's Constitution cannot change, but the ability of one's jīng shén to take charge of the Officials (zàng fŭ) can certainly be weakened by any of the etiological factors recognized in Chinese medicine. Most commonly, pathological factors interfere with this natural control by the Essence Spirit over the functioning of one or more of the zàng fŭ (or their associated Meridians). Thus I believe that to strengthen a person's Essence Spirit means to restore control of one's functions to one's original nature or Constitution. Contrary to the beliefs of many commentators on Chinese medicine, I believe that in many cases this can be accomplished by dispersing techniques rather than by supplementing techniques. It depends on the inherent tendencies of each Constitution to gravitate towards repletion or depletion when under uncontrollable stress from pathogenic factors.

> *Sù Wèn* 15, Jade Tablet Discourse on the Essentials: "The Spirit turns around, it does not go backwards. When it goes backwards, then it no (longer) turns around. In this case one has lost (one's vital) mechanism."

Most commentators interpret this as a reference to the clockwise circulation of the Five Elements (wŭ xíng) as metaphorically reflected in the orderly turning of the seasons. Wŭ xíng theory is of major importance in CCA, as it is involved in formulating all acupuncture treatment planning, so this passage is important to understand. The normal pattern of wŭ xíng relationships with respect to degrees of solidness and hollowness of each Element is fixed for each individual, and is encoded in their Constitution. When one of the Organs of any Element is functioning outside the bounds of its Constitutional situation, that means the Spirit has lost its original control. In such a case, it is understandable that the healthy 'vital mechanism' has been lost. Moreover, the practitioner must start planning treatment from the perspective of the Spirit or Constitution in order to regain a healthy 'vital mechanism' regardless of which Organ is manifesting symptoms. These symptoms are reflective of what I have been calling the Condition, and while it is perfectly appropriate to treat these Conditional symptoms, it should be done in such a way as to restore control to the individual's Original nature (Constitution). Zhāng Zhìcōng (張志聰) commented, "The fact that the periods of resting and dominance of the wŭ xíng in the qì of the human body can parallel, and do not oppose, the progression of the wŭ xíng in the four seasons depends on the correct adjustment of the

'Spirit power' ruling the interior of the human body. Hence phenomena indicating opposition, backward movement, or reverse circulation demonstrate that the 'Spirit power' has lost its proper function, which, in turn is a sign of life coming to an end."[44]

> *Sù Wèn* 26, Discourse on the Eight Cardinal (Turning Points) and on Spirit Brilliance.[45]

Unschuld, in his footnote to this chapter title, states that, "This treatise discusses piercing, but (its contents) have nothing to do with piercing", citing the opinion of previous commentators. Clearly there has historically been a divide between acupuncture practitioners who based their treatments on the fundamental role of the Spirit, and those who dismissed such a notion as mystical nonsense. This chapter can only be understood in relation to those just discussed which precede it. These chapters were undoubtedly focused on the importance of the Spirit in treatment. If we look at an excerpt from chapter 26, we read,

> "The Spirit, ah! the Spirit! The ears do not hear (it). When the (physician's) eyes are clear, his Heart is open and his mind goes ahead, he alone apprehends (it as if it were) clearly perceivable. But the mouth cannot speak (of it). Everyone looks out, (but) he alone sees (it). If one approaches it, it seems to be obscure, (but to him) alone it is obvious (as if it were) clearly displayed. As if the wind had blown away the clouds. Hence one speaks of a 'Spirit'".[46]

What I understand from this passage is that there is a clear link between the Heart/mind of the physician and his/her ability to perceive the Spirit. The

44 Kaptchuk summarizes the import of this chapter thusly, "for all needling to be reliable, one must first regulate the Spirit." "The placebo Effect and the Penetrating Divine Illumination", PCOM Symposium, San Diego, 10/28/2018.

45 It is commonly assumed that the Eight Cardinal Points refer to the cardinal and sub-cardinal compass directions. These are the origins of the 'eight winds' that play a role in *Sù Wèn* cosmology. I find that there is a hidden pun in this title. While the 'eight winds' can be the source of the various kinds of pathology in human beings (when they are "unseasonal"), it is also the 'wind' that can blow away the clouds (pathology) that obstructed the now illuminated Spirit (神明 shén mìng) of the chapter title.

46 Henry Lu has a different translation than Unschuld of the lines that immediately follow. While Unschuld does not even mention Spirit, Lu writes, "The Spirits are detected through taking pulses…." In fact, the text does not mention the pulse explicitly in this sentence, however Lu's interpretation is not without merit, as the whole chapter discusses both the practice of pulse diagnosis and the primacy of the Spirit.

key word I believe is "open" describing the state of the physician's Heart. This is a more or less direct parallel to the Daoist teachings about the correct state of the Heart as being empty.[47] Another comment I'd emphasize is that the text implies that the ability to perceive the Spirit is not based on every day normal use of the sensory faculties, but rather comes from training the senses to attend to the more subtle signals that would naturally be expected to emanate from the subtle (obscure to use Daoist imagery) realm of the Spirit. This idea was certainly emphasized in my training in Five Element style acupuncture by Worsley, and I find it equally true in developing skills in the pulse techniques in CCA. The pulse diagnosis of Constitution has been in the hidden or obscure realm of acupuncture for thousands of years, and only recently have methods been described for piercing this veil of obscurity.

> *Língshū* 8, To Consider the Spirit as the Foundation. The very first sentence in this chapter provides a summary of the ideas I have been developing in the preceding citations. "All norms of piercing (require one) to first of all consider the Spirit as the foundation."

The text goes on to state that life begins with the 'clashing of the two Essences, calling Spirit'. This new life has its unique identity now provided, and that is what I have been referring to as the Constitution. The text clearly indicates that all the norms (rules or laws) of acupuncture must first consider this Constitution (Essence Spirit), which is the fundamental premise of CCA. That is why the first task in CCA diagnosis is to properly identify the patient's Constitution. Only afterwards is it appropriate to focus one's attention on the present Condition (irrespective of their Western or Eastern medical diagnosis), and to proceed to the more complex questions of how this Condition came about, and how one might use acupuncture to help the patient's Condition revert back to conformity with their Constitution.

I might add one additional point here that has been inspired by Ted Kaptchuk's discussion of "Penetrating Divine Illumination (通神明 tōng shén mìng)" in a 2018 lecture[48]. Divine illumination is an alternate translation of the last

---

47 See *Huáinánzǐ*'s chapter 6 discussion, as analyzed in *Grasping the Donkey's Tail.*

48 . Kaptchuk, T., op. cit. I have shared this chapter with Ted, and he has graciously consented to my presentation of his thoughts on this topic.

two characters in the title of Sù Wèn chapter 26, which I rendered above, per Unschuld, as Spirit Brilliance. Without going into detail about Kaptchuk's thinking on this matter, which he has yet to publish, it is my impression that his basic contention is that healing comes about, at least in part, from the interpersonal relationship between practitioner and patient at the level which I have been calling Spirit. This concept is not confined to Chinese medicine as was pointed out to me by my teacher of Ayurvedic pulse diagnosis. It is believed in Ayurveda that when the practitioner concentrates on the patient's pulse signals, the awareness of both participants is drawn to the same point of focus on the etheric level (Spirit), and that this shared awareness in itself is enough to initiate a healing response in the patient. Kaptchuk cited several passages from pre-modern Oriental medical texts that essentially make an equivalent claim:

> "Feel better before taking medicine is the most direct method" from Xǔ Shūwēi (許叔微 1132 CE), and "By using this method the process begins even before the medicine reaches the mouth" from *Precious Mirror of Oriental Medicine* (Dongui Bogam) by Heo Jun, originally published in 1613 CE during the Joseon Dynasty of Korea.[49]

Penetrating Divine Illumination was first mentioned in the *Sù Wèn* in chapter 3, which presented a disquisition by Huángdì (the Yellow Emperor) on how the ancients were able to maintain health and longevity. Here is my translation of this passage, "The sages of former times transmitted Essence Spirit, complying (with) the Qi of Heaven, and so (used) Penetrating Divine Illumination."[50] Wáng Bīng, whose edition of the *Sù Wèn* comprises our oldest extant copy, noted that "To be able to transmit Essence and Spirit, only the sages who have attained the Way are able to do that." At least to Wáng Bīng, Penetrating Divine Illumination represented the epitome of the art of healing. I beg to differ with Wáng Bīng on the specifics of this assertion. I believe Penetrating Divine Illumination is being activated every time

49 There is a fictionalized Netflix series from Korea (with English sub-titles), *Live Up To Your Name*, that revolves around this Dr. Heo Jun and other historical Oriental medical figures. The emphasis on Spirit, and the importance of the doctor/patient relationship is striking in its presentation of the "miracles" that can occur only when the doctor is capable of "changing the patient's thoughts and Spirits". It is also the most impressive depiction of acupuncture in a cinematic work that I have encountered.

50 故聖人傳精神,服天氣,而通神明.

a practitioner connects with their patient in a way that reaches to the root of their problems. To the degree that this root is discovered and touched in such a way as to restore its proper role in guiding one's life, healing will occur and Penetrating Divine Illumination will have taken place. This etheric method was addressed again in *Sù Wèn* chapter 13, which I have already discussed. As part of a longer presentation on the most ancient methods of healing, Qí Bó mentions Jiù Dàijì (a legendary figure) as the originator of perhaps the most profound methods of healing in Chinese medicine, both of which form an important aspect of how I understand and practice CCA.[51] The first of Jiù's contributions was to establish the relationship between individuals' complexions and their pulses. Complexion in this context should be interpreted as color, because this passage about Jiù specifically enumerates the Five Elements (wǔ xíng) that are to be correlated with the pulse. This is a most interesting passage for me because it ties together the color aspect of Worsley's CF diagnosis with the pulse focus of CCA, both of which systems are essentially based on Constitutional diagnosis. This canonical passage fails to inform us, however, about just how to examine and classify colors and pulses, so it has been left up to later generations of practitioners to discover, comment on, and transmit the specifics of these diagnostic methods. The second contribution by Jiù was to tie together this color/pulse diagnostic schema with the methodology of Penetrating Divine Illumination, reconstructing the nature of the highest art of healing practice prior to the development of both herbal medicine and acupuncture/moxibustion.

There is one final mention of Penetrating Divine Illumination in the *Sù Wèn* in chapter 69. I'll footnote my translation (amateur level though it is)[52], but I include it because it brings in what I have proposed as the fundamental axiom of Chinese medicine in *Grasping the Donkey's Tail*, i.e. resonance or correspondence theory, mentioned here right in the middle of the passage

51 上古使僦貸季，理色脈而通神明，合之金木水火土四時八風六合.... My translation: "High antiquity employed Jiù Dàijì, to put in order the colors, pulses and penetrating divine illumination, joining them with metal, wood, water, fire and earth, the four seasons, the eight winds and the six junctions...."

52 善言天者，必應於人，善言古者，必驗於今，善言氣者，必彰於物，善言應者，同天地之化，善言化言變者，通神明之理 "Those who can speak about heaven, must know the correspondence to man, those who can speak about the ancients, must know about the present, those who can speak about qì, must know about material things, those who can speak about correspondence (yìng), they are like the transformations (huà) of heaven and earth, those who can speak about transformation and change (biàn), know the inner essence of penetrating divine illumination."

as yìng 應. The more common compound term for resonance, gănyìng 感應, was not used in the *Nèijīng*, but yìng by itself carries the same connotation. Thus we can see that the concepts of Constitution, Essence Spirit, and resonance are all tied together in the practice of pulse diagnosis, and find one method of practical expression via Penetrating Divine Illumination. Lest the reader misunderstand me, I want to make it perfectly clear that I believe that the ethereal nature of Penetrating Divine Illumination is but one part of the rationale for the practical efficacy of CCA. The other part, which forms the major component of my writings about CCA, consists of the nuts and bolts of the different techniques of pulse diagnosis and the various methods of acupuncture treatment that depend on factual knowledge gained from persistent study, practice and experimentation. These form the core of this application of Oriental medical science. Each Point, Meridian, pulse finding and Official (zàngfŭ) has a unique individualized nature that CCA aims to explore and utilize in the practice of healing through acupuncture. On the other hand, I insist on the importance of the role played by the relationship between practitioner and patient in healing through acupuncture practice. Often times my patients have told me that the questions I asked them and the suggestions I gave them as homework exercises was what impressed them the most, and changed the way they looked at their issues in life, including but not limited to their struggles with illness. I do not claim that what I use is identical to the Penetrating Divine Illumination of the ancient sages, but neither do I deny its possible role in my work.

CHAPTER 2

CCA Imagery in the Theory of Acupuncture and its Mechanisms

Without doubt, the traditional understanding of how acupuncture works is based on the concept of qì.[53] Sinologists, historians, philosophers, medical practitioners and other scholars have suggested numerous English translations for the term qì, and yet there is no universally accepted standard translation, and I believe there can be none because the term qì has multiple meanings depending on context. For this reason I have mostly left qì untranslated in this treatise. However, there are times when I believe readers can better understand the propositions of CCA by choosing a more evocative English term. George Soulié de Morant, who initiated the first lineage of acupuncture practitioners in the West, used the term energy (l'énergie) as a translation of qì, and it has been widely adopted by practitioners up to the present time, if not by Sinologists. Although many scholars are critical of his choice, I find it to be quite useful, as long as one recognizes that this "energy" is not one of the concrete forms recognized in Western scientific disciplines. That said, there is a useful analogy I shall present between the thermodynamic kinetic energy we experience as heat, and the behavior of qì (Fig.4).

53 There are, of course, non-traditional ideas about the mechanisms underlying the effects of acupuncture, but that is not germane to this presentation.

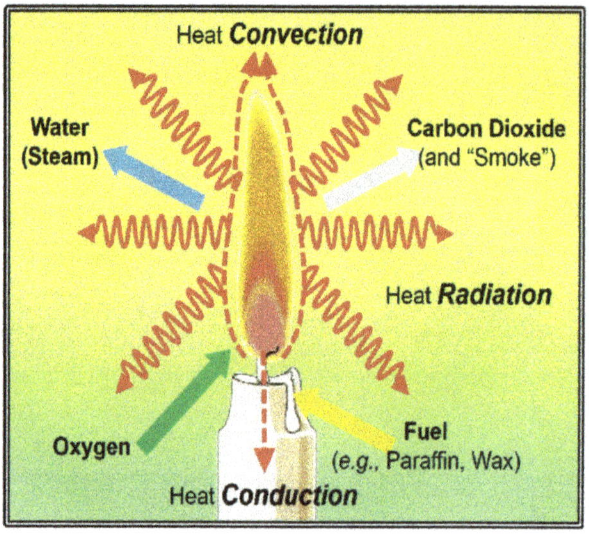

Figure 4: The Thermodynamics Model [54]

Physics describes three ways in which heat is distributed, for example by a radiator in your office or a simple candle on your dining room table. The first is called conduction (Fig. 5), which takes place by direct contact either with the heat source itself, or with a highly efficient intermediary means such as a metallic bar or wire. Early on, we all learn not to stick our hand in a flame, and also to be careful picking up a metal spoon when it is sitting in a hot bowl of soup that just arrived at our table. The second is called convection, which takes place more diffusely when a mobile medium such as a liquid or gas conveys a stream of heat in a particular direction (Fig. 6). Convection ovens use fans to create such a focused flow of hot air. Finally heat can be spread by radiation, a process that can occur even in a vacuum (picture our sun), and which is transmitted in all directions of space unless artificially restricted. I would like to offer the following analogy for how to understand the way in which qì can account for the various mechanisms that form the basis of CCA theory.

54 Source: Henry Norman. The candle is an apt image for depicting the concept of qì and of its related technical terms, as expressed by Rhonda Chang: "主 zhǔ means a master, a host, main, owner. The older form of the character 主 is a pictograph of a lighted candle. A candle flame gives light to the darkness. Light is yang, darkness is yin, yang is the power source, without yang, yin is inactive, so 主 is the main and master of activity and life." Yinyang Wuxing Yi Facebook Group post, 6/6/21. Ms. Chang is clearly associating the characters for qì and zhǔ as a means for understanding yīn and yáng, as well as the more complex concepts they engender.

Figure 5: Conductive Circulation is Linear[55]

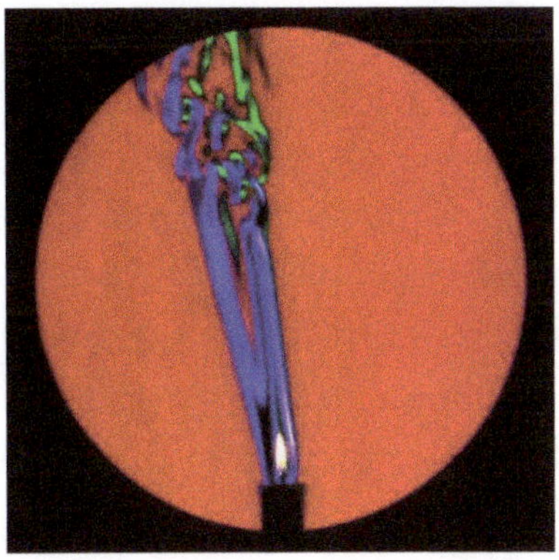

Figure 6: Convective Circulation is Semi-Linear[56]

Figure 7: Radiant Circulation is Non-Linear[57]

55 Source: Wikimedia Commons. Power lines are highly efficient energy conductors.

56 Source: Wikimedia Commons. The thermal currents from the candle are highlighted by this temperature/color image display.

57 Source: Wikimedia Commons. Because radiant energy is emitted in all directions, it can have the widest sphere of influence.

Some of the effects of acupuncture have been shown to be mediated by the human nervous system. Injecting local anesthetics into the tissue where acupuncture needles are inserted, or into the nearby nerve tissues can block the pain relieving effect of acupuncture in experimental human and animal subjects. What are nerve tissues but specialized intermediary structures serving the function of "conducting" forms of energy, whether electrical, chemical or simply informational? They have a well-defined anatomical structure that allows them to function as energy conductors. Much research has been done to investigate whether or not the traditional acupuncture Meridians follow neural pathways, and I believe it has been established that nerves and Meridians follow separate trajectories. In fact, there has as yet been no established anatomical tissue that matches the Meridian pathways, although there are speculations that the fluids in the more amorphous interstitial cleavage planes between anatomical structures might correlate with the traditional Meridian trajectories. Such a hypothesis is evocative of the convection mechanism in which the directed flow of energy through a fluid or gaseous medium transmits heat. I like this analogy because it can explain much of the semi-linear relationships employed as part of Meridian strategies in acupuncture. A nice example would be the use of Entry/Exit Point treatments to unblock the connection between sequential Meridians. The use of Points at one end of a Meridian to treat problems at the other end is another example of this semi-linear mechanism, which I will call a kind of convection effect. Since the Principal Meridians are believed to extend to their associated Organs, even the use of Points on a Meridian to treat problems of its Organ can be understood as "convection" effects. These two mechanisms, conduction (neural) and convection (fluids in cleavage planes), have really been the only ones seriously studied by researchers intent on discovering how acupuncture works. Unfortunately, they've left out the most important mechanism by which qì, the subtlest of energies, is operative in effectuating the majority of phenomena that constitute the traditional methods of acupuncture described in the classical texts, and that also form the rationale for much of CCA. I am referring to the radiant transmission of energy (qì), which can occur even in a vacuum and has no necessary anatomical substrate. In the previous chapter I mentioned resonance (gǎnyìng 感應), which shares this delocalized attribute with other radiant phenomena. In Five Element (wǔ xíng) theory, as it is applied in acupuncture, there are no linear trajectories to explain how the

Liver controls the Spleen/Pancreas or how the Urinary Bladder supports the Gall Bladder, for example. I believe this lack of a physical trajectory in radiant transmission of qì is the main reason that Five Element theory has been the most heavily critiqued part of traditional Oriental medicine. However, the fact is that the essential methods of effective acupuncture treatment as a holistic approach to healthcare simply cannot do without it (Fig. 8). This thesis about radiant effects is true, not only for CCA, but for all the classically based styles of practice I have encountered, reminding me once again to emphasize that I have great respect for the other classical traditions and lineages. CCA is but one branch of a vital organic tree, a field that I call Traditional Oriental Medicine.

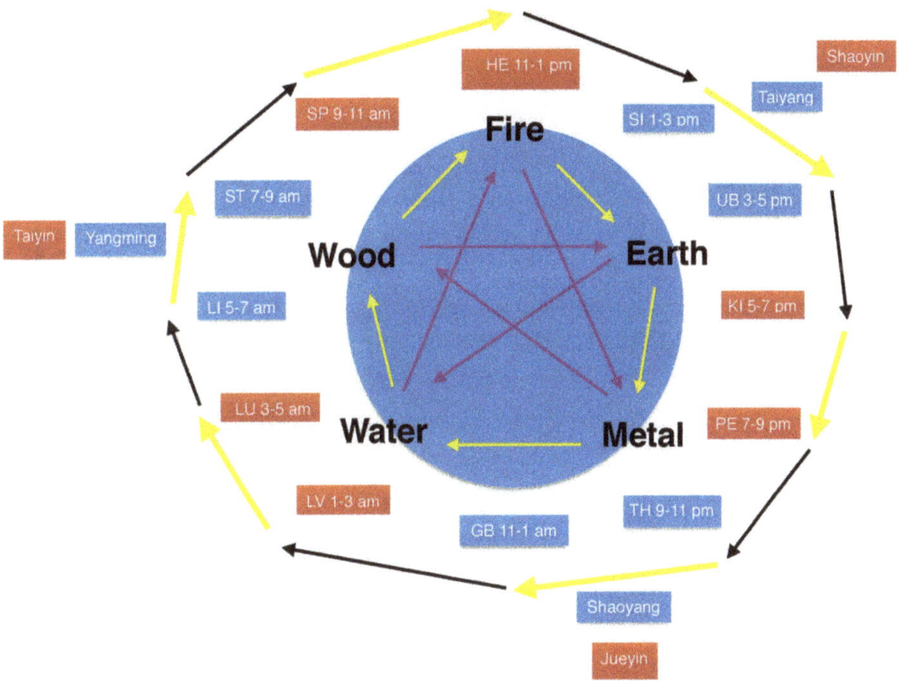

Figure 8: Meridian based (peripheral) & Resonance based (central) relationships [58]

58 Source: Adaptation by Peter Eckman.

I wish to reiterate that the analogy of (heat) energy to qì is simply that: an analogy. Oriental medicine is inherently based on an idealistic philosophical set of beliefs, while Western medicine (and its definition of science in general) is inherently based on a materialistic philosophical set of beliefs. For this reason alone, it is not possible to fairly reduce propositions from one of these systems of healthcare to assessment based on knowledge established in the other's domain. Neither system's knowledge base contains more truth than the other's. They are complementary approaches where each has much to offer individuals with challenges to their health.

CHAPTER 3

Expansion of the Constitutional Conditional Paradigm

Figure 9: Our Donkey puzzling over the Nèijīng [59]

59 Source: Neal White. The curious reader might wonder, "Why is Donkey up a tree?" Perhaps because some mischievous primates were monkeying around with his precious scroll, and he had to break with conventional norms in order to retrieve it? Or, he's "up a tree" as in being "stumped about the meaning" of something in the classical text. Or, he simply wants a quiet place to ponder.

I recognize that the new material to be introduced in this text is rather dense, including aspects of scholarly research into the Classics, technical minutia of how to examine many different pulse manifestations, protocols for constructing acupuncture treatments, and much more. To encourage readers to stay the course, here is a short list of some of the simpler issues I will be covering, and I hope that at least some part of my discussion of them will resolve questions that both beginners and seasoned practitioners typically find themselves wondering about.

1. What are the natural standards for locating the radial pulses of the 12 Organs/Officials? (Not what you learned in school.)
2. Does the laterality of treatment points matter, and how can the optimal choice be made?
3. Practically speaking, how does the pulse indicate whether the hand or foot branch of one of the Six Great Meridians is imbalanced?
4. How can one best distinguish an excess from a deficient pulse?
5. Can one detect Akabane imbalances from the pulse?
6. Just what is Ministerial Fire (in the Classics), and why is Fire included as one of the Six Climatic Factors?
7. How can one clearly distinguish problems due to malfunctions of the Extraordinary Meridians from those of the Regular Meridians?

Of course much more complicated issues will need to be addressed in addition, but I hope I have sparked enough curiosity to encourage practitioners and lay folk of every level of experience to read on. For the more experienced, and possibly more adventurous, there is a unique interpretation of the circulation of yíng and wèi qì, demystifying the very obscure passages in both the *Língshū* and *Nánjīng* on this topic. A new approach to understanding, diagnosing and treating the Extraordinary Meridians, not as isolated Vessels, but as forming an integrated system, is presented. Finally, my most important clinical discovery to date: how to use CCA to diagnose and treat the many types of cancer, not just to alleviate the symptoms or the side effects of conventional biomedical treatments, is explained in detail.

So let me start with a question that I have tried to answer by reflecting on a passage in the *Língshū* that puzzled me. From an acupuncturist's point of view, what are the different types of etiology and pathology that can be identified by examination and addressed by acupuncture treatment? Traditionally, etiology in Chinese medicine has been divided into three classes: exogenous factors (weather), endogenous factors (emotions) and miscellaneous factors (lifestyle, diet, trauma, etc.). Pathology was never systematized so clearly, as far as I can tell, however various schools of thought have divided problems into those of the Meridians or Channels (jīngmài) and those of the Organs or Officials (zàngfǔ). The former group can be further subdivided into the various types of Meridians, i.e. Principal (zhèngjīng), Connecting (luòmài), Extraordinary (qíjīng bāmài), Divergent (jīngbié) and Sinew (jīngjīn) Channels. While all these subjects have been discussed in the classics, practitioners of different styles of acupuncture have freely chosen to focus attention on one or more of these groupings. Since CCA is a unique approach which is still in the developmental stage, I have tried to find guidance in the classics to help with narrowing down and systematizing the appropriate types of treatment for each patient's presentation, while still relying on classical foundations. One idea that may prove useful can be found in *Língshū* 9, and I have tried to understand its transmission of teachings in the following passage[60]:

> "The scope of all piercings encompasses three categories of piercings to make the grain qì [61] arrive.

60 凡刺之屬，三刺至穀氣，邪僻妄合，陰陽易居，逆順相反，沉浮異處，四時不得，稽留淫泆須鍼而去 from *Língshū* 9.

61 *Língshū* 9 specifies that by "grain qì", it is referring to the "proper qì". 所謂穀氣至者，已補而實，已寫而虛，故以知穀氣至也. "When it says, the grain qì have arrived (it is like this): When acupuncture is successful at draining excesses or supplementing depletions, then the qì has returned to a normal state. This is called the arrival of grain qì" (author's translation). Some authors claim that acupuncture (as opposed to herbal medicine) can only disperse, not supplement, but that is not the position of the classical texts. Moreover, the passage quoted here clearly states that acupuncture is employed for the purpose of facilitating the arrival of the normal (grain) qì. Thus, as a corollary, acupuncture is not focused on the different forms of perverse qì, but simply on returning the circulation of the proper, or normal qì. Such an interpretation of the classical understanding of acupuncture's goal is wholly consonant with the approach underlying CCA.

> (1) Whether evil (qì), having united (with proper qi), and if so, yīn and yáng have changed their places. [62]
> (2) Whether movements, against or with the flow, have turned into their opposites, and if so, what is in the depth and what floats at the surface have changed their locations.[63]
> (3) Whether adaptations to the four seasons failed to occur, and so (evil qì) were able to settle internally, and if so they were present excessively and have spilled over.[64] All this requires needles in order to be eliminated."

Why have I presented my (rather unorthodox) footnoted interpretation of this passage from the *Língshū*? As with all situations dealt with by CCA, the three mechanisms mentioned here present with different pulse findings, as

---

62 (1) Yīn and Yáng Meridians within a given Element are separating (moving apart) rather than uniting, resulting in a RYCK imbalance, manifesting as Yáng Excess with Yīn Deficiency (Yáng Condition) or Yīn Excess with Yáng Deficiency (Yīn Condition). Notice that in this circumstance, it is not a question of "with or against the flow," but rather, is either Excess or Deficiency of Yīn or Yáng present at the moment (the Condition). To treat these Conditions per the *Língshū*, one Meridian is supplemented while its coupled Meridian of the same Element is dispersed. The type of evil qì involved was not a factor in the *Língshū*'s protocols for addressing these types of RYCK imbalances. The healthy Meridians and their Organs, following treatment, apparently neutralize the evil qì (author's interpretation). The RYCK doctrine stipulates that if one Meridian of a coupled pair in the same Element is in Excess, then the coupled Meridian must be in Deficiency. While this situation is often the case, as illustrated by the efficacy of the classical RYCK treatment protocols, in my experience it is not always true. Both Korean Constitutional Acupuncture and Worsley's Five Element Acupuncture approaches are based on treating coupled pairs of Meridians with the same polarity of intervention. Both of the coupled Meridians may potentially be in need of either tonification or of dispersion. See also Simonov's contention below in category 3, that Yīn Deficiency (e.g. LU Deficiency) is accompanied by Yáng Deficiency (i.e. LI Deficiency). That is because the Yīn Organ is Yīn Deficient while the Yáng Organ is Yáng Deficient. My experience indicates that there is no absolute rule for determining the polarity of coupled Meridians of the same Element. Only a complete pulse diagnosis, such as the examinations codified in CCA, will allow a correct diagnosis in each individual case. Because the Yīn and Yáng Meridians within the same Element are treated with opposite polarities of needling, I believe this situation is a reasonable interpretation of the text's statement that Yīn and Yáng have changed places.

63 (2) Meridian/Organ imbalances between Elements as described in *Nánjīng* 5. These represent excesses and deficiencies which must be dispersed or supplemented, but which reveal themselves by abnormal depths of the radial pulses at the affected locations (more floating or sinking than is proper). They also display RYCK abnormalities (author's observation), and should be expected to present that way, as they are described in this *Língshū* chapter on the RYCK methodology.

64 (3) I believe we must take the reference to the 4 seasons symbolically, similarly to the way the text uses terms like "wood" to refer to Liver, East, green, anger, etc. The seasons are examples of the changing environment to which people are exposed over time, and this can include both smaller time units (minutes, for example) and also years (for example), not just seasons. Chapter 7 will elaborate on Simonov's application of this aspect of physiology with regard to the effect of solar radiation on the Meridians and Organs, but a significant portion of the *Língshū* is devoted to establishing a basis for the later codification of zǐ wǔ liú zhù (子午流注) theory. The grammar in the above passage from the *Língshū* does not clearly indicate which of categories 1, 2 and 3 are due to evil qì having united with normal qì. Perhaps each of these categories results from this situation. Evil qì may be the result of penetration from the exterior, but it may also be an interiorly generated situation if normal qì becomes damaged beyond the organism's ability to adapt and recover its original state. What differs between the 3 is the change in the individual's physiological state and presumably its reflection in the pulses.

will be explained further along. Additionally, the three mechanisms I'm proposing all represent classical pathological conditions recognized in Chinese medicine as follows: yīnyáng imbalances, Five Element imbalances and loss of harmony between the macrocosm and the individual's microcosm. These pathological Conditions are all deviations away from the normal physiological state. Are these Conditions caused by pathogenic factors, or are they the Conditions that allow pathogens to penetrate and obstruct normal functions? That's a "chicken or egg - which came first?" kind of question. As long as normal physiology can be restored, it doesn't matter in my opinion. Any of these mechanisms could potentially lead to obstruction by retained pathogenic factors. Herbal medicine may be superior to acupuncture in directly acting on these pathogens, while it is my suspicion that acupuncture may be superior in reestablishing the normal physiological functions. At least the *Nèijīng* states that "All this (*these three types of pathology*) requires needles in order to be eliminated." I will provide explanations of my understanding of the diagnosis and treatment methods for each of these three types of pathology, which I have found to be most effective. Treating any one of the three may at times correct all of these pathological situations, but it may also be necessary to combine them in other clinical situations where two or all three are present. The possibility of gradually expanding the range of diagnostic and treatment options in CCA depends on careful study of the pulse phenomena found to reveal themselves in each case. While I have initiated the CCA approach, it is ripe for further development by any observant practitioner.

There is one fundamental tenet of CCA that I wish to emphasize here, as it speaks to the issue of pathology and pathomechanisms. CCA posits a unique "constellational" approach to these topics in which Organs and Meridians play a role that is not simply governed by their intrinsic functions, but additionally play roles that are governed by their relationships to the Constitutional Organ. This revolutionary "constellational" thesis, borrowed from Korean Constitutional Acupuncture (KCA), posits an analysis of pathology that takes some aspects of knowledge from both Eastern and Western systems of medicine. Experience shows that specific types of pathology identified with disease names in Western medicine may share a common constellational pathomechanism in CCA. The most startling example of this is in diagnosing and creating effective methods for treating the disease labeled as cancer.[65] I will discuss this topic in detail later, but it is enough to state here that all cases of cancer I have encountered appear to share the same constellational pathomechanism, whose treatment differs only according to the individual's Constitution, not the type of cancer nor the organs and tissues involved. However this approach is not limited to cancer. Many malfunctions of the human organism are well treated and understood by discovering the common constellational pathomechanism they share. The treatment of pain is possibly the most common use of acupuncture, and constellational CCA treatment of pain has been remarkably effective in my practice. As far as I have been able to determine, traditional Oriental medical theory has not previously investigated this constellational analysis of pathomechanisms, but has rather focused on the specific responsibilities of each Organ and Meridian to simply carry out its intrinsic functions, regardless of the individual's Constitution. I regard this issue as one of the most promising areas for major breakthroughs in healthcare if future investigators learn CCA theory and practice. The outcome will be a much closer coordination between Eastern and Western medicines.

There are some other fundamental aspects of Chinese medicine that I have previously more or less ignored (not incorporated in CCA), and which I still view as a subject in need of further research. The most prominent of these is the importance, or lack of importance, in identifying the specific type of pathogenic factor (xié qì) responsible for an individual's illness. Several reasons underlie my original decision not to address this aspect of pathology. Let me start by noting that my main inspiration for developing CCA, the work of Kuon Dowon, did not include such considerations in his pulse based system of Korean Constitutional Acupuncture (KCA). Traditionally, the basis for the pulse diagnosis of pathogenic factors in Chinese medicine has largely depended on the identification of pulse quality abnormalities, while CCA is based more on both pulse quantities, as reflected in the size (and solidity or hollowness) of the pulse wave, and in the pulse locations along the radial artery, with a specific focus on the depth at which the pulse is widest at each location. In addition to the complexity and difficulty of differentiating 28 pulse qualities, I have felt that something was missing from a model of

65 Eckman, P. "The Diagnosis and Treatment of Cancer with Constitutional Conditional Acupuncture" Journal of Chinese Medicine #125, 2021. Pps. 3-16.

pathogenesis that emphasized exogenous pathogenic factors (the liù xié, 六邪) over the endogenous ones. In Chinese medical history, I have no doubt that exogenous factors such as wind, cold and dampness, for example, were more frequently able to disrupt normal physiological functions, due to the less than adequate material resources of the largely rural agrarian Chinese population in earlier ages. However, as was first pointed out to me by J.R. Worsley, nowadays most people have adequate food, shelter and clothing, at least in the developed countries where I have been a practitioner, so that the illnesses encountered in the present times are much more often due to endogenous factors, i.e. psycho-emotional disturbances. For now I propose this thesis: that restoring the patient's Condition to the pattern of their Constitution should be effective treatment for dealing with most, if not all, abnormalities due to the disruption by xié qì. Pulse quality diagnosis may be best able to differentiate these various types of xié qì, and yet still not be necessary for their treatment by acupuncture. Properly functioning Organs and Meridians may be inherently capable of successfully eliminating xié qì. A new pulse method of detecting abnormalities due to retained xié qì, and the complementary acupuncture protocols for eliminating it will be introduced in Chapter 10. It is certainly possible that some herbal medicines could work at this level in a different manner than acupuncture, introducing new substances into the human body that might have a direct effect on the various types of xié qì, as has already been mentioned, rather than acting indirectly by first rebalancing the Organs and Meridians. These ideas are only hypothetical, and it should be recognized that the *Língshū* passage cited above explicitly states that retained pathogens (xié qì) require acupuncture treatment (piercing) for their resolution, but there is no notion that needles target xié qì directly. If the normal (grain) qì is induced to arrive, acupuncture will have initiated the healing process and allowed the organism to resolve the pathogenic qì disruption. Thus the question of whether or not identification of specific pathogens by pulse quality diagnosis is necessary in order to formulate effective treatments to resolve these pathogenic factors remains unanswered, in my opinion, and for the present is not included in CCA methodology. However, there is a difference between identifying pathological pulse qualities on the one hand, and devising strategies for detecting where pathogens might be disrupting normal functions on the other hand. In fact, it is my present understanding that pulse qualities may also express states that are not issues related to xié qì at all. Among these issues are the determination of whether the Hand or Foot branch of a Great Meridian is expressing itself. Additionally, Chapter 8 on the Luò Vessels will discuss recent CCA findings that bear on the issue of diagnosing and treating "stuck" or "latent" pathogens, which in the words of the *Língshū*, "have spilled over", and thus disrupted the normal relationship of microcosm to macrocosm.

I would like to also address the topic of pathogenic factors by examining Worsley's teachings in 5 Element acupuncture (FEA), as they leave some questions unresolved. His teachings contradict common observations by the vast majority of acupuncture practitioners of all other traditions. In FEA it is taught that the illnesses provoked by pathogenic factors often present as Aggressive Energy (AE), which can only be removed by a specific method of treating the Back Shū Points of the zàng Organs. The method he used for accomplishing this he described as "draining Aggressive Energy" (AE), which I believe represents the translation of 泄邪氣 (xiè xié qì) from Chinese to French to English. The specific protocol Worsley taught involves shallow needling of the Back Shū Points of the yīn Organs (zàng) until the skin around the needles shows an absence of erythema (redness). Lavier had passed aspects of this methodology on to Worsley, but had later written about other techniques for expelling pathogenic energy. Although Worsley claimed to have taught the classical approach to acupuncture dating back thousands of years, there is no obvious mention of this Back Shū Point technique in the *Nèijīng*. To add even more confusion to the picture, the vast majority of acupuncturists throughout the world have pursued their practice for ages without employing this protocol for draining AE, and there is no evidence that such an approach has been injurious. Quite the contrary, acupuncture without draining AE has been recognized worldwide for its beneficial effects in numerous conditions, even by the World Health Organization. Worsley claimed that between 6 and 10 percent of all chronic illnesses were due to AE, and that if left untreated by this specific protocol, would invariably lead to death within one year.[66] In my own practice of over 45 years I have mostly treated patients with chronic illnesses, and they routinely live for many decades without having had specific treatment for AE. It should

66 Worsley, J. Class notes of Peter Eckman

be noted that the *Nèijīng* contains many passages warning about the rather immanent fatal outcome associated with various pulse diagnostic findings, but to my knowledge none of these have been verified by prospective studies, nor are they promulgated by contemporary teachers and practitioners. Thus Worsley's hyperbole in this regard was perhaps merely an unconscious nod to a traditional Chinese manner of exposition.

This problem has led me to reexamine the origin of Worsley's teachings about AE. As previously reported in *In the Footsteps of the Yellow Emperor*, Worsley was introduced to the concept of AE by Jacques Lavier at a class he attended in 1964. As mentioned, Lavier subsequently wrote several books that discussed AE, a term he coined in French, in which he proposed various alternate methods of expelling AE, however at no time did he propose a fatal outcome if AE was left untreated. What he did claim was that it was preferable to treat AE prior to balancing the Five Elements (a doctrine shared by FEA)[67], however he also wrote that the very first phase of acupuncture treatment should be purely symptomatic, quite the opposite of Worsley's teachings. Following this line of inquiry, I might propose that Worsley's elaboration of Lavier's initial remarks about AE might have come from his own early experiences as a practitioner, as I have never encountered another tradition than Lavier's that would have led Worsley to expand AE teachings in the way that he did. However, another possibility is that this was something he learned from one of his Chinese teachers, Hsiu Yangchai. Hsiu, a Taiwanese acupuncturist, was Worsley's source for the Seven Dragons for Seven Devils treatment of Spirit Possession (purportedly a Táng dynasty protocol), and for the protocol for treating Husband/Wife imbalances. Both of these conditions were also described by Worsley as being fatal if left untreated, and were the other two blocks that needed to be cleared before initiating Five Element treatment, along with draining AE. Worsley might conceivably have grafted the fatality teachings of Hsiu onto the AE teachings of Lavier. Given these variant teachings of Worsley and Lavier, I believe we are on safe ground to propose

67 Dispelling pathogenic factors (xié qì) before balancing the normal ones (zhèng qì) is supported by this quote from *Sù Wèn* 20, "One must first remove (an evil from) the blood vessels, and only then one harmonizes any (excess and deficiency)." 實則寫之，虛則補之. 必先去其血脈而後調之. The character 調 (tiáo or diào) could be translated as harmonize, as I have done, but could also be translated as transfer, which reasonably could be interpreted as validating the Worsley method of transferring qì from where it is excess to where it is deficient.

that xié qì, or AE, is not necessarily a particularly fatal type of pathology if left untreated. At the same time, restoration of health might be hastened if a reliable method of identifying and treating xié qì by acupuncture could be ascertained. The passage I quoted from *Língshū* 9 could reasonably be interpreted as proposing three general methods for eliminating xié qì, with the AE protocol being merely one way of addressing mechanism number 3 (failure of adaptation leading to overflow and retention of xié qì.

Does the *Nèijīng* contribute anything to our understanding of the basis for Worsley's AE protocol? Perhaps *Ling Shu* chapter 51, titled "The Back Shu Points" is a good place to start. This chapter only discusses the Back Shu Points of the 5 zàng Organs. In one of it's more controversial sentences, this chapter states, with regard to these Points, "One may cauterize, but one must not pierce them."[68] Now cauterization by moxabustion is a very superficial treatment comparable to the extremely shallow needling used in treating AE. Both techniques lead to local skin erythema (redness), reflecting the body's diversion of qì and Blood flow to the surface. It is commonly believed that the surface of the body is richly supplied with Defensive Energy (wèi qì, 衛氣), and that as such it is a primary site where xié qì can either penetrate the body or be expelled from it, depending on the relative strengths of xié qì and wèi qì. Chapter 51 seems to thus parallel some ideas embodied in the teachings about AE. For one thing, only the Back Shū Points of the zàng Organs are dealt with, and secondly, only the very surface layer of the body is specified as appropriate to treat. Perhaps the biggest objection to seeing these as validation supporting concepts is the contraindication of needling these Points in *Língshū* 51. Although not as often cited, there is an early mention of the Back Shū Points in *Su Wen* chapter 24 (血氣形志), which appears to be the first appearance of the term Back Shū Points in that text. Similar to the chapter in the *Língshū*, *Su Wen* chapter 24 also talks only about the Back Shū Points of the 5 zàng Organs. However, it says these points <u>may be cauterized and pierced</u>.[69] These discrepancies within the *Nèijīng* are not unusual, as the text is believed to have been a compilation by a number of different authors, but it begins to look like there is at least a possibility that Worsley's AE teachings may have originated in, or at least have been inspired by, *Nèijīng*

68 灸之則可刺之則不可.

69 是謂五藏之腧，灸刺之度也.

concepts. My present belief is that using Worsley's Back Shū Point protocol as an early step in acupuncture treatment is certainly a reasonably safe procedure, and in some cases can lead to dramatic improvements in a patient's health. I would interpret it as one method for adjusting the balance between yíng and wèi qì, a topic to be covered in Chapter 7. Another chapter will also present my views on diagnosing and treating cancer, which is certainly a condition where AE (or H/W or Possession) might be expected to be close to universally present, according to Worsley's teachings. In my approach to treating cancer, I have used a combination of several CCA techniques, none of which involved AE, H/W or Possession treatments, and I have seen remarkable improvements in the health and longevity of these patients with cancers that run the gamut from stage 1 (in situ lesions) to stage 4 (metastatic lesions throughout the body with poor 5 year survival rates). For now I will simply state that it may be possible in other styles of acupuncture practice to directly treat xié qì, but that such direct approaches are not necessary components of treatment formulation, even in cases of lethal diseases, and that CCA has been successful without incorporating such an approach.

Turning to a separate topic, I have recognized that it is important to reflect on the role embryogenesis plays in the sequential development of the human organism, as this progression has implications for the order in which abnormalities detected via the pulse might be best addressed. I will elaborate on this topic in the chapter on the Extraordinary Meridians, but I wish to emphasize here that processes which become active in the fetal stage of life do not necessarily disappear later on, but may remain dormant until called back into action by specific forms of trauma later in life. Most practitioners of Chinese medicine diagnose and treat the Extraordinary Meridians based on symptomatology alone, perhaps with the addition of palpation for tenderness in specific related areas of the body. I have not found this to be sufficiently accurate, and will reexamine the protocols for diagnosing and treating the Extraordinary Meridians in CCA by attending to pulse diagnostic findings. One aspect of Extraordinary Meridian theory demands consideration here: There is no direct Five Element mechanism involved in their functioning, however there is clearly a division of the Extraordinary Meridians into yīn and yáng categories. If these Extraordinary Meridians do play a significant role in embryological development, then not only Five Element considerations, but also yīn yáng considerations must be incorporated into a more complete understanding of the Constitution. It is reasonable to posit that individuals have an inherent tendency towards either a yīn or yáng nature, as was described by Li Jema in his development of Sasang Constitutional Medicine (SCM), based on the four yīn yáng images (sì xiàng) from the *Yì Jīng*. This issue will also be addressed in the ensuing chapters.

Finally I wish to refine my previous remarks about the correct standards for examining the pulses at cùn, guān and chǐ in both the yáng and yīn positions for finger placement.[70] I previously published an article in the Journal of Chinese Medicine on this topic, the substance of which I am reprising here, as I would like to add back the mathematical reasoning that was deleted in the Journal's editorial process.[71] The importance of that article is partially due to the disagreement it proposes with the finger position standards of the popular Shen/Hammer methodology. The following remarks pertain to what I call the yáng pulse positions. It is a simple matter to locate the yīn pulse positions afterwards by keeping the same relative distances between the examining fingers, but moving them proximally on the patient's wrist so that the examiner's index and middle fingers are placed equally on either side of the styloid process rather than placing the middle finger exactly above it. The mathematical reasoning is important in my opinion, as it supports the assertion that acupuncture is based on an Eastern scientific foundation, rather than simply being a collection of anecdotal recipes. The pulses represent wave phenomena, which carry information that logically will vary depending on exactly which part of the wave is located under the examiner's finger. Thus every aspect of CCA diagnosis depends on placing one's fingers as close as possible to their ideal location, including, but not limited to, the diagnosis

70 In yáng position the examiner's middle finger is placed over the radial styloid process. In yīn position the examiner's index and middle fingers are on either side of the styloid process. Although I am not aware of any Chinese textual sources that describe separate yáng and yīn pulse positions (my personal innovation), the idea is not without precedent in TCM. Lu Yubin makes the following comment in his book *Pulse Diagnosis*, published by Shandong Science and Technology Press, page 48: "The so-called above and below the vessels refers to feel the anterior part, the middle part and the posterior part of the Cun, Guan and Chi respectively. It is said that the anterior part (of the pulse) can be used to examine Yang Qi, while the posterior part (can be used) to examine Yin Fluid." Lu describes this procedure as part of the pulse examining method named "pushing," although he does not include citations of sources for this idea, nor the Chinese character for this method. Pushing would appear to be a searching method rather than a distinct repositioning of the examiner's fingers.

71 Eckman, P., "Precision in Finger Placement for Pulse Diagnosis", The Journal of Chinese Medicine (England), Issue 109, pp. 40-55, 2015.

of the Constitution and the deviation pattern discussed in the *Màijīng*. The latter determination is necessary in order to be able to identify Akabane imbalances via the pulse, a new addition to CCA that I will explain further along in this treatise. Obviously the *Nánjīng* Elemental depths of the pulses reflecting the 12 Organs (zàngfǔ) will also be unreliable if the fingers are not positioned correctly.

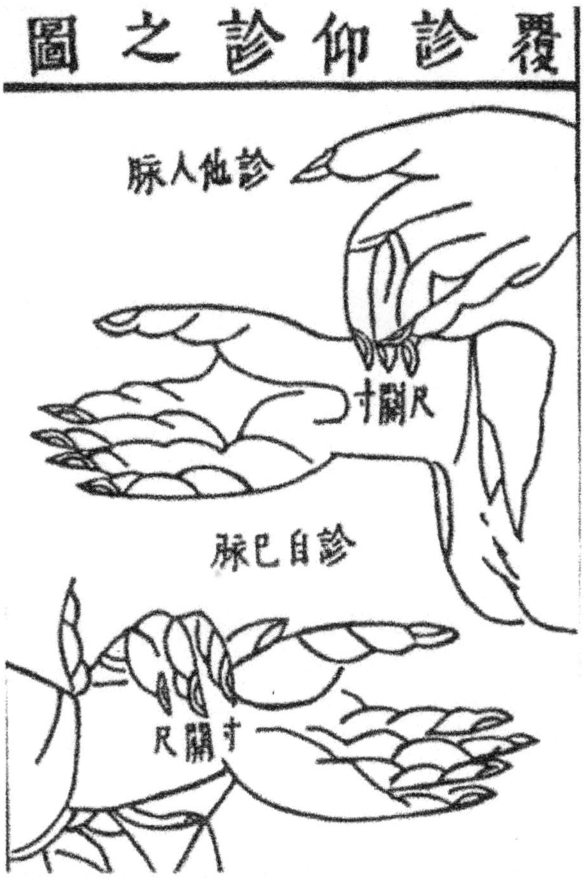

Figure 10: Examining the pulses of others vs. self [72]

72 Image from 三才圖會 (*Sāncái Túhuì* - 1607) by Wáng Qí (王圻), an illustrated encyclopedia that serves as a good source for early representations of classical doctrines, which were not given graphic examples in the original classical texts. This image was reproduced by Professor P. Huard (1953) and later published again by G. Soulié de Morant (1983). The top image shows taking someone else's pulse, while the bottom image shows taking one's own pulse. In the source text it seems clear that the radial pulses during the Míng dynasty were examined with the index finger just proximal to the wrist crease, which I designate as the yáng position pulses in CCA.

My thinking is as follows: The originators of radial pulse diagnosis most likely spent a lot of time palpating their own radial pulses (illustrated in the bottom half of Fig. 10), looking for meaningful clues to the energetics of their own Organs and Meridians. I know I have spent considerable time examining my own pulses, and the majority of the teachings in CCA having to do with pulse diagnosis have been triggered by findings I perceived in my own pulses. Thus it is reasonable to suggest that any individual's index, middle and ring fingers will naturally fall into place at cùn, guān and chǐ if the middle finger is situated over the crest of the styloid process. The exact position of this anatomical feature might not be perfectly clear in some individuals, however if the examiner is close, then centering the middle finger so that it at the peak of the pulse wave will produce even more reliable results. In this case, the index finger will naturally abut the scaphoid bone, with no gaps or overlaps between one's fingers. A difficulty arises however when we recognize that patients have both different size fingers and different anatomical characteristics of their radius bone than those of their examiners. Thus if the correct locations of cùn, guān and chǐ on the patient's wrist is important to determine with precision, then a correction of the examiner's finger positions is necessary in order to end up in the same place for each of these three locations as the patient would find if they were examining themselves. This sounds complicated, but actually it is a rather simple mathematical problem with an easy solution. Let us assume that we always start by placing our middle finger above the crest of the styloid process. Please refer to figures 11 and 12 to visualize the following descriptions.[73]

73 The following passage describes how to locate the yáng position pulses. It is necessary to determine the finger positions for the yáng position pulses before relocating one's fingers to examine the yīn position pulses. The yīn position pulses are then located by keeping the same relative spacing of the examiner's three fingers, but moving then proximally so that the index and middle fingers are placed on either side of the crest of the styloid process. The yáng position pulses reflect the functioning of the fŭ Organs, while the yīn position pulses reflect the functioning of the zàng Organs. It is almost universally believed that the superficial pulses reflect the fŭ Organs, while the deep pulses reflect the zàng Organs at any given location. I believe this is an error that is based on a misunderstanding of the *Nèijīng* text, as I will explain. The proof of my alternative belief, however, was established for me by the discovery of how one can perceive Akabane imbalances via the pulse, also to be explained later. What's important here is to note that these Akabane imbalance signals are independent of pulse depth checked by the examiner, but instead depend on the yīn or yáng finger position employed by the examiner. For example, a Lung Akabane imbalance only shows in the right cùn yīn finger position, regardless of superficial or deep pressure, while a Large Intestine Akabane imbalance only shows in the right cùn yáng finger position, regardless of superficial or deep pressure.

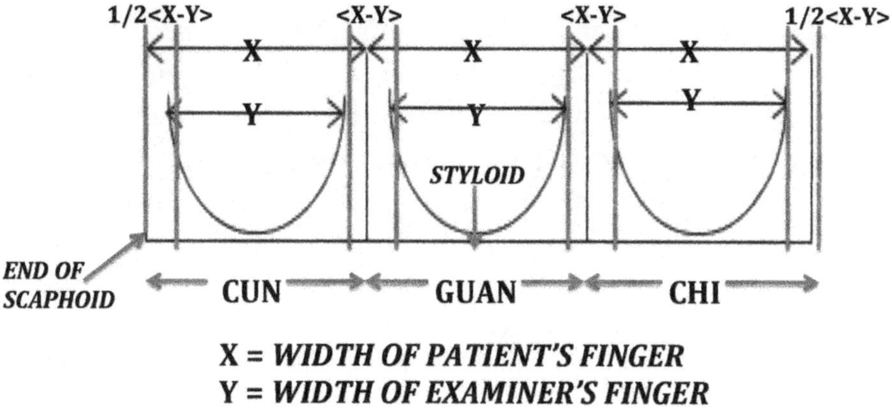

Figure 11: Patients with wide fingers

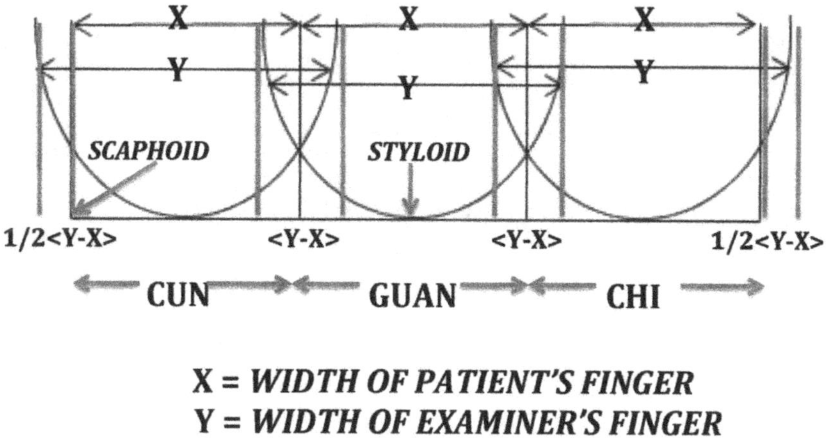

Figure 12: Patient with narrow fingers

Let us call the width of the patient's middle finger x. Similarly, let us call the width of the examiner's middle finger y. Figure 11 shows that for a patient whose fingers are wider than the examiner's fingers, there will be a gap between the examiner's three fingers if they are to end up where the patient's fingers would have been, if they examined their own pulses. Mathematically, this gap is exactly (x-y) in length, and additionally, the gap between the edge of the scaphoid and the examiner's index finger is exactly one half (x-y) in length. This illustrates why the standard for pulse examination cannot mandate that the index finger abut the scaphoid bone, as is taught in the Shen/Hammer tradition. This difference is however usually quite small. There is no need to use any measuring tools to correctly place one's examining fingers. One only needs to place one's fingers so that the space between index and middle fingers is twice as wide as the gap between scaphoid and index finger. A similar solution exists for patients whose fingers are narrower than the examiner's (Fig. 12). In this case, the examiner's fingers need to overlap each other, but by how much? The overlap is shown to be (y-x) in length, but in this case the examiner's index finger will overlap the scaphoid edge by a length of one half (y-x). The overlap between fingers is twice the length of that between index finger and scaphoid. These differences in finger placement are, as mentioned, usually quite small, but they can result in the difference between an accurate pulse diagnosis and one that results in erroneous conclusions. As mentioned, one reason that I wanted to present the mathematical analysis given above is my strong belief that traditional acupuncture, with its emphasis on pulse diagnosis, is based on natural law, and fully merits being included in the natural sciences. I should also remind the reader that these standards have been formulated specifically for acupuncture pulse diagnosis. I leave it as an open question whether the rules for herbal medicine pulse diagnosis might differ, as I do not have the necessary experience to take a position on that matter. From what I know of Dr. John H. F. Shen's practice, his herbal formulations were quite sophisticated, whereas the few acupuncture treatments he described in his writings were rather unsophisticated by comparison, in my opinion, of course.[74]

CCA is largely based on material that is not ordinarily studied by most acupuncture practitioners. Certainly the theories and pulse exams developed by both Koreans (e.g. Korean Constitutional Acupuncture and Korean Hand Acupuncture)[75] and Indians (e.g. Ayurveda) are largely unknown

74 Shen, J. *Chinese Medicine*, Educational Solutions, New York, 1980.

75 In a May 2, 2021 podcast of an interview of influential teacher James Flowers by Daniel Gelman of the Purple Cloud Institute, there was discussion of how little most practitioners of acupuncture know about the unique aspects of Korean medicine. Flowers, who had himself studied at Kyung Hee University in Korea, was even unaware of Korean Hand Acupuncture, a discipline that has contributed significantly to the development of CCA.

in the profession, but my observation is that additionally, practitioners knowledgeable about developments in Japanese acupuncture (e.g. Akabane theory) and even traditional Chinese methods (e.g. the Rényíng Cùnkǒu comparison) are also distinctly in the minority. The result is that learning CCA involves a lot of material and procedures that will undoubtedly present some difficulties at first, but the ideas and methods of CCA are intrinsically simple and straightforward. Learning CCA is a bit like learning a foreign language. With practice, this approach to acupuncture becomes almost intuitive, and practitioners can expect to become fluent after the initial period of bewilderment, analogous to the experience in learning any new language.[76] To help ease this process, I will briefly summarize material I have presented in my prior publications, and also organize this material in what I believe to be a logical approach to treatment planning. Although there are many different treatment strategies in CCA, they are all based on encouraging the return of Original Nature (Constitutional energetics) to each patient, at every step of the way.

When Kuon Dowon presented KCA to the world in 1965, he originally gave an explanation of how he chose the Root treatments for the 8 different Constitutions. There was a certain degree of theoretical reasoning, coupled with trial and error feedback from patients based on using different forms of the 4 needle technique. When Kuon later introduced the Auxiliary Formulas in 1973, he unfortunately gave no description of the way in which he had formulated this significantly more complex development. However, based on personal observation of Kuon's practice, and reports from other Korean students who learned his methodology, it is apparent to me that Kuon kept changing the details of his approach, but

76 Here is the text of an email I received 6/25/19 from the organizer of my classes in Spain, "A few of us got together to practice your method of CCA…We were particularly proud of managing (collectively) to diagnose one patient with Yang Wei Mai imbalance on one side. After the treatment (EEV first, then Condition) his symptoms of tendinitis improved dramatically, and the pulses went to their correct position. We also treated an elderly man who has cancer. His son is (one of our group's) teacher of osteopathy, and who was (himself) so impressed with the results that his own Constitutional formula produced in him, that he brought his father. We diagnosed the father as having a SI Deficiency Constitution, and treated him following the cancer procedure you taught us. His face color changed immediately, and he started walking around the clinic without his walking frame (in fact, he almost left it behind when leaving). In a couple of days we got a message from his son that the man has been (finally) eating and sleeping very well." I'm including this note simply to document that the methodology of CCA is transmissible from teacher to student, even including the most advanced aspects, such as the treatment of cancer.

never revealed the thinking behind these changes. It is almost inescapable to conclude that his approach was essentially one of trial and error, albeit with a significant amount of inspiration based on the fundamentals of yīn yáng, wǔ xíng, zàng fǔ and sasang theory. My own approach in developing CCA has been quite different, as I feel that there needs to be some rational basis for guiding its development. This rational basis is provided by a much more elaborate approach to pulse diagnosis.

Specifically, each component of diagnosis, i.e. the Constitution and the Condition, must be corroborated by at least two different pulse findings. Some of these have been described in the Chinese medical classics, such as the RYCK system from the *Nèijīng*, the Meridian deviations from the *Màijīng* and the depth analysis from the *Nánjīng*. Other corroborative methods for the Constitution include Kuon's KCA pulse, the Sasang Constitutional Medicine (SCM) pulse originally proposed by Puramo Chong, the Ayurvedic Prakriti pulse and the 4 foods O-Ring test (see Chapter 16, Figures 46 and 47). For the Condition, the additional corroboration (supporting the *Nánjīng* depth profile) is via the Ayurvedic Vikruti pulse as manifested in the Subdoshas, the zào versus jìng pulse quality (see Chapter 6), plus the radial pulses associated with Excesses and Deficiencies of Yīn or Yáng and qì or Blood, topics that are still in need of further research before incorporation into CCA. Even the width ratios of the RYCK pulses can disclose quite specific information about the Condition. I have not previously sufficiently emphasized the <u>relationship between</u> the Constitutional and Conditional pulse findings. Only certain Conditional states are possible in an individual of a given Constitution. Thus, not only must the Constitution and Condition be corroborated by at least two different pulse exams, but the Condition must be one that is possible for that Constitution. The relationship between these two states is actually a reflection of the type of pathology exhibited by the patient. We can use this information to help validate our diagnoses of Constitution and Condition. For example, if we have a patient with severe arthritis, in my experience, and that of my mentor Kuon Dowon, they will have a relationship between Constitution and Condition that is called "zàng inflammation," and only one pair of Organs and their Meridians is indicated for addressing this pathology in any given Constitution. If we do not find abnormalities in the pulses of these two Organs, then we must seriously question whether or not our Constitutional and Conditional pulse interpretations were correct. Ultimately, I am hopeful that the pathomechanisms underlying all diseases can be identified in this way, and the effective acupuncture treatments should follow. This approach has proven itself extremely reliable in the diagnosis and treatment of cancer in my practice, so I am optimistic about the future development of this approach.

A number of practitioners who have learned some of the methods Kuon incorporated in KCA have asked me if I had any idea why he changed his treatment formulas over the years. I will offer my opinion, but while Kuon is still alive, perhaps a different explanation will be forthcoming from him. As I have pointed out previously, in my opinion each of KCA's 8 pulse types do not strictly identify a specific Constitution, but rather a family of either 2 or 3 closely related Constitutions. Since Kuon did not recognize these 12 additional Constitutions, he had to formulate Root treatments that would work for all family members with the same KCA pulse type. Let me use LV Excess as an example, a Constitution that Kuon originally called Jupita, but later renamed Hepatonia. His original Root treatment published in 1965 used the following Points: (LU 8, LV 4+ and HE 8, LV 2-). The family member that shares the same KCA pulse type with LV Excess is SP Deficiency. If one treats SP Deficiency by the above formula, which includes dispersing the HE, it will further weaken the SP, which is not desirable. Instead Kuon changed the Root treatment to LU 8, LV 4+ and KI 10, LV 8-). These 4 Points are effective for both dispersing the LV and Supplementing the SP due to their Creative and Control Cycle relationships. I am reminded of Chapter 75 in the *Nánjīng*, which recommends a somewhat similar indirect approach to treating a situation of LV (East) Excess by focusing treatment on HE (South) and KI (North). While these various protocols may all show some efficacy, the original Jupita treatment recommended by Kuon has been clearly the most effective choice in my experience. It also has the merit of being the exact 4 needle combination discovered by the Korean monk Sa Am in the 1600s, which is the single most profound and influential contribution in the history of Korean acupuncture. Each of the other 7 Constitutions' newer Root Formulas that were updated by Kuon can be similarly accounted for by consideration of their effect on the other Constitutions that share the same KCA pulse pattern.[77] In CCA however, all the Root formulas follow Sa Am's original method exactly, producing the 20 unique Root formulas. Kuon also changed the Root formula in people with Fu Organ based Constitutions from treating their own Meridians to treating the Meridians of their coupled Zang Organs of the same Element. He never introduced a separate formula to treat coupled Meridians, but as many styles of acupuncture are based on treating both together, it is my impression that Kuon probably made this change to accommodate such a clinical finding.

Although many aspects of CCA were inspired by Kuon Dowon, I should emphasize that I have had a number of other influential Korean teachers and colleagues, as well as teachers from other backgrounds, whose ideas have helped shape CCA. One topic that has intrigued and puzzled me for many years, and still does, involves the question of laterality in the choice of treatment Points and their associated protocols. There are passages in the *Nèijīng* which state that one should choose Points on the right to treat problems on the left, and Points on the left to treat problems on the right.[78] Many styles of acupuncture have made use of this principle, including the well-known approaches of Master Tong and Richard Tan. I can mention another Korean practitioner, Byoung Soon Kim,[79] who recommends treating male patients on the left side and female patients on the right side, unless the symptoms are strongly unilateral, in which case he also chooses the opposite side for treatment. Thus it does not surprise me that Kuon also favored an opposite side laterality approach in formulating KCA, but of course it is not an invariable rule, and must be adapted to the particulars of each clinical case. Kuon's general method in this regard was to treat

77 Kuon's KCA approach included dietary advice in addition to acupuncture treatment. Unfortunately, we do not know how the Constitutional diagnosis was confirmed in the research study that led to the specific diets. If the basis was simply the KCA pulse pattern, then individuals with different CCA Constitutions would have been grouped together, and the diets might be a compromise that was reasonable for each individual. If however, the research included only individuals whose diagnosis was confirmed by repeated positive responses to acupuncture treatment, then the diets would be more accurate for the original eight Constitutional types, but possibly quite inaccurate for the other 12 Constitutional types that share the same KCA pulse pattern. Much more research is needed to rectify this aspect of therapeutics.

78 *Sù Wèn* Chapter 64 and *Língshū* Chapter 7. These discuss the needle methods of miùcì (謬刺) and jùcì (巨刺), both of which involve treating right side for left side problems and vice-versa.

79 Kim, B. *The Silver Bullet*, KOSA of the America, Buena Park, 2017.

individuals of fǔ Organ Constitutions on the left side and those of zàng Organ Constitutions on the right side, but he treated all cases of pain on the left side, regardless of the Constitution. I believe Kuon was accurate in this observation regarding laterality. In CCA I call this primary treatment side the Constitutional side and the other (symptom side) the Conditional side. Other Korean teachers did not come to the same conclusion about the choice of laterality of treatments. David Lee, a contemporary practitioner of Sasang medicine, believes that in using his "Bi-horary" guidelines for acupuncture, it makes no difference at all whether the treatments use Points on the left or right sides, as long as all the Points are on the same side in a given treatment. Lee bases his treatments on Sasang Constitutional diagnosis, but although he originally was inspired by the work of Kuon, he has formulated an entirely different style of acupuncture treatment, and his rules for choosing treatments naturally are different. Not only does Lee believe that treating either side is equally effective, but he also believes that treating either the zàng or fǔ Meridians provides another equally efficacious choice, as long as one does not mix the two in the same treatment. Lee's experience in this regard may help explain the efficacy of Kuon's later choice of always picking the zàng Meridian of the Constitutional Element for the Root Formula, regardless of whether the patient has a zàng or a fǔ Constitution. My own experience in developing CCA is that Kuon's original idea about the Root Formula, i.e. treating the fǔ Meridian for people with a fǔ Organ Constitution and the zàng Meridian for people with a zàng Organ Constitution, was correct, and should not be changed. I have great respect for these other Korean styles of acupuncture, knowing that they have helped many people. What I miss in them, however, is a rational connection between the symptoms people experience, and the combinations of Meridians and Points that can alleviate their suffering. Pulse diagnosis is the way I have personally found to be capable of making this crucial connection, and is the primary diagnostic method in CCA for that reason.

I have also previously mentioned another of my Korean teachers, Puramo Chong. I was quite surprised by his laterality choices when I spent time observing his practice. Virtually every one of his patients was treated with the 4 needle technique for the Root Formula on one side of the body, while needles for the Auxiliary Formula, also using the 4 needle technique, were inserted on the opposite side of the body.[80] However, in the cases I observed (almost all cases of painful complaints), only the zàng inflammation Auxiliary Formula was employed, regardless of the symptoms or pathology. I was favorably impressed with the results I observed in his clinic, and even more so by the results he described in an unpublished theory and case history manuscript that he shared with me, where he used the combined treatment of both left and right sides with multiple Auxiliary Formulas for more complex cases. In my own practice, however, I was not able to duplicate his results when I initially tried this treatment strategy. Of course his diagnostic methodology was quite different from what I use in CCA, so I was not completely nonplussed. My approach had been to use the Meridian associated with the Subdosha pulse pattern on each side as my laterality guide, which seemed the logical way to allocate treatment formulas to both sides of the body. I must admit that I was puzzled by the conflicting results between Puramo, myself, and Kuon in this regard. One day it dawned on me that the Subdosha pattern corresponding to the Constitution was usually found on the Conditional side for that person's Constitution (i.e. the left side for zàng Constitutions and right side for fŭ Constitutions). Since Kuon routinely treated the Constitutional side almost exclusively, I wondered if perhaps the Subdosha patterns usually reflect the energetics on the opposite side of the body from where they appear. When I tried reversing the sides from those used in my previous cases, the results were generally favorable, but I don't think they have been nearly as reliable as Kuon's unilateral approach. To reiterate, both the Root treatment and the Auxiliary treatment should be given on the side with a RYCK imbalance (usually found to be on the Constitutional side), thus reinforcing the idea that the most important aspect of treatment was indeed supporting the Constitutional Organ/Meridian. The Conditional Subdoshas typically appear on this treatment side, as long as only one other Organ is out of balance with the Constitutional Organ.

Given all the different treatment recommendations I've discussed, how might one make a more objective decision about laterality for the different parts of

80 This laterality approach has been a fundamental tenet of versions of the Japanese Meridian Therapy styles, which are based on Constitutional theory, similar to CCA. Robert Hayden has written, "Toyo Hari makes use of a unique concept known as Sokoku Control, based on unilateral patterns of needling. Briefly, the primary pattern is needled on one side of the body (usually the stronger or asymptomatic side), the secondary on the other." From "Honchi Ho, Hyochi Ho: Root and Branch Treatment Staging in Contemporary Japanese Acupuncture," Robert Hayden, 1996, Midwest Center for the Study of Oriental Medicine, Racine, WI and Chicago, IL.

any CCA treatment, particularly the more complex ones? While I do not have a dogmatic answer, I believe that pre-testing treatments with gold and silver colored Korean Hand Acupuncture press pellets on the corresponding points of the hand microsystem are best at elucidating this issue, with the practitioner monitoring the pulse changes as the pellets are applied separately on each side. The evoked changes in the pulse have proven to be a reliable guide to laterality choices in my experience. Using this investigative approach, I have confirmed that the Subdoshas and the RYCK pulse quality always agree with each other on each wrist, but sometimes this reflects that side of the body and sometimes it reflects the opposite side of the body. The rules for determining which of these possibilities is correct in any situation will be described later in this text, but I should add that this use of the RYCK pulse quality matching the Subdoshas can make the pulse diagnostic process much more accurate.

CHAPTER 4

The Basics of Pulse Examination

Fig. 13: A bewildered doctor checking the pulse of a lovesick young woman [81]

81 Her concerned mother comforts her. In the background Cupid is pointing to one of his arrows. Note the doctor's three fingers properly placed at the cùn, guān and chǐ positions! Colored mezzotint by William Ward (1802). Courtesy of Wikimedia commons, donated by the Wellcome Trust, photo no. V0017086. Source: Creative Commons Attribution 4.0 International License.

There are several different methods of pulse examination used in CCA, and each one is carried out according to specified standards, so that practitioners can confirm each other's findings. Each type of pulse examination requires its specified technique, often employing different parts of the examiner's fingers than other types of pulse examination, for the most accurate information. It is for this reason that I believe that CCA is best learned in a hands-on manner from one's teacher. Books and articles can communicate didactic information, but pulse diagnosis depends on perceptual training, so written works like this one are necessarily incomplete. The reason for different methods of examination is that they each reveal a different aspect of the functional dynamic (qì jī) occurring simultaneously in the patient. The most familiar examination is taking the radial pulses at the wrist (Cùnkǒu). As mentioned, I divide this well known method into two separate maneuvers, called yáng position pulse examination and yīn position pulse examination[82]. These reflect the functioning of the fǔ and zàng Organs respectively. I am not aware of any other publications or claims that the differentiation of the pulses of the fǔ and zàng Organs is based on a longitudinal, as opposed to a vertical (depth), assessment. It is more or less universally taught in Chinese medicine that the superficial pulses reflect the fǔ Organs while the deep pulses reflect the zàng Organs at each position. While this belief appeals to common sense (the fǔ are more superficial Organs than the zàng), it is rather difficult to reconcile such an idea with the depth teachings in Chapter 5 of the *Nánjīng*. Moreover, I confess that the standard teachings about the differing pulse depths of the fǔ and zàng Organs have never impressed me as being of much clinical use in my acupuncture practice. This conventional idea can be traced back to passages in the classics, but surprisingly, there is a certain degree of uncertainty about how to interpret two key terms in this regard, 外 wài and 內 nèi. The orthodox view is that wài refers to superficial because it is most often translated as "exterior", while nèi refers to deep because it is most often translated as "interior". A recent English translation of one of the key passages in *Sù Wèn* Chapter 17[83] by a Chinese author[84] may explain

82 Although briefly mentioned in Chapter 3 of the present text, a more complete discussion can be found in the author's prior text, *The Compleat Acupuncturist*, pages 169-170.

83 *Sù Wèn* 17: 尺內兩傍，則季脅也，尺外以候腎，尺裏以候腹中。附上左，外以候肝，內以候鬲；右，外以候胃，內以候脾。上附上右，外以候肺，內以候胸中；左，外以候心，內以候膻中。前以候前，後以候後。上竟上者，胸喉中事也；下竟下者，少腹腰股膝脛足中事也。

84 Kong, Y. C. *Huangdi Nèijīng; A synopsis with Commentary*. The Chinese University Press, Hong Kong, 2010. This is a translation of *Nèijīng Zhiyao* by Li Zhongzi (1588-1655).

the reason for the possibility that my variant interpretation is potentially supported regarding this "Biblical" text. Here is Kong's translation with my annotation in italics: "On both sides *(left and right wrists)* of the chi site, [all organs lying as far as] the hypochondrium can be detected. The distal *(wài)* part of the chi site [on the wrists] reads the kidney, proximal the abdomen. *(Next to and above)* on the left [wrist], distal_*(not wài)*_to the chi site [is the guan site], its distal *(wài)* part reads the liver, proximal *(nèi)* (closer to the chi site) the diaphragm. On the right [wrist], the distal *(wài)* part [of the guan site] reads the stomach, proximal *(nèi)* the spleen. Uppermost [furthest] from the chi site [is the cun site]; on the right [wrist], the distal *(wài)* part reads the lung, proximal *(nèi)* the chest. On the left [wrist], the distal *(wài)* part reads the heart, proximal *(nèi)* the pericardium." In Kong's view, there are different structures or functions at cùn, guān and chǐ on both wrists, depending on whether the distal or proximal part of the pulse site is being examined. With the exception of the Spleen, all the distal parts *(wài)* reflect zàng Organs, so there is an apparent contradiction with my experience that the zàng Organs' pulses (yīn position) are felt more proximally than those of the fǔ Organs (yáng position). Again I must thank Igor Simonov for pointing out that in classical Chinese grammar, proximal and distal were always used with reference to the observer! Given that astuce, the pulses of the zàng Organs, being distal from the observer, are actually proximal from the point of view of the patient, which is exactly as I observe them to be.

So where exactly do we feel the yáng position (fǔ) and yīn position (zàng) pulses? Yáng position pulses are felt with the middle finger located at the crest of the radial styloid process and the index finger located as described in the previous chapter. Yīn position pulses are felt with the crest of the radial styloid process located between the examiner's index and middle fingers. On each wrist there are three yáng position pulses and three yīn position pulses. All together this defines 12 pulse positions, and each of these positions are examined at five depths (which I call the *Nánjīng* depths, after Chapter 5 in that text where they were initially described), to determine where the pulse is widest. These 12 pulses correspond to the 12 zàng fǔ and their Meridians. The depths at each position correspond to the Five Elements in the order Metal, Fire, Earth, Wood and Water, from superficial to deep (as specified in the *Nánjīng*). In a healthy individual, each of the

12 pulses will be widest at the depth corresponding to the same Element as its Organ, with the exception of the Triple Heater and Pericardium, both of which should be felt as widest at the Water depth in healthy people.[85] Other examples appear throughout this text demonstrating that the Ministerial Fire functions do not obey the usual Five Element rules that govern the other zàng fǔ, an important distinction that I have not seen in other English language sources.

A second type of pulse examination is performed at both the yáng and yīn positions just described, but in this case it is not the depth that is examined, but rather the longitudinal alignment of the radial pulse along its proximal/distal pathway. In healthy individuals (specifically those who have neither an activation of one of the Extraordinary Meridians nor any Akabane imbalances), these three pulses on each wrist will show the same deviation, radially or ulnarly, under either the index, middle or ring finger. This finding should be the same on the left and right wrists, and in both yīn and yáng finger positions. I call this the *Màijīng* deviation, as it was first described in Book 10 of that text, although I have expanded upon and adjusted a few of the *Màijīng*'s specifications, based on my clinical experience. When Extraordinary Meridian signals or Akabane imbalances are present, these *Màijīng* deviations are more accurately identified at positions 4, 5 and 6, further proximally up the arm. At these proximal locations, neither Extraordinary Meridian activation nor Akabane imbalances distort the *Màijīng* deviation. Radial deviations correspond to Taiyang, Yangming and Shaoyang Meridians at the index, middle and ring fingers respectively. Ulnar deviations correspond to Jueyin, Taiyin and Shaoyin Meridians under the same three fingers. Everyone has one of these six deviations, which are Constitutional indicators, as they do not change throughout an individual's life.

A different method of pulse examination involves comparing the carotid and radial pulses at the neck and the wrist, known as the Rényíng Cùnkǒu

85 This order (Metal, Fire, Earth, Wood, Water) differs from those of the Creative and Control cycles of the Five Elements. I feel it is crucial to point out that the relationships of the Elements are not fixed to any sequence, but depend entirely on context. This statement is true for all theoretical ordering theories in Oriental medicine, including yīnyáng sequences, trigram and hexagram sequences, etc. While this opinion may seem self-evident to some readers, I have been surprised at how often it confounds otherwise seasoned practitioners who insist that the ordering system is immutable.

(RYCK) examination, so named after these acupuncture locations. The left side pulses are compared to each other, as are the right side pulses. A wider carotid pulse indicates yáng excess or yīn deficiency. A wider radial pulse indicates either yīn excess or yáng deficiency. In addition to the size of the pulse at the neck and wrist, there is also an important difference to be diagnosed concerning their overall shapes or waveforms. I believe this may be an unstated implication of what the *Nèi Jīng* was referring to in describing the ratios of these pulses. The components of each pulse may represent the three yáng Great Meridians contributing to the Rényíng pulse and the three yīn Great Meridians contributing to the Cùnkǒu pulse. The ensuing Chapter 6, on RYCK diagnosis, will present an analysis of the different theories about RYCK diagnosis, and propose unique standards for comparing these pulses. From that perspective, I will hypothesize why the RYCK methodology has not remained a prominent part of contemporary TCM theory and practice. That chapter will also explain how the pulse quality is evaluated in RYCK examination, in order to determine if the hand or foot branch of one of the six Great Meridians is imbalanced.[86] I have previously asserted that the correct location for feeling the ST 9 pulse in women is at the level of the crico-thyroid membrane, rather than the apex of the thyroid cartilage, as women do not have an Adam's Apple.[87] I learned this teaching from the originator of Korean Hand Acupuncture, Yoo Tae Woo, but have often wondered whether it is correct or not. I now feel comfortable in confirming the accuracy of this location in women, as I find that the *Màijīng* deviations at Cùnkǒu, just discussed, are identically represented at the Rényíng location in both men and women, and in women this *Màijīng* deviation can be felt most clearly where Yoo claimed it to be. A discussion of the location of the Cùnkǒu pulse is best left for Chapter 6, as it is a complex and controversial topic.

The Ayurvedic pulses are examined further proximally on the forearms at positions 4, 5, and 6 (with respect to those just discussed at positions 1, 2,

86 Alert: I have previously stated that the RYCK ratio is an unvarying reflection of an individual's Constitution, manifesting whenever any imbalances are present. While this statement is a basic tenet of CCA, the typical RYCK pattern may appear on either the left side, right side or bilaterally, depending on both the Constitution and Condition of the patient.

87 Eckman, P. *Grasping the Donkey's Tail,* p.92.

and 3). Ayurveda is not based on yīn and yáng, but rather on three Doshas named Vata, Pitta and Kapha. My research has established a correspondence of these three Doshas with specific patterns of the Five Elements as defined in Chinese medicine. The deep pulses at this proximal location indicate Constitutional information (Prakriti), while the superficial pulses at the same location indicate Conditional information (Vikruti). These superficial pulses are more commonly referred to as the Subdosha pulses. Another finding from my research is that at the Subdosha level, the locations on the examiner's finger, from radial to ulnar edge, can be divided into four places at which an impulse associated with a specific Subdosha might appear. The order from radial to ulnar edge of these four superficial places corresponds to the Metal, Wood, Earth and Water Elements (of Chinese Medicine) respectively, on all three fingers. If any finger has impulses at two of these places, the correspondence is to the Fire Element.[88] These Subdosha pulses provide invaluable information about the type of pathological disturbance affecting the patient, so although it is one of the more advanced and subtle examination techniques in CCA, it should not be omitted. However, most of the information from the Subdosha examination can (and should be) be corroborated by the *Nánjīng* depth examination already mentioned, as these are both Conditional pulse evaluation methods for the most part. Comparing the findings of these two methods will help in developing skill at interpreting both of them. Warning: Be aware that the Subdosha findings are not strictly correlated with the side of the body on which they are felt. Sometimes they reflect the same side (ipsilateral), but often they reflect the opposite side (contralateral). The same warning applies to the pulse quality association in RYCK diagnosis just mentioned. The actual rules for laterality appear to depend on the type of pathology (bìng lǐ 病理) being reflected, but more study is needed to confirm the rules pertaining to all types of pathology.

Keeping the fingers in the same 4, 5 and 6 positions also allows one to examine the Korean Constitutional and Korean Sasang pulses. Both of these exams involve the very deep impulses, but different parts of the examiner's fingers are used in palpation, as well as different methods of exerting

88 Of note, the Fire Element has twice the number of associated Officials and Meridians as any of the other Elements.

pressure via the fingers. The Korean Constitutional pulse is felt with the flat of the distal phalanx of each finger pressing downwards after pulling the flexor tendon over the radial artery, while the Sasang pulse is felt with the fingertips pressing in the same direction, from ulnar to radial sides of the pulse. The pattern of where the impulse is first felt as pressure is released after occluding the radial artery provides the diagnostic information, which is solely about the Constitution. These findings should match what has been found by the Ayurvedic deep pulse exam. All of the information in this Chapter can be found in greater detail in *The Compleat Acupuncturist*, but these summary descriptions may provide a helpful orientation for beginners at CCA pulse diagnosis.

What I have described above includes most of the pulse examinations that I have been studying for many years, along the way to developing CCA. At this stage of my own development I can now determine an individual's Constitution with a high degree of accuracy by using only the *Màijīng* deviation and both the Ayurvedic Prakriti and RYCK patterns. For practitioners only interested in Constitutional diagnosis (perhaps practitioners of the Worsley Five Element style), these three examinations may be performed without regard to the other methods. As I was trained in the Worsley style, I know how often different practitioners find different Constitutions (CFs) on the same patient, so these simple exams may provide tools for achieving greater accuracy and confidence in diagnosis. In a recent blog post[89], Nora Franglen, an influential teacher of Five Element style acupuncture, stressed how frequently the CF is not definitively identified on the initial examination, but only becomes apparent during the series of treatments that serve as a guide to correct diagnosis. CCA may make this process proceed much more rapidly. I might add that over the years I have treated a number of patients who were diagnosed by Worsley himself. My observation has been that those patients who had done very well with treatment based on the CF diagnosed by Worsley had this same Element confirmed as the Constitutional Element by CCA, whereas those patients who had inconsistent results based on Worsley's CF diagnosis usually had a different Element identified as Constitutional by using CCA pulse criteria. My current belief is that in these latter cases,

89 norafranglen.blogspot.com 3/9/20

the color, sound, odor and emotion method (CSOE) taught by Worsley did not always point directly to the Constitution, but in some cases pointed to the present Condition, whether it was an Element out of balance with the Constitutional Element, or possibly an Element that was simply chronically distressed by perverse factors (AE or xié qì), topics to be attended to further in this treatise.

CHAPTER 5

Five Element Balancing: The Meridians (Jīng Mài) and Organs (Zàng Fǔ)

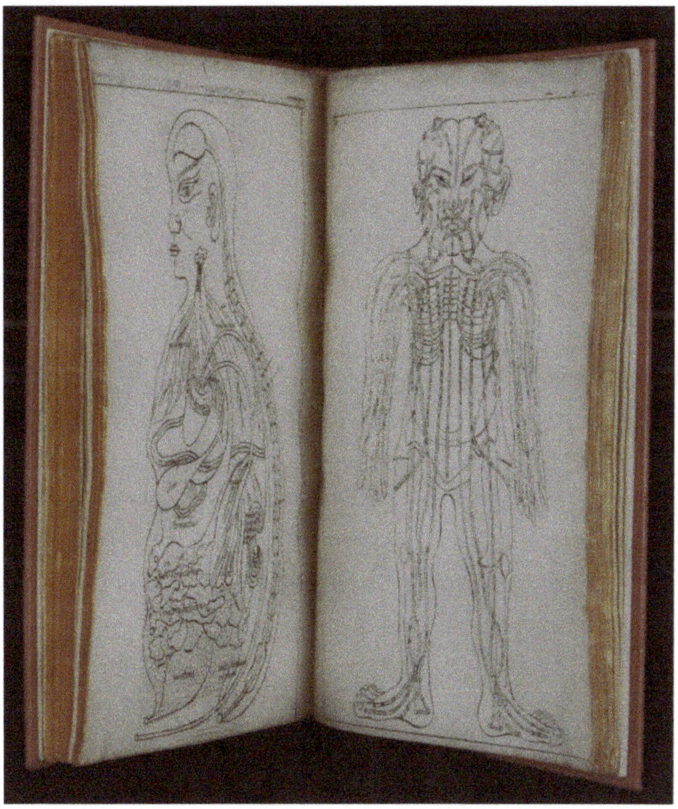

Figure 14: The Meridians and Organs [90]

90 Source: National Library of Medicine: Specimen medicinae Sinicae: sive, Opuscula medica ad mentem Sinensium: cum figuris aeneis et ligneis / edidit Andreas Cleyer ... [Public domain] asserted by commons.wikimedia.org.

Although balancing yīn and yáng would appear to be the most important goal of treatment from a philosophical perspective, as well as from certain passages in the *Nèijīng*, in practice I have followed the method used by virtually all of my teachers, which is to begin treatment[91] by addressing the Five Element (wǔ xíng) related issues.[92] One theoretical justification for prioritizing this step in the treatment process is that all the functions of the organism are regulated by the zàng fǔ, and if these Organs (Fig. 14) are not able to perform properly then the individual will no longer be guided by their Constitutional Organ, meaning they have lost touch with their Original Nature. The overall balance of yīn and yáng in an individual is subject to continual flux, and is therefore a Conditional rather than a Constitutional matter, although this statement is an oversimplification, as will be discussed further along. The zàng fǔ relate to each other via the radiating effects of their qì, governed by the Five Elements. This notion was previously described in Chapter 2 as a delocalized interaction that is independent of any linear substrate including Meridian connections. The deep coordination of the Organs, necessary to maintain the expression of the individual's Spirit Essence (shén jīng) is thus a logical place to begin acupuncture treatment, for as the *Nèijīng* cautioned, all needling should be based on the Spirit[93]. Actually both the *Yì Jīng* and the *Sù Wèn* have an identical statement that supports this approach, "What is beyond measuring by yīn and yáng is called Spirit."[94] Since the zàng fǔ house the various aspects of the Spirit, treating to balance them is obeying the dictum to base treatment on the Spirit. Literally all the patients I treated in my three workshops in China in 2017, 2018 and 2019 presented with imbalances involving the Five Elements, and treatments designed to address that issue were uniformly effective. Additionally all the case histories detailed in *The Compleat Acupuncturist* were of patients also treated in this manner. For that reason, only a brief review is presented here.

---

91 After clearing blocks, a subject to be addressed later.

92 Pathomechanisms 1 and 2 previously cited from *Língshū* 9 are both due to Five Element imbalances. The first type classically refers to imbalances (of yīn and yáng) within a given Element (Rényíng Cùnkǒu imbalances to be described in the next chapter) while the second type refers to imbalances between different Elements. Extraordinary Meridian imbalances however (see Chapter 12), are primary yīn yáng imbalances that should be treated before addressing Five Element issues.

93 *Língshū* 8, 凡刺之法, 先必本于神. "All norms of piercing (require one) to first of all consider the Spirit as the foundation."

94 陰陽不測謂之神. *Sù Wèn* 66 and *Yì Jīng: xì cí zhuàn*. Translated by Lorraine Wilcox, who noted its identical appearance in both classics. Personal communication 12/5/20.

Before CCA treatment can be formulated, it is first necessary to determine the individual's Constitution and Condition. The Constitution always reflects the potential for either hyperfunctioning or hypofunctioning of one of the 10 formed Organs, and thus is correlated with one of the Five Elements. This Organ's pulse should be widest at the depth associated with its Element, but this is not typically found to be the case in people who feel less than perfectly healthy. The actual depth found indicates the Element with which the Constitutional Organ is most out of balance. Since there are both a zàng and a fǔ Organ that resonate with the Conditional Element just identified, the next step is to examine the pulses of these two Organs. In most cases one of these Organs will have its pulse depth shifted to that of the Constitutional Element. In many patients, this is enough to make an initial diagnosis and treatment plan, but it is always better to confirm these findings by checking the Subdoshas on both left and right wrists. Again, in a simple case, one side will show the Subdoshas of the Constitutional Organ (usually the symptomatic side), while the other side will show the Subdoshas of the Conditional Organ (usually the side with the RYCK imbalance). These should be the same two Organs that were identified as having abnormal pulse depths.

In my experience, such simple cases always involve Elements that have a Control Cycle (xiāng kè) relationship. This situation determines that if the Constitutional Organ is in Excess, the Conditional Organ will be in Deficiency. Likewise if the Constitutional Organ is in Deficiency, then the Conditional Organ will be in Excess. There are two fundamental methods for addressing this kind of imbalance: either a transfer protocol can be selected in the manner taught by Worsley and others, or several Four Needle protocols can be selected in the manner taught by Kuon for Korean Constitutional Acupuncture (KCA). Let me first illustrate the use of the transfer technique. Worsley taught that these transfers can be done using either Creative Cycle (xiāng shēng) or Control Cycle (xiāng kè) relationships, with these latter involving only the yin Organs. Many authors have differed with Worsley about the correct usage of the Control Cycle, asserting that it operates from yīn Organ to yáng Organ and yáng Organ to yīn Organ. My experience is that this latter interpretation is the correct one, i.e. that the Control Cycle transfers work most effectively from yīn Organ to yáng Organ and vice-versa. This teaching happens to be widely shared among Asian traditional practitioners and was even the approach initially transmitted by Worsley's teacher of this material, Jacques Lavier. If one of the two Organs being harmonized in this way is a zàng Organ and the other is a fŭ Organ, then it will alternatively be possible to incorporate the Connecting Meridian (luò mài) in the treatment instead of using the Control Cycle. If the Control Cycle is not being used, then the first Point to be needled is always the tonification (mother) Point of the Deficient Meridian identified. This needle is inserted in the direction of flow of the Meridian, but all subsequent needles are inserted perpendicularly. Needles are added at the tonification Point of the Meridian that just gave up some of its qì, one after the other until the Excess Organ's Meridian has been emptied of its surplus qì. If it is necessary to involve a Connecting Meridian, then the last Point to be needled is the luò Point that draws qì out of the Excess Organ. If the Control Cycle is used instead, then the last point to be needled might be the Grandparent point, rather than the tonification point, although the Grandparent Point might also be selected as the first Point in this type of transfer. When all these needles are in place the first needle is stimulated with a clockwise rotation and removed. Then the other needles are removed in the order they were inserted. All the Points that were needled should be closed with finger pressure. It is not necessary to leave these needles in place for any length of time, however my experience is that leaving them for up to 20 minutes does not affect the outcome, and some patients prefer to have this 20 minute period to experience the change in their internal state brought about by the treatment, undisturbed. I routinely perform this kind of transfer treatment only on one side. In this usage, the treatment side should be the side that shows the RYCK imbalance, to be described in the next chapter. Worsley taught his students to do these types of transfer treatments bilaterally, but also mentioned that an advanced practitioner would know how to perform the treatment unilaterally. As a recipient of treatment from J.R. Worsley himself in the early days of my study with him, I was always treated by him with unilateral needling. Unfortunately he never elaborated on the criteria for choosing laterality, other than one remark I heard, in which he stated that yáng Meridians could be treated on the left side and yīn Meridians could be treated on the right side, but I never witnessed a demonstration of this idea in any of the clinical cases he supervised while teaching his students.

Here's a practical example of my use of the transfer method. The patient is found to have a Liver Excess Constitution with a Large Intestine Deficient

Condition. On the right side, CK is wider than RY. The Liver pulse is widest at the Metal depth (superficial) and the Large Intestine pulse is widest at the Wood depth (fairly deep). The left side Subdoshas show Pitta/Vata of the Wood type (spikes in the second position on the index and middle fingers) and the right side Subdoshas show Vata/Kapha (spikes in the first position (Metal) of the index and ring fingers). Since the RYCK imbalance is on the right side, that will be the treatment side. This is also the side showing the Conditional Subdoshas rather than the Constitutional Subdoshas, and this finding can be used to corroborate the correct side for treatment. The Large Intestine is Deficient in this example, so the first needle is inserted in LI 11 towards the shoulder. LI 11 is an Earth Point, so the next needle goes in the tonification Point of the Stomach Meridian, ST 41. Similarly, the next needle goes in SI 3, the tonification Point of the fŭ Organ of the Fire Element. Finally a needle is inserted in GB 37 (luò Point) to empty the Excess qì from the Liver. LI 11 is stimulated and all the needles are removed. The pulse depths on both the Large Intestine and Liver should now be normalized and the Subdoshas on both wrists should be Pitta/Vata of the Wood type (the Constitutional state). An alternative treatment using the Control Cycle, also on the right side, would use the following Points: LI 11 inserted towards the shoulder followed by ST 43[95] inserted perpendicularly. LI 11 is stimulated and then both needles are removed.

A different approach for treating this Liver Excess patient utilizes the Four Needle technique. The first step in planning such a treatment is to identify the order of strengths and weaknesses of the Elements making up this person's Constitution (Fig. 15). For Liver Excess, the Liver is strongest, its daughter the Heart is next strongest and its mother the Kidney is third strongest. The weakest (most Deficient) is the grandmother the Lung (which is in the control position), and finally the less weak Organ is the granddaughter the Spleen. This ordering must be kept in mind when choosing the two parts of treatment as follows. First the Liver is dispersed by tonifying LU 8 and LV 4 plus dispersing HE 8 and LV 2. Next the Large Intestine is supplemented by tonifying ST 36 and LI 11 plus dispersing GB 41 and LI 3. Note that these last two Points would not have been chosen if Large Intestine Deficiency had

95 Grandfather Point on the ST Meridian. The Wood Element is the Grandparent of the Earth Element.

been the Constitutional diagnosis (SI 5 and LI 5 would need to be dispersed in that case). As noted in my prior publications, all of these Points should be treated on the right (Constitutional) side only, with the needles all inserted in the direction of Meridian flow to tonify, or against the direction of Meridian flow to disperse. They can be stimulated and withdrawn one after the other, or they can all be left in place for up to 20 minutes, as in the transfer protocol. A variation of this technique was introduced by Kuon, who used very superficial needle insertion at every Point in a repetitive sequence that recurred a specific number of times (hereafter referred to as "touch treatment"). His treatments rarely took longer than a minute or two, and were highly effective. I have used his style of treatment on many patients with excellent results. Choosing which protocol and style of needling to employ is dictated by the patient's comfort, as they all result in comparable outcomes in my experience.

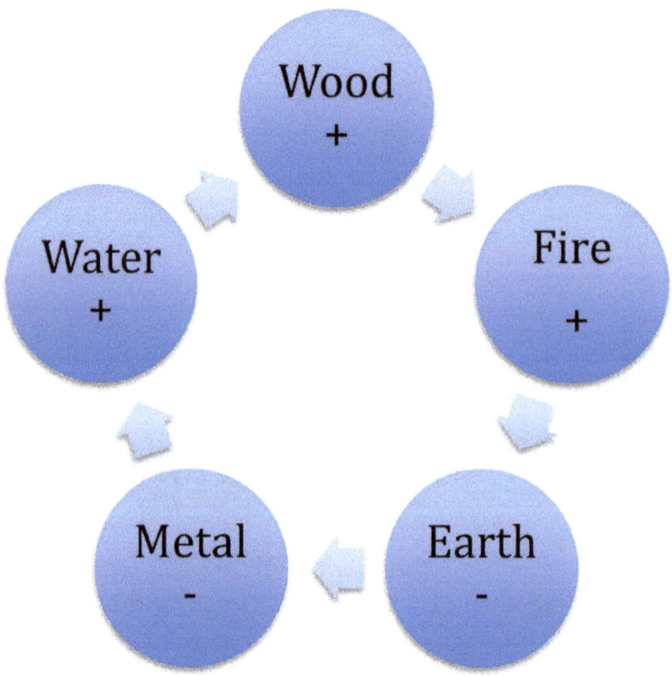

Figure 15: Liver Excess' Order of Elemental Strength & Weakness
Wood (5)>Fire (4) >Water (3) >Earth (2) >Metal (1)[96]

96 Source: Peter Eckman. In a Liver Deficiency Constitution, the order from weakest to strongest would be Wood, Water, Fire, Earth, Metal. This is not exactly opposite to the Liver Excess order, so care is needed in Point selection based on Five Element considerations.

I'd like to present one further example of using the Control Cycle in the qì transfer method, as it demonstrates several unique features which are both attractive due to their simplicity, while at the same time quite complex in terms of demonstrating a deep understanding of some of the advanced concepts incorporated in CCA. This is a case where I performed a treatment on my wife during the Covid-19 pandemic. Ordinarily I don't like to treat family members because it is difficult to approach things objectively, but in this case we were under quarantine, so it was not possible for her to seek treatment elsewhere. As background I should say that although my wife believes strongly in acupuncture, she is quite "needle phobic," so I approach treating her with the least number of needles and with minimal stimulation. She complained of abdominal pain from the right upper quadrant to the epigastrium, with no obvious precipitating cause. I diagnosed her Constitution as UB Excess and her Condition as HE Deficiency using all the pulse exams described above. Her RYCK imbalance was on the left side, so that became the treatment side. Only one needle was used, HE 3 in tonification, without leaving the needle in place for more than a few seconds. This corrected all the pulse abnormalities, but made no immediate change in her symptoms. However, the next day she reported that not only was the pain gone, but that for the first time in quite a while she felt like herself again. Apparently she had been feeling tired and uninspired for some time without recognizing that something was wrong. HE 3 is the Water Point on the Heart Meridian that pulls qì from the Urinary Bladder to the Heart. Worsley used to talk about a "Su Wen treatment" that consisted of only one Point. Although I believe that terminology is a misnomer, this is a dramatic example of how powerful such a simple intervention can be. The question remains, however, what could explain her abdominal pain, as there is no direct relationship to either the Heart or Urinary Bladder, and HE 3 is not generally described as a Point for treating abdominal pain.[97] This is where taking a complete medical history proves invaluable. My wife has had a decades long history of digestive problems with imaging and blood tests showing chronic low level inflammation of her pancreas. Knowing this history, it is reasonable to suspect that

97 There are rare citations of HE 3 for subcostal pain, however I have never seen it used for that. One could also make the argument that by body image mapping, the elbow is homologous with the abdomen (both being in the middle of their respective terrains). Such an explanation however lacks the rigor of that provided by CCA theory, and would not be treated by needle insertion for just a few seconds in any case.

the abdominal pain was coming from pancreatic inflammation. Advanced students of CCA might recognize that the Zàng inflammation formula for UB Excess Constitutions involves tonifying the Heart. Tonifying HE 3 using the transfer protocol has exactly this effect, and in this clinical situation it is superior to the use of HE 9 or any other Point on the HE Meridian. There are many theories in Chinese medicine that could explain this effect, such as invoking the "Mother-Child" law which indicates that tonifying the Heart will tonify its child, the Spleen/Pancreas, but the more specific theories underlying CCA impress me with the most accurate understanding of pathophysiological mechanisms in cases such as this one, and provide several options for successfully treating them with acupuncture.

CHAPTER 6

Yīn Yáng Balancing Lost Secrets of Rényíng Cùnkǒu Diagnosis in the Língshū: Pulse Shape, Width and Quality

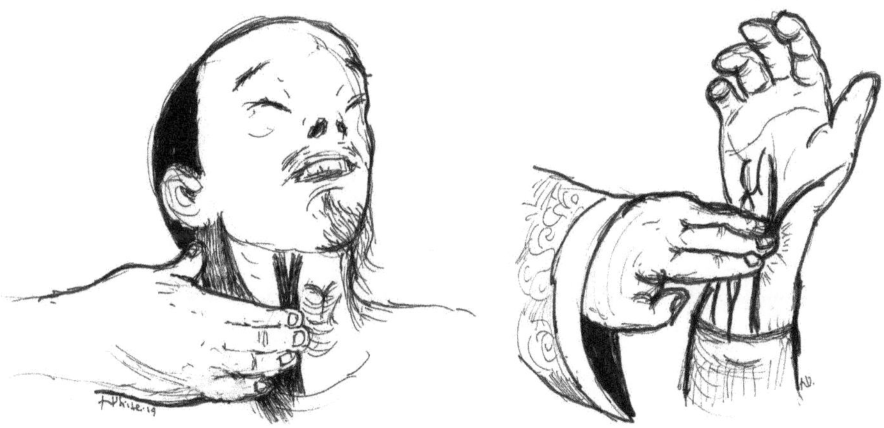

Figure 16: The Rényíng and Cùnkǒu Pulses [98]

I have discussed the *Nèijīng*'s method of pulse diagnosis, which compares the carotid and radial arterial pulses, in my prior books. It is commonly known as the Rényíng Cùnkǒu (RYCK) method, and as I have pointed out it was the first systematic description of how pulse diagnosis is performed, interpreted, and carried forward in the progression from diagnosis to treatment and finally to prognosis in the Chinese classics. And yet, it is rare to find an acupuncturist who examines these pulses, let alone bases their treatment on

98 Source: Neal White.

its findings. Why is that, considering the almost universal acclaim for this "Bible" of Chinese medicine? The *Nèijīng* clearly assigns the RYCK methodology to imbalances of the Six Great Meridians (liùjīng 六經) with its focus on the balance between same Element coupled yīn and yáng Meridians. Thus historically, it was primarily a method for identifying and then correcting imbalances of yīn and yáng within a given Element. I shall put forward the thesis that the Element revealed by the RYCK pattern is what I have been calling the Constitutional Element, and it is only this Element whose yīnyáng status is revealed by this method of pulse diagnosis.[99] The individual's global balance of yīn and yáng is a Conditional matter that is entirely different. For example, someone may have a Liver Excess Constitution, where Cùnkǒu exceeds Rényíng in width, and yet they might still be yīn deficient overall. The RYCK pattern should disappear (i.e. RY becomes equal to CK) when an optimal acupuncture treatment is given, as indicated in the *Nèijīng*, but whenever it does appear as other than equal in any given patient, it shows the characteristic ratio of these pulses that is reflective of that individual's Constitution, with the Extraordinary Meridian exception noted above. Note that this doctrine is based on my own experience, and is at odds with the ideas of my teachers, many of whom used RYCK pulse diagnosis. Several of them taught that the RYCK ratio changes as the treatment process proceeds, and sequentially reflects different stages associated with the Six Great Meridians, before reaching normality (RY equal to CK). One of these teachers however, the late Master Doeki Ogura, who was the lineage leader of the RYCK (Jingei in Japanese) approach to diagnosis in Japan following World War II, taught that normality is reflected by having a carotid pulse that is four times the size of the radial pulse, thus directly contradicting the text of the *Nèijīng*! No wonder that practitioners have mostly ignored this controversial aspect of the *Nèijīng*'s teachings.

Before proceeding further, the issue of laterality in RYCK diagnosis needs attention, as it is not a simple matter, and may additionally explain why it is

99 As noted, this dictum only applies if no Extraordinary Meridians have been activated. In cases of Extraordinary Meridian activation, the RYCK width imbalance may not reflect the mature Constitution, similar to the unreliability of the *Màijīng* deviation when Extraordinary Meridian pulse signals at cùn, guān and chǐ are present. However with Extraordinary Meridian imbalances, the RYCK pulses also become imbalanced bilaterally rather than unilaterally, and additionally they display a typical pulse quality signature that will be elaborated upon in Chapter 12.

so challenging to interpret these pulses accurately. The *Nèijīng* did not state explicitly which side of the body was being examined or treated with regard to this topic. In fact, it did not explicitly state that the two pulses should be compared to each other separately on each side of the body, but I think such an interpretation is so obvious that the authors must have felt no need to mention it. My most coherent teacher of this method, Yoo Tae Woo, taught that one should compare left RY to left CK and right RY to right CK, and treat any imbalance found on that same side of the body. In my opinion, the unequal size ratio of these pulses provides information about the primary side of the body in need of treatment, which is often what I call the "Constitutional" side, but that is not an invariable rule. Sometimes, as will be explained, the "Conditional" side is the one needing treatment, as it may be the side with a RYCK imbalance. The pulse quality to be discussed later in this chapter can differ on the two sides, and my suspicion is that this quality often reflects information about the opposite side of the body. This finding is less surprising than one might expect, as I have also found that the Ayurvedic Subdoshas also frequently reflect information about the opposite side of the body. In fact, as I already stated, the RYCK quality and the Subdoshas on any side of the body always agree in terms of indicating a Hand or Foot branch of a Great Meridian. I will explain this further along in this chapter. As far as I can tell, there is very little discussion of laterality in the *Nèijīng* regarding either diagnosis or treatment, with the exception of times when opposite side needling is specifically indicated. That exception will be discussed in the chapter on Akabane imbalances, but in general, my impression is that even in modern times, most acupuncturists treat the majority of Points bilaterally, although there are clearly exceptions, such as Master Tong's style of family lineage practice. From my many years of preceptorship with Chae Lew, I can hardly recall a single instance of unilateral needling, although the intensity of stimulation occasionally differed on the two sides. Similarly, all the treatments I observed at the Leamington College, supervised by J. R. Worsley, were also bilateral, except when treating Akabane imbalances. We do know, from modern studies of anatomy and physiology, that the brain primarily receives and sends messages from and to the opposite side of the body, so it would surprise me more if there were no indications of such polarized phenomena revealing themselves in the pulses.

The preceding remarks regarding conflicting and ambiguous teachings might reasonably account for why this classical RYCK method of pulse diagnosis was abandoned by the vast majority of acupuncture practitioners and practitioners of herbal medicine in China (i.e. as part of TCM), as well as in the West, due to its complexity, controversiality and divergence from the more common TCM approaches. The following is pure speculation on my part, but I believe that some of these problems in RYCK diagnosis most likely arose due to the presence of competing schools of thought about this method in ancient times. I suspect that there were at least two different schools of thought concerning RYCK diagnosis during the time the *Nèijīng* was compiled, as the passages in both the *Sù Wèn* and the *Líng Shū* were almost certainly composed by contributions from different authors, and all the information discussed in these classics regarding RYCK examination cannot be fit into a single model. Bear in mind that the classics being discussed were not accompanied by graphic illustrations, so the exact meaning of some passages is open to varying interpretations.

The first level of analysis is the simpler one: Rényíng reflects yáng, while Cùnkǒu reflects yīn.[100] If one is carefully observant in examining these pulses, they will find that in most cases, one of these four pulses (two on the left and two on the right) will not appear as a single uniform pulse wave, but instead will be separated into two distinct components. In the examination being discussed now (see Fig. 17), it is imperative to use the size (width) of all components of the pulse wave together for comparison in judging the size (width)

100 Rényíng is a point on the Stomach (foot yangming) Meridian, which is a yáng Meridian. The Point is also located at a superior (yáng) position in the body. In CCA, Stomach Excess is classified as belonging to Soyangin (Korean terminology), which is the hottest or most yáng Constitutional grouping. Stomach Excess is characterized by having a Pitta Vata makeup. All five Soyangin Constitutional types have the Ayurvedic Pitta Dosha in their makeup, and Pitta is indicative of heat (yáng). In contrast, Cùnkǒu is located on the Lung (hand taiyin) Meridian, which is a yīn Meridian. This Point is located lower (more yīn) in the body than Rényíng. When checking RYCK pulses, the patient should ideally be in an upright posture (sitting or standing), and the Cùnkǒu pulse should be examined with the patient's wrist at their heart level, so as to conform with these yīn yáng associations. In CCA, Lung Excess is classified as belonging to Soeumin, which is the coldest or most yīn Constitutional grouping. All five Soeumin Constitutional types lack Pitta Dosha in their makeup. Lung Excess is a Vata Kapha type, emphasizing its lack of Pitta. One other comment: Rényíng is one of the "Windows of Heaven" which are associated with the Spirit (shén) that is distributed from the Heart, while Cùnkǒu is often said to be the Source Point of the Lung (yīn) Meridian. Source Points are where the Original Energy (yuán qì and jīng) is distributed from the Kidneys (yīn). This constitutes another fundamental association to yáng and yīn respectively. Yoo Tae Woo, the initiator of Korean Hand Acupuncture, used the simple yīn yáng associations of CK and RY to choose the polarity of Extraordinary Meridian treatments, a matter that will be highlighted in Chapter 12.

of the overall pulse. My observation is that this overall width, as illustrated in Fig.18, is usually found to closely parallel the component (shape) comparison of Fig.17, so both of these methods yield comparable conclusions. In my opinion, the information gathered in this way reflects the balance between yīn and yáng in the individual's Constitutional Element. However, any such imbalance found indicates that the Constitutional state has been replaced by an altered Conditional state. According to the *Nèijīng*, these pulses should change to having equal width following optimal treatment. At least according to Wáng Bīng, when acupuncture is applied correctly, these pulse widths will become equal, like a continuous rope drawn through both locations.[101] It is my impression that this statement is clinically accurate, and may be used to confirm that a given treatment was both effective and has accomplished its goal.

There is an additional aspect of RYCK pulse analysis, found in a number of passages in the *Nèijīng*, which can often be helpful in diagnosing the Constitution of each individual. In this method the numerical ratio between RY and CK pulses is discriminated, not just which is greater, yáng or yīn, but the text is unclear about the parameter that is being compared (size, strength. or shape/pattern), and even the interpretation of the specified ratios is ambiguous and controversial.[102] One possible reason TCM and other modern approaches have largely abandoned RYCK diagnosis is that they usually are not focused on separately diagnosing both the Constitution and the Condition of patients. Rather, these contemporary styles of acupuncture are essentially focused on the present Condition, whether by syndrome

101 The *Sù Wèn* 7 teachings about RYCK diagnosis were glossed by Wáng Bīng as follows, "If the (movement in the) vessels is equally big or small (at both locations), as if a rope of equal diameter (were pulled through both locations), one speaks of a healthy person. Hence (the text) states: This is what is called one". Cited by Unschuld (2011) in Footnote 8 on page 139.

102 The text in *Sù Wèn* 9 uses the term 盛 (shèng) to compare pulses, which means flourishing or vigorous. Something might be displaying more vigor by its size, force, shape or other measure. I have favored size over force due in part to Wáng Bīng's reference cited above, that used "big" and "small" regarding pulses that become equal in size, like a rope pulled through both locations when health is restored. *Língshū* 48 would appear to clarify this ambiguity: "寸口大於人迎一倍，病在足厥陰，一倍而躁，在手心主." (With) Cùnkǒu bigger than Rényíng one-fold, the disease is in the foot Jueyin (author's translation). The contemporary French teacher Jean-sylvain Prot chooses the other interpretation: "Le Méridien touché est Jueyin si le pouls de Cunkou est deux fois plus fort que Renying" posted on his Facebook site: *Língshū* Medicine Chinoise Classique & Acupuncture Optimale on 2/12/2018. "Plus fort" means stronger, not bigger. Unschuld translates shèng as "fullness". Henry Lu translates the same term as "big". Lavier chose "excess (en excès)" in his French translation. In my opinion, only clinical practice can point towards the correct interpretation.

differentiation or other methods of analysis, and I do not believe that the RYCK ratio directly corresponds to the Conditional state being diagnosed. I believe that the RYCK ratio, when it is unequal, is a fixed characteristic of each person, and is therefore a Constitutional indicator. Although the *Nèijīng* did not explicitly differentiate the concepts of Constitution and Condition, it does stipulate that by treating RYCK imbalances via Six Great Meridian methodology, all manner of illnesses can be alleviated. This statement, assuming that RYCK imbalances are Constitutional aspects of the pulse, is similar to the propositions found in Worsley's Five Element Acupuncture approach regarding treatment of the "CF" (Causative Factor or Constitution) and Kuon's Korean Constitutional Acupuncture approach. In both of these latter systems it is crucial to determine and treat the Constitution. As these systems have strongly influenced my own thinking, I have determined that RYCK diagnosis must be reflective of the Constitution of the individual, since this diagnosis was formulated in the *Nèijīng* as a basic guide to acupuncture treatment regardless of the presenting symptoms or Conditions. The RYCK ratio can only indicate either a healthy state when balanced (RY equals CK), or the underlying Constitution when imbalanced (RY does not equal CK). When the balance is restored, RY should become equal to CK again. In developing this interpretation of RYCK diagnosis I have been forced to challenge the interpreted meanings of the RYCK passages in the *Nèijīng* as they have been understood by commentators in both China and elsewhere, from post-classical to modern times, but I believe there are both linguistic and experiential grounds for doing so.

At first, in my clinical practice, I was not able to reliably confirm any of the teachings about RYCK diagnosis, especially as my teachers did not agree with each other, but subsequently I began to feel that the information from this exam would be useful if it was parsed in ways that I had not seen discussed in the available literature. It is also important to recognize that there are differences of opinion on where to feel both the Rényíng and the Cùnkǒu pulses. The location for Cùnkǒu that I found most useful in RYCK diagnosis in the past was specified as being at the radial styloid process, approximately where point LU 8 is located.[103] Many practitioners however, claim to use LU

103 Wáng Bīng's gloss: "The qi opening is located one inch behind the fish-line (separating wrist and palm), cited by Unschuld (2011) in Footnote 8 on page 139.

9 as the location of Cùnkǒu.[104] In actuality, I believe they mean the pulse at the cùn position (slightly more proximal), as the pulse directly over the wrist crease is typically much narrower than the pulse at Rényíng. Because the pulse at the cùn position often shows similar findings as the pulse at the styloid crest (guān position), I think that the cùn position itself often provides similar data for comparing size with the pulse at Rényíng. The respected teacher of a purely *Nèijīng* based approach to acupuncture, David S. White, has written that following his teacher's advice he uses two fingers to palpate the pulse at Cùnkǒu (presumably covering both cùn and guān locations), as one finger at any location in the vicinity was less informative in his experience.[105] My recent findings lead me to select the cùn pulse in applying the RYCK method whenever the cùn pulse is the one that manifests a different size from the pulse at Rényíng, but most often I still use the guān position for this examination, as stated in my prior publications. It seems to me that this part of the RYCK system could use further investigation. I have stated that RYCK imbalances reflect Constitutionally determined <u>ratios</u>, as specified in the *Nèijīng*, but that after successful treatment the ratio of RY to CK becomes one to one, implicating this diagnostic system as also providing Conditional information of a sort. A finding that appears to be reliable is that if the RYCK ratio (defined as either a matter of components, waveform shapes or overall width) is not one to one on either the left or right side, then that side of the body (at least) needs treatment to deal with resolving one's present (or potentially future) health problems. [106]

---

104 It is crucial to recognize that the term Cùnkǒu has denoted different places and phenomena in the course of Chinese history. Igor Simonov has made a strong argument for it being the pulse at LU 9 in "old Chinese," which predated the classical Chinese of the *Nèijīng* and subsequent Han dynasty texts. Other sources use the cùn position in the RYCK examination, as noted above. The later *Màijīng* once again redefined Cùnkǒu as only referring to the radial pulse on one of the two wrists. This historical variability in the meaning of Cùnkǒu may explain why practitioners who use this method say they are comparing the pulse at LU 9 with the carotid pulse, even though their actual finger palpates the pulse at the cùn position more proximally. It would be surprising if there were no real difference between the terms cùn and cùnkǒu in the minds of the originators of these terms. On another related anatomical matter, it is noteworthy that both the Adam's Apple and the styloid process of the radius are similar in their anatomical form, and would likely have suggested using their comparative pulses as meaningfully connected by early Chinese investigators.

105 Personal communication, 10/16/21.

106 This dictum refers specifically to the rebalancing of the Five Elements. In almost every case where I have found a RYCK imbalance, there has been an inter-Elemental imbalance present. The *Nèijīng* does however present the RYCK system as specifically addressing intra-Elemental imbalances, such as between Liver and Gall Bladder, for example, thus implying its ability to treat imbalances between yīn and yáng within the same Element. Most likely these intra-Elemental imbalances are responsible for the development of subsequent inter-Elemental imbalances.

In addition to the issues that I have just discussed, I believe there is an even more significant reason for the abandonment of the RYCK methodology historically, due to what I believe to be a misinterpretation of the *Nèijīng* text. I previously discussed the issue of the ratios of radial and carotid pulses, and my opinion that the standard translation of shaoyang and jueyin imbalances as having ratios of 2:1, taiyang and shaoyin as having imbalance ratios of 3:1 and yangming and taiyin as having imbalance ratios of 4:1 is not correct. Rather these ratios should be 1.5:1, 2:1, and 3:1 respectively. When I first described my interpretation of the *Nèijīng* text concerning this subject, I remarked that I was not convinced that the ratios were meant to refer to relative width, strength or distribution of components (shape) present in the pulse. Having now had much more experience checking these pulses, I believe that the correct interpretation has to do with the observation that these pulses, when imbalanced, present as "broken up" into separable components. The components together must be included in assessing the width of the pulse, as they together create different pulse shapes. A normal RYCK pulse feels like it has only one coherent and symmetrical waveform (Fig. 17A), while imbalanced RYCK pulses display a variety of more complex waveforms as shown in Fig. 17 B, C and D. These variant waveforms appear to me to indicate that either the first component of yáng or yīn (shaoyang or jueyin) has become imbalanced leading to a ratio of 1.5:1 (Fig. 17B), or the second component of yáng or yīn (taiyang or shaoyin) has become imbalanced leading to a ratio of 2:1 (Fig. 17C), or the third component of yáng or yīn (yangming or taiyin) has become imbalanced leading to a ratio of 3:1 (Fig. 17D). Granted, it is not easy to differentiate 1.5 from 2, or even 2 from 3 times wider, when assessing the size of pulse ratios, so that difficulty may have contributed to both the ambiguity and the decline in the historical usage of the RYCK method in acupuncture practice. I have thus added another illustration to show the edges of each pulse wave (compared to each other) in the normal state and the three imbalanced states (Fig. 18). Of course, as shown in Fig. 17A, the normal pulse wave does not really have edges, but tapers on its medial and lateral sides. Be that as it may, practically speaking it clearly feels like there are these edges as shown in Fig. 18, probably due to the sharp decrease in the pulse wave on both sides. These two different assessments of the RYCK pulse ratios agree with each other in all individuals I have examined. I think my interpretation of the RYCK ratios is additionally in keeping with the other ideas presented in the *Nèijīng* regarding the importance of the six divisions of yīn and yáng, which do not exist independently, but rather coordinate into a harmonious mechanism when functioning in a healthy manner. At least that is what I believe Wáng Bīng was indicating in his footnoted remark regarding *Sù Wèn* Chapter 7, in his image of a continuous rope of equal size through both Rényíng and Cùnkǒu. It is possible that the "broken" waveform felt in an imbalanced pulse represents the absence of a healthy energetic contribution from one of the three yáng or three yīn Great Meridians (liùjīng 六經). This finding may relate to the three circuits of Meridians that are important in understanding the Meridian system as a whole, but are not named as such in the classics.

The first circuit is composed of Pericardium, Triple Heater, Liver and Gall Bladder. It is associated with the Chinese concept of Heaven (tiān 天) as one of the three powers (sāncái 三才). The second circuit is composed of the Heart, Small intestine, Bladder and Kidney. It is associated with the Chinese concept of Man (rén 人). The third circuit is composed of Lung, Large Intestine, Stomach and Spleen/Pancreas. It is associated with the Chinese concept of Earth (dì 地).[107] These three circuits present the same ratios of RYCK pulses in all of the Meridians in each circuit, and the order for these ratios fits the circuit order of Heaven, Man, Earth. This order is a frequently encountered one in traditional Chinese metaphysics, where Man stands between Heaven and Earth. It is significant that the Heaven circuit contains both Ministerial Fire Meridians, reflecting functions without form. As such, I associate these Ministerial Fire functions with the equally insubstantial balance of yīn and yáng, noting that yīn and yáng are not things (nouns), but rather qualities (adjectives). I tend to think of such issues as relating to pure Heaven or Anterior Heaven (xiāntiān 先天), while the Earth circuit contains both Stomach and Spleen Meridians that are not only Earth Element associated, but are also the root of Posterior Heaven (hòutiān 後天). In a similar way, I associate these Earth Element Organs with the more material substances qì and Blood, which are properly understood, in contrast, as things (nouns). What then is the Man circuit about? Comprising Fire and Water Elements, I imagine that the connection is

107 Suger, P. "The Three Circuits: The Missing Link in Channel Theory" North American Journal of Oriental Medicine, Vol. 27, No. 81, 2021. pp. 12-14.

to Man's Spirit and Essence, stored in the Heart and Kidney respectively, and reflecting the uniqueness of each individual being (their Constitution) between Heaven and Earth.

There is one further remark I wish to include pertaining to the RYCK ratio findings. I am not aware that previous commentators have discussed this subject with regard to the Eight Extraordinary Meridians. In my experience, to be elaborated in Chapter 16, imbalances of these Extraordinary Meridians are another possible reason for RYCK disparities. In fact, RYCK imbalances appear to always occur bilaterally when Extraordinary Meridians are malfunctioning, although further research is needed to confirm the reliability of that statement. If it is accurate, it would provide one more reason why the RYCK system lost popularity. Trying to always associate RYCK imbalances only with the Six Meridians wouldn't explain the clinical picture or guide the necessary treatment when Extraordinary Meridian imbalances are present.

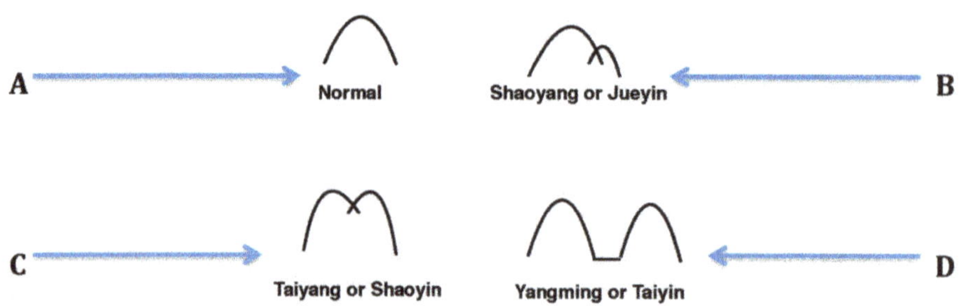

Fig. 17 (A,B,C,D)[108] *RYCK Pulse Shapes*

108 Source: Peter Eckman. The orientation of the pulse waves is from radial to ulnar for CK and lateral to medial for RY.

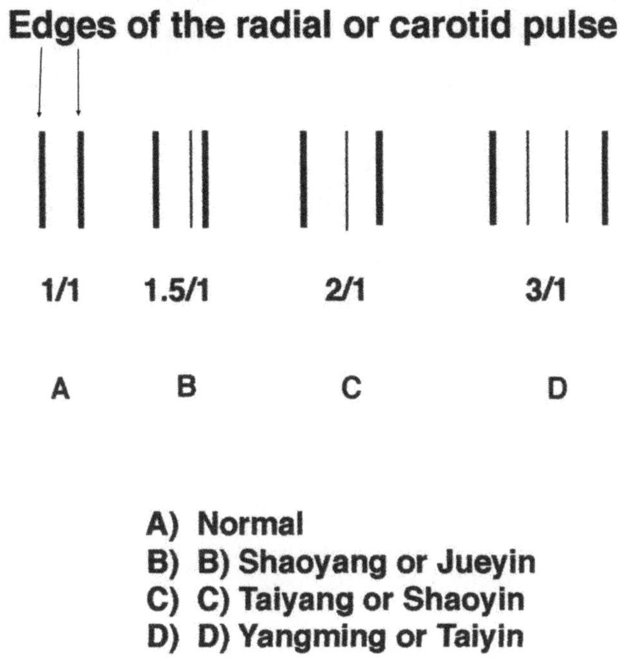

Fig. 18 RYCK Pulse Sizes [109]

A question that I have often been unable to answer to my own satisfaction is, "What is the difference, if any, between abnormalities of the Meridians versus abnormalities of their associated Organs, from the perspective of examining the pulse?" If only the superficial pathways of the Meridians were important to take into account, then the question might be more easily answered, as abnormalities along the superficial trajectories of the Principle Meridians are amenable to inspection and palpation, thus local abnormalities could be detected by these examinations. However, the deep pathways of the Principal Meridians cannot be examined in this way, so it is not obvious whether any symptoms or pulse findings are reflecting deep pathways or the Organs themselves. Clearly the RYCK system was developed in regard to the 3 yīn and 3 yáng system of Meridians. The *Nèijīng* hardly mentions any Organs in the chapters that describe RYCK diagnosis. Rather, it focuses mainly on the 6 levels of shaoyang, yangming, taiyang, taiyin, shaoyin and jueyin, and whether the foot or hand branch

109 Source: Peter Eckman

of each Meridian was imbalanced. From this beginning it is understandable that many practitioners believe that RYCK findings specifically reflect only Meridian abnormalities. But can we confidently say that Organ pathologies are not also reflected in abnormalities of the RYCK pulses? That is not a question that can be answered definitively from a philological perspective (i.e. textual analysis), as it clearly deals with clinical matters that depend on clinical findings for resolution. My conviction is that the RYCK findings definitely include information about both Meridians and Organs, since many treatments that have corrected RYCK imbalances in my patients have also corrected Organ pathology, as reflected in both laboratory tests and clinical symptoms. The most dramatic example of this conclusion can be found in the treatment of cancer, which I will discuss and provide numerous case histories as examples. These obvious Organ level pathologies always present with RYCK imbalances, and the treatment that appears to reverse the cancer process also eliminates the RYCK imbalance.

The RYCK system has three aspects: The first is the shape images (waveforms) displayed by these two pulses on each side of the body, as has just been discussed. The second is the size comparison also alluded to in the prior discussion of yīn and yáng deficiency and excess. The third is the pulse quality that determines (among other phenomena to be discussed) involvement of the hand versus foot branches of the Great Meridians. My belief, consistent with the *Nèijīng*, is that the size and shape ratio aspect typically reflects the six levels of yīn and yáng, while it is the quality aspect instead that reflects information about the hand vs. foot Meridians and their associated Organs. Contrary to the parameters I have been discussing so far, which reflect energetics on the same side of the body as the RYCK pulse findings (which is why it is selected as the treatment side), the pulse quality displayed appears to me to reflect energetic information about either the same or the opposite side of the body, depending on the type of pathology present.[110] There has been no prior mention of such an association, as far as I know, so I encourage readers to investigate this claim for themselves. I might reiterate here that the RYCK pulse quality is not the only energetic signal that frequently reflects the energetic state of the opposite side of the body, as the Ayurvedic Subdosha pulses appear to always share this contralateral energetic signal when it is present in the RYCK pulses. These findings, discovered in my clinical practice, are perhaps the most confusing aspect of pulse diagnosis, but they can be learned, and actually the RYCK quality and the Subdosha findings, which frequently share this contralateral nature, serve to confirm one's diagnostic interpretation, and as a result actually make pulse diagnosis easier! I'll return to this topic shortly. Let's examine the pulse quality aspect first, bearing in mind that my interpretation is once again decidedly different from any other that I have come across, but it is one that does not appear to conflict with the *Nèijīng*'s teachings.

In the classics, the character zào 躁 is first introduced in chapter 5 of *Sù Wèn*. "陰靜陽躁" (yīn is tranquil, yáng is turbulent), where zào is contrasted as the opposite of jìng 靜 or calmness, tranquility, stillness. Like yīn and yáng, this preliminary statement implies neither health nor pathology to either quality. They are just reflections of the yīn and yáng categories. Zào does however

110 This laterality issue is perhaps the biggest reason that RYCK diagnosis has been largely abandoned. My tentative belief is that one of the determining factors for which side of the body is expressed in the RYCK pulse quality is the number of Meridians/Organs that need to be treated to restore Constitutional energetics. If only one Meridian needs treatment (see the later discussion of Akabane imbalances), then the RYCK quality is expressed ipsilaterally (on the same side as the imbalanced RYCK width ratio, i.e. the side that needs treatment). If two Meridians are out of balance, e.g. cases of zàng or fǔ inflammation to be discussed later, then the RYCK quality gets expressed contralaterally (on the opposite side from the RYCK width ratio). These inflammatory conditions are quite common, so they would prevent a satisfactory understanding of what the zào and jìng qualities feel like to anyone not familiar with this phenomenon. If three Meridians need treatment, e.g. many cases of pain, the RYCK quality is once again expressed on the same side as the abnormal width ratio, i.e. the treatment side. Finally, in Extraordinary Meridian imbalances (eight types), the pulse qualities reflect the Opening Point's Meridian on the opposite side of the body, to be detailed in a later chapter. These variations in laterality appear to reflect numerical influences determined by natural law. One and three are yáng numbers, while two, four, six and eight are yīn numbers. As will be described later, cancer also happens to present with this opposite side pulse quality for zào and jìng, and the number of pulse depth abnormalities varies from four to six. It seems that laterality issues are reflections of basic yīn and yáng theory, expressed via odd versus even numbered phenomena.

clearly belong in the yáng category.[111] The next mention of zào is in *Sù Wèn* chapter 17. The text says that pulses near the surface (superficial) indicate (evil qì) is in the yáng (Meridians), which causes heat. If in addition the zào quality is present, it indicates the (heat is in the) hand (yáng Meridians). Wáng Bīng commented that if the zào quality was absent in a superficial pulse, the imbalance must be in the foot yáng Meridians. Here we see that zào indicates location, but not degree of pathology[112]. The next sentence discusses deep pulses, and indicates that these mean there is evil qì in the yīn Meridians. If the opposite of zào quality is also present (i.e. jìng 靜), the evil qì is in the foot (yīn) Meridians. This confirms that likewise, jìng does not indicate health, but only location, the same as for zào. Thus we can interpret this passage as indicating that superficial pulses represent a yáng Meridian/Organ condition, while deep pulses represent a yīn Meridian/Organ condition, and zào is the quality present in the pulse if the Hand Meridians are affected, while jìng is the pulse quality if Foot Meridians are affected.[113] It would be nice if all the subsequent passages using the character zào followed this simple locational paradigm, however such is not the case. Many passages in the *Sù Wèn* and *Líng Shū* employ zào specifically when discussing the identification of pathological conditions. What I propose to introduce at this point is the idea that zào has a general meaning, (turbulent is a reasonable choice) but that depending on context, such turbulence can be either a physiological indicator or a pathological one.[114] It is only by making this

---

111 *Sù Wèn* 20 groups zào with other pulse qualities classified as yáng. These include 盛, 喘 and 数. Thus this section of *Sù Wèn* distinguishes zào from 盛 shèng (flourishing), 喘 chuǎn (panting) and 数 shuò (frequent). Based on this passage, it would appear that a frequent, rapid or overly fast pulse is not the primary meaning of the zào quality, although there are practitioners who classify any pulse as zào whose rate is greater than 80 beats per minute. What we do know at least, is that zào indicates something that should be classified as yáng.

112 I am not aware of any passage in the *Nèijīng* that classifies hand Meridian imbalances as indicating a greater degree of pathology than foot Meridian imbalances.

113 The associations of zào and jìng with hand and foot Meridians has always been in reference to the Regular Meridians, as far as I can tell. In Chapter 12 I will present the thesis that they are likewise associated with the Extraordinary Meridians that are treated by an "Opening Point" on either the hand or foot Regular Meridians, but in that case manifesting on the contralateral side. This idea could not have been stated in the early classics, as the Opening Points were first identified more than a thousand years later. This discovery is another strong argument for viewing classical acupuncture theory as an evolving body of scientific knowledge, rather than as a discipline restricted to what is stated only in ancient texts.

114 A similar double meaning can be attributed to jìng (靜). Kroll defines jìng as stillness, motionless, or immobile. Clearly this cannot be a universally healthy state, since movement is also a part of any healthy life; immobility is often a pathological sign. This must be analogous to the couple yīn and yáng, which are also characterized by stillness and movement respectively. However, Kroll also notes the (later?) association of jìng with Daoist and Buddhist meditation practices, where it has then taken on a positive connotation.

distinction that clinicians can avoid interpreting hand Meridian imbalances as generically more serious than foot Meridian imbalances. I should emphasize that aside from its technical use in the RYCK method of diagnosis, zào (turbulent) generally has a negative connotation, while jìng (tranquil) has a positive connotation. These terms, which were introduced in the *Nèijīng* did not appear in the *Nánjīng*, and are not generally listed in tabulations of the classical pulse qualities in later works on Chinese medicine.[115] How might one account for such an eventuality? On a recent visit to Guangzhou, I had a discussion with a teacher at a tài jí quán (太極拳) school there who had a cup with the character jìng (靜) on his desk (Fig. 19). He explained that tranquility was the goal of training in this art. I asked him what word in Chinese expresses the opposite state, and he readily replied zào, so that this general sense of desirable and undesirable states represented by these terms has lasted in Chinese culture down to modern times, but we cannot rule out the possibility that in pre-classical Chinese these characters were not originally associated with desirable and undesirable states of being, similar to the way we think of the terms yīn and yáng.[116] An analysis of the Chinese character might give us some insight into the original formulation of zào, and how its early technical use in medicine (in the RYCK system) might have morphed into a more generalized term for unfavorable psycho-physiological states that are not so different from the connotations of "turbulent" in English. Post-*Nèijīng* commentators have routinely missed the distinction between these separate usages of zào, and have thus missed a key aspect of the RYCK methodology. Such a hypothesis would possibly add to the rationale for the rarity of this method in the contemporary practice of pulse diagnosis in Chinese medicine.

115 There is philological evidence that the *Nánjīng* may actually be a pre-Han dynasty text, and that its terminology may be representative of an older stratum of characters than that of the other extant classical medical texts. In this regard, the *Nèijīng* may represent the transitional period when both meanings for zào found their way into separate chapters of the compiled text. My speculation on the earlier dating of the *Nánjīng* is based on personal communications from Dr. Igor Simonov.

116 Kroll (page 220) gives dòng (動 movement) as the antonym for jìng (靜) in pre-modern Chinese. Movement does not have the negative connotations associated with turbulence, bolstering my contention that originally both zào and jìng were generically neutral terms as regards desirable vs. undesirable states.

Figure 19: Jìng (靜), the state of tranquility [117]

Sù Wèn Chapter 5 says, "the 6 Meridians (Channels)[118] are streams." Talking about 6 Channels is in reference to the 3 yīn, 3 yáng classification, amplified in *Sù Wèn* chapter 6, where the beginning points (roots) of all 6 Channels are said to be located on the feet at the jǐng /well points. This fits very nicely with the idea of a river or stream of qì flowing from the feet upwards from well to spring to stream to river to sea and onward, a major concept in understanding the nature of the 5 shū points. There is no discussion here of hand and foot branches of these 6 Channels, nor is there in *Sù Wèn* chapters 9 and 40 where the RYCK system is introduced. Essentially this system is simply based on the idea that 6 parts of yīn and yáng flow up the body from their origin

117 Source: Peter Eckman.

118 There are many possible translations of jīngmài into English, with advocates arguing for their preferred choice. My own opinion is that the nature of Chinese characters is such that any given usage depends on context for deciding on the best translation. Personally, I was originally taught to use Meridian as the standard, but contemporary scholars tend to prefer terms like Channel or Vessel. I believe these are all correct, as long as the reader recognizes the wider range of possible interpretations. I tend to use Meridian as my default translation, but when there is a clear analogy to the flow of water I often will switch to Channel or Vessel.

in the feet. It is of interest to speculate on why the ancient Chinese thought these "streams" flow upwards in the human body, but downwards in inorganic nature. I imagine the answer to be the recognition that living beings function by opposing entropy, thereby creating order rather than dissipating it. This explanation fits my understanding of Chinese medical thought as being based on scientific observations and conclusions, even if the terminology is quite different than that found in the later development of the Western scientific worldview.

The role the zào quality plays in the RYCK diagnostic system is not explained until *Líng Shū* chapters 9 and 10. Chapter 9 is entitled "(From) Beginning (to) End" (終始). A 1909 text by Sun Dingyi glossed this title as referring to "the locations where the jīngmài appear and where they end." This is of course reminiscent of *Sù Wèn* 6, where the roots of all the Meridians are said to be located on the feet. Now *Líng Shū* chapter 9 reintroduces the way of parsing the 6 Channels indicated by RYCK pulse diagnosis briefly mentioned in *Sù Wèn* 17. The pulses of each of these 6 Channels can now be divided into a hand branch, which is associated with the zào pulse quality, and a foot branch, which lacks the zào pulse quality. Neither branch is said to indicate a more severe degree of pathology, only a difference in location. Let's keep the image of a flowing stream in mind as we examine the character zào 躁[119] that has now been introduced into the RYCK method.

Why was zào chosen as the term for the indication used to differentiate pulses of the hand versus the foot Meridians? I don't know if prior commentators have ever considered this question, and proposed an answer, but it would seem that in this context, zào must refer to something that is not inherently pathological. The radical on the left (#157) indicates the foot,

119 The original meaning of the character zào 躁 (as a specifically defined type of pulse finding) is nowhere described in the *Nèijīng*, other than its juxtaposition as the opposite of jìng 靜 (tranquil). Schuessler (2007) in his *Etymological Dictionary of Old Chinese* does not include this character for zào, but does cite a perfect homonym (same tone, page 601) 造, that meant to "to proceed to" or "reach to," as attested in the *Hòu Hàn Shū*, where it was used to describe mountains reaching to the sky. Kroll gives several definitions for this character in classical Chinese, including restless, unquiet, unsettled and agitated, but he does not refer to its use as a technical term in medicine. It makes sense that a technical term will often have a relationship to the same word as used in common language, but often with a more restricted and nuanced definition. Schuessler's homophonic zào is suggestive of a possible route by which the technical meaning of zào reflected both its homonym and its more general meanings, as I propose in the remainder of this paragraph. Admittedly, my interpretation has not been previously mentioned by other commentators, at least to the best of my knowledge.

while the phonetic on the right (series 764)[120] indicates birds singing in a tree. On first impression there isn't any obvious connection to any of the commonly proposed meanings of zào (agitated, turbulent, rapid or some variation of these adjectives), nor to any claimed pulse signature for zào. Let us keep in mind however, both the *Sù Wèn* chapter 6 assertion that the 6 Channels have their origin as streams originating on the feet at the jǐng / well points, and Schuessler's homonym of zào that means "reaching to". If we think about this pulse related zào character, there is one striking finding: the radical points to the lower part of the body, while the phonetic points to something in the upper region (analogous to foot and hand). So we have indications of two separate and opposite parts, bottom and top. And we already are oriented towards a flow of qì from bottom (foot) to top (hand). In a body of water flowing along, is there any natural analogy that makes sense of the terms rapid and agitated/turbulent? I think there is actually a rather obvious one: the rapids that are found in rivers where the flow is narrowed or obstructed by geographic factors such as mountain terrain, rocks, trees, or any kind of debris. Such rapids are characterized by both turbulent (agitated) and faster (rapid) flow due to the same volume of water having to pass a smaller dimension of space. This is exactly how I experience the 躁 quality associated with the hand Channels, when the pulse is examined at the LU 8 location. The zào quality has a wider presence at a deep level (its origin) on palpation, but as the pressure of the examining finger is released the pulse becomes narrower and might almost disappear before widening out again nearer the surface (its most distant manifestation). To me this suggests the image of something obstructing the flow from below to above. We know that the foot branches of the 6 Channels connect to the hand branches, but these connections are not part of the Principal Channel trajectory. Rather they are secondary branches that make an ideal place for the flow of qì to be subject

120 These numbers refer to Wieger's listings.

to change (restricted), leading to the appearance of this zào quality[121]. Figure 20 illustrates this hypothesis.

Figure 20: River Rapids [122]

I have used this interpretation of zào in diagnosing my patients for quite some time and have found it to be accurate and clinically very useful. I diagnose the quality from the radial pulse alone, as I never compress the carotid artery strongly enough to look for this quality difference. Earlier, when I mentioned methods of differentiating excess (solid) from deficient (hollow) pulses, a similar (though not identical) pulse image often appears, but there it was not described as zào 躁 versus jìng (靜), but rather as shí 實 versus xū 虛, and the reader is encouraged to be alert to the differences in these pulse images. My experience is that by assessing zào or not zào

121 It has been suggested that the interpretation of classical Chinese characters should not go beyond those offered in the long history of Chinese textual commentary, and that interpretation by reference to the radical and phonetic components is not a sufficient basis (personal communication from Dan Bensky). I offer these 2 quotes in rebuttal (from *Classical Chinese Medicine* by Doctor Liu Lihong, pages 75 and 76), "When we look at a character, we need to do more than just look it up in a reference text. We need to analyze its construction. This means not only discerning the "radical" or "signific" part of the character (i.e. the semantic part, the component of the character that indicates its range of meanings), but also the phonetic part, as both are highly significant for its overall meaning." "To fully understand Chinese characters, we must of course rely on secondary resources, as well as on an intuition about the inherent structure of the character. Both of these are indispensable." I would be remiss not to emphasize the clinical efficacy resulting from, and inspired by this analysis of the character 躁.

122 Source: https://www.newzealand.com/us/lake-taupo/

(i.e. jìng 靜) as the quality of the pulse on the side that is opposite to the one that has a RYCK imbalance, one can often (whenever only one other Organ is out of balance with the Constitutional Organ) determine if the Constitution and the Condition are each separately associated with either a hand or a foot Meridian.[123] Patients with zào quality pulses on the side without a RYCK imbalance in such cases (2 Meridians out of balance) usually have either Fire or Metal Element (hand branch) Constitutions, which focuses my diagnosis on these Elements, while patients without the zào quality pulses, in the same location and situation, usually have either Earth, Wood or Water Element (foot branch) Constitutions, and their differential diagnosis follows accordingly. Again, the Constitution usually presents with a RYCK size imbalance on only one side of the body, however it can occasionally be bilateral, apparently reflecting the need to address an Extraordinary Meridian activation, as mentioned. Further research is needed in that regard (See Chapter 12). Interestingly, with unilateral RYCK imbalances, the pulse quality on this same side as the RYCK imbalance, as in the examples just cited above, appears to reflect the Condition, so checking for the presence or absence of the zào quality on this ipsilateral side can be helpful in confirming the Conditional diagnosis as well. To repeat what I said earlier, the Subdoshas and the RYCK pulse quality on each side of the body always agree with each other about whether a hand branch or a foot branch of a Great Meridian is expressing itself. Pitta Dosha always corresponds to a hand branch. Kapha always corresponds to a foot branch. Vata (Fire type) always corresponds to a hand branch. Vata (Water type) always corresponds to a foot branch. Vata/Kapha always corresponds to a hand branch. Pitta/Vata always corresponds to a foot branch. Pitta/Kapha (Fire type) always corresponds to a hand branch. Pitta/Kapha (Water type) always corresponds to a foot branch. I must reiterate that this

123 To be clear, one first identifies a RYCK abnormal ratio on one wrist (which informs about the yīn yáng balance in the Constitutional Element) and then one identifies the pulse quality on both the same side and on the opposite wrist as zào or jìng (which informs about the Elements currently involved in the individual's energetic state). In many cases the quality found on one wrist reflects the energetics on the opposite side of the body, which might pertain to the Constitutional Meridian, but not always. One exception, and not a rare one in my experience, is in more complex illnesses where the Constitutional Organ is calling on either its Parent or Child Organ for assistance. In such cases the pulse quality reflects the energetics on the RYCK imbalanced side ipsilaterally. This finding is a confusing phenomenon, and is best clarified by studying the case histories to be presented. The reader is alerted that similar contralateral energetic information conveyed by zào and jìng qualities also manifests when Extraordinary Meridian imbalances are present (see Chapter 12).

interpretation of zào elaborated above, and of the RYCK system in general, is not described in the *Nèijīng* or any other classical texts, to my knowledge.

Using the RYCK pulse quality on both sides can often narrow down the choice of which of the 6 Great Meridians need treatment for the Condition, as just explained, but I usually include treatment of the Constitutional Meridian to ground the treatment in the individual's sense of self or Spirit. In such cases the Constitutional Meridian is treated on the side with the RYCK imbalance, as is the Conditional Meridian. In my experience, the choice of hand or foot branch needing treatment for both the Constitutional and Conditional aspects can be confirmed based on which ones of the 12 pulses of the zàng fŭ have been displaced from their proper depth (per *Nánjīng* 5). Both the Constitutional and Conditional Organs/Meridians will show abnormalities in their pulse depths. If there are no depth displacements for the pulses of any of the Organs, but there are still RYCK ratio differences and symptoms indicating imbalance being present, then such a finding suggests that while the Five Elements are balanced with each other appropriately, the yīn and yáng Organs or Meridians of the Constitutional Element are separating, rather than working together; obviously an issue that needs to be addressed in treatment. This situation corresponds to the pathological category 1 (yīn yáng imbalances), which I cited earlier in the passage from *Líng Shū* 9. If however, there are depth displacements of any Organs noted on the pulse, this finding suggests that the Organs of the Five Elements are not correctly balanced with each other (wŭ xíng imbalances representing pathological category 2 in *Líng Shū* 9), and that such RYCK findings might indicate a need to treat Elemental imbalances first or rather to treat both the Constitution and the Condition simultaneously, which is the typical approach used in CCA.

As mentioned earlier, I generally use Kuon's method of treating both Constitutional and Conditional Meridians on the same side of the body. That strategy has been effective, as nicely validated in the case histories cited in the article I published on the cases I treated in Kunming, China in 2017.[124] In summary, my present understanding is that the presence or absence of the

124 Eckman, P. "Constitutional/Conditional Acupuncture: Basic Theory and Clinical Practice", The Journal of Chinese Medicine (England), Issue 117, pp. 45-53, 2018.

zào quality might differ on left and right sides of the body. It appears that the presence of the zào quality in healthy individuals with balanced pulses correlates with their having a hand Meridian based Constitution (Fire or Metal), whereas if the zào quality is absent their Constitution is based on a foot Meridian (Water, Wood or Earth). That is one reason I think the RYCK pulses do contribute both Constitutional and Conditional information. The ratio of the RYCK pulses however reflects whether or not the Meridians and/ or Organs are functioning in an optimal manner. Again, these statements are challenging to grasp at first if they are only written descriptions, but careful attention to the case histories should make them clearer. Figure 21 may help in visualizing my pulse quality thesis. Zào pulses are concave in shape, while non- zào pulses are convex.

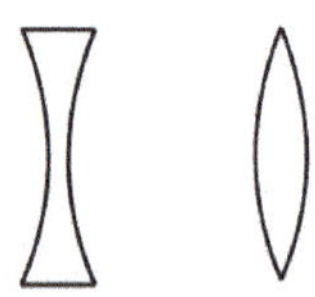

Figure 21: Concave (Zào) and Convex (Jìng) Pulse Shapes [125]

There was a time when I was taught there was such a notion as "the Chinese anatomical position", with upraised arms [126] (See Figure 22). If that is historically verified, it would certainly support my hypothesis of how the ancient Chinese imagined the zào quality to manifest the transformation of yīn to yáng in the human body as we move from the feet (Earth) to the hands

125 Source: Peter Eckman.

126 This Chinese anatomical position seems almost implicit in classical Meridian theory. All the yīn Meridians flow from below upwards (Earth to Heaven), as seen in the numbering of their points. All the yáng Meridians flow from above downwards (from Heaven to Earth), as seen in their point numbering. The only position of the body that reflects this yīn yáng dynamic is with the hands raised above the head. The following quotation shows that this terminology has been part of the oral Chinese medicine tradition, even if no images or explicit Chinese textual citations have been found to document it: "The … Chinese anatomical position, the position in which a person stands to accurately display the energetic aspect of anatomy, is in a hands-up position…with arms extended above the head with palms facing forward." From *Chinese Holistic Medicine in your Daily Life*, Steven Cardoza, Llewellyn Publications, 2017. Also note, "In western anatomical position, we are looking at a person standing who is facing forward, with their legs slightly apart, with arms by their sides, and their palms facing forward. In eastern anatomical position the feet may be positioned together and the hands may be held above the head rather than by the sides of the body," accessed from the Internet 10/12/19 at: https://3dmusclelab.com › Anatomy And Physiology.

(Heaven). The common belief is that yíng qì flows inside the Channels, while wèi qì circulates outside them. Yíng qì is yīn with respect to wèi qì, so it makes sense that the pulses (yíng qì) felt emanating from the jīngmài reflect this flow from the most yīn part of the body (resonating with Earth) towards the most yáng part of the body (resonating with Heaven).

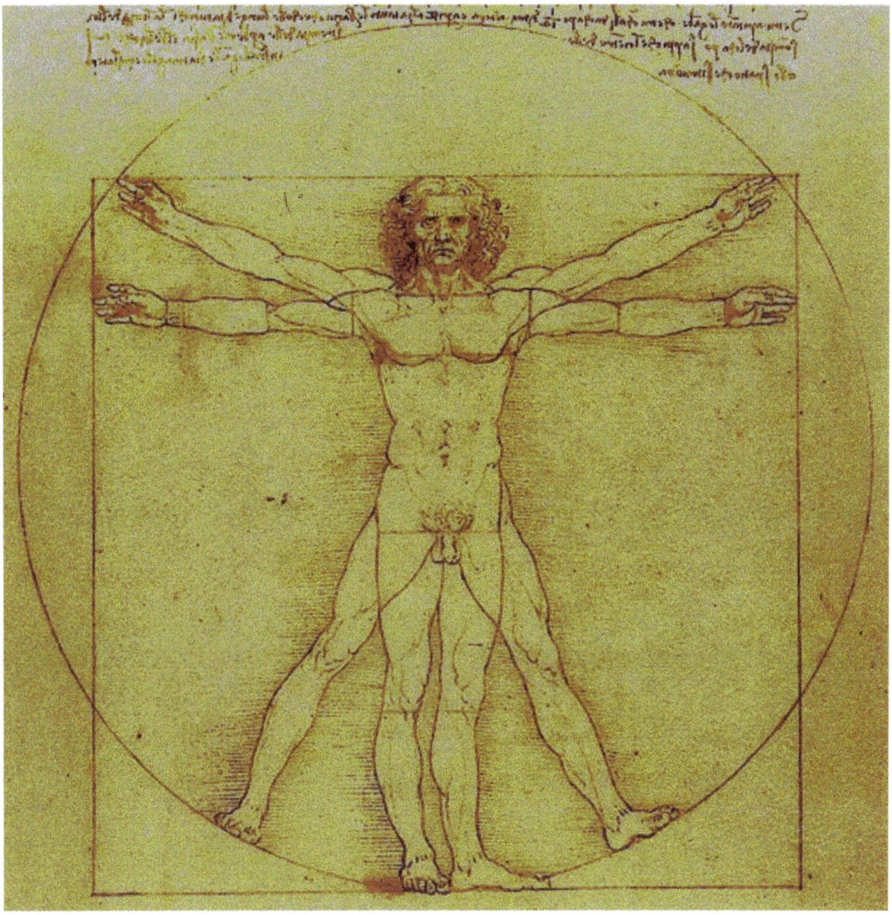

Figure 22: Leonardo da Vinci's Vitruvian Man
The Chinese Anatomical Position? [127]

127 Although I have been unsuccessful in locating a Chinese illustration of the anatomical position in question, it is interesting how similar da Vinci's ideas were to those of Chinese medicine. In addition to the upraised arms, his drawing shows the "Three Powers" of Heaven, Earth and Man as represented by the circle, the square and the body, superimposed to indicate their resonant relationship, with the navel at the center. The dimensions all illustrate numerical ratios reminiscent of Chinese numerology. The drawing is in the Public Domain per Wikimedia Commons. The original drawing is dated c. 1492 and is located in the Gallerie dell'Accademia in Venice, Italy.

CHAPTER 7

Macrocosm/Microcosm Balancing: A Reinterpretation of the Língshū *and* Nánjīng *Theory of the 50-Fold Circulation of Qì*[128]

128 This chapter mainly presents material developed by Dr. Igor Simonov. He discovered phenomena in the pulse that were consistent with teachings in the *Nánjīng*, which had not previously been reported in modern times. It is Simonov's contention that these pulse findings were among the earliest discoveries in ancient China that allowed for a rational approach to acupuncture treatment. This approach Simonov named "Archaic Acupuncture". In 2019 I carried out an experiment, using first year students at the Beijing University of Chinese Medicine, to test one of Simonov's principal theories about the true meaning of passages in the *Nánjīng*. I will describe this experiment in detail, but I want to emphasize here that the findings were surprisingly consistent with Simonov's interpretation. Simonov has not written any books on his work, but is currently focused on developing a new translation of the *Nánjīng*. It is with his gracious personal approval that I am reporting several of his findings in this chapter, and if I have misunderstood any of his ideas, it is my mistake entirely. Thus I would highly recommend that interested readers study with him in a Skype class, as I did, for a very educational experience. His website www.pulsediagnosticum.com includes contact information for this purpose, in addition to many other interesting articles. The slide rule pictured in Fig. 24 can be purchased from him for a nominal fee, and is invaluable for designing treatments to synchronize macrocosm with microcosm in any clinical encounter. I have added some of my own literary research (principally from the *Língshū*) in this chapter, so references to that text should be assumed to reflect my own ideas rather than those of Simonov.

Figure 23: The Four Seasons [129]

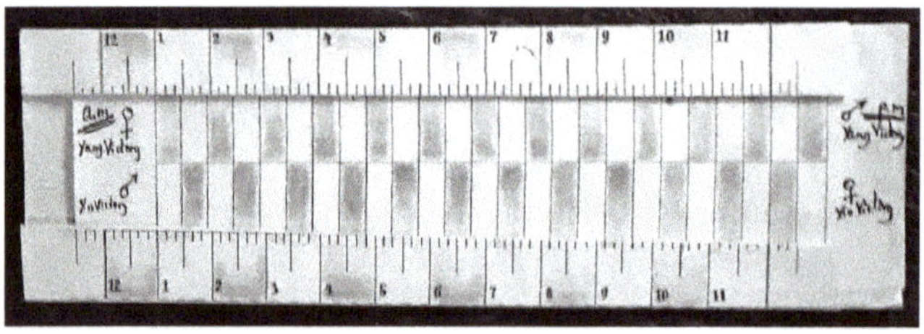

Figure 24: Simonov's Slide Rule [130]

The third pathomechanism presented in *Líng Shū* chapter 9 has to do with the failure of the individual to adapt to the macrocosmic changes exemplified in the seasonal cycles of each year. As I indicated earlier, I believe the choice of seasonal adaptation is in this case a general reference term implying all cyclical phenomena that result from the interaction of macrocosm and microcosm, and perhaps more specifically those between the sun and the earth. Before going into detail, I would like to say a few words about the relationship of the two most important classical texts regarding the practice

129 "Allegory of the four seasons with Chronos". Oil painting by Frenc Plachy, 1874. Source: wikimedia.commons.org, in the public domain. I particularly like the inclusion of Chronos, implicating time as the progenitor of the seasons.
130 Author's photograph of the device devised by Simonov to depict the 50 daily cycles of yīn and yáng. I have set solar noon at one o'clock, just as a possible example.

of acupuncture, namely the *Língshū* and the *Nánjīng*. My understanding is that each of these texts are compilations from multiple authors, presenting material from different time periods as well as different schools of thought, so it is not surprising to find conflicting material in each. Some historically prominent commentators, such as Liao P'ing, have strongly argued that the *Nánjīng* is not really about difficult issues in the *Nèijīng*, as is commonly taught, but is rather a heretical break from the *Nèijīng*'s teachings, instigated by Daoist theoreticians. I will try to refute this argument by showing the important similarity, including almost identical statements, in both texts about the topic covered in this chapter.

Língshū chapter 5 is titled "Roots and Outcomes".[131] Its very first line is, "Heaven and Earth mutually affect each other." I think we can interpret this as shorthand for the well-known trinity of Chinese medical philosophy: Heaven/Earth/Man. While these three powers interact, they are also part of a hierarchy in which Heaven is the initiator and Earth is the recipient. Man reflects the outcome of their interaction, and is influenced by both. Thus this early chapter sets the stage for a presentation of the relationship between macrocosm (Heaven and Earth) and microcosm (Man). Following this introduction in the *Língshū*, the four seasons are mentioned, along with their Heavenly nature in terms of the balance of yīn and yáng qì emitted by the sun (which determines the seasons). Since all life on earth is ultimately dependent on the sun's radiant energy, the details of how the yīn and yáng qì vary during any time segment (day, month, season, year, etc.) become important in devising treatments aimed at altering the state of yīn and yáng in patients. The next passage of chapter 5 brings up the question of where to drain and where to supplement. Obviously, for acupuncture, this involves knowing the details of the zàng fǔ and their Meridians, which is the next topic discussed. The text states, "He who fails to know root and connection, as well as the 5 zàng and 6 fǔ will… let the proper qì go away." I should point out here that there is an alternate translation for this chapter's title, i.e. Root and Connection[132],

131 根結. Author's translation, as is the chapter's first line 天地相感. Here the term xiāng gǎn is used instead of the more familiar term gǎn yìng, often translated as resonate, with which it is roughly synonymous. This and the other translations of brief Chinese quotations in this chapter are typically paraphrased from Unschuld unless otherwise noted, with the substitution of Meridian for conduit, or vessel at times.

132 That is Unschuld's translation of the chapter title.

which has just been cited as essential knowledge. Then the text does something strange. Instead of keeping the focus on root and connection, it switches to a focus on the end and beginning (終始, shǐ zhōng) of each Meridian. This is not only a conundrum, but it is crucial to some of the treatment recommendations that Simonov proposed for balancing microcosm to macrocosm, and which I have found clinically potent in doing acupuncture (which is why I am going into such a detailed textual analysis here). The final sentence in this segment of the text states, "The subtlety of the nine needles lies in the importance of the ends and beginnings (of the Meridians)". The important phrases here are "nine needles" and "ends and beginnings". The nine needles include several that are not meant to penetrate the skin, a technique of Simonov's that I will describe, and in using this technique it is important to utilize as many Points along the chosen Meridian, from beginning to end, as possible.

At this point however, the text reverts to a presentation of each Meridian's root and connection, with imagery proper to the theory of opening, closing and pivoting of the 3 yáng and 3 yīn Great Meridians. The order is taiyang (opens), yangming (closes) and shaoyang (pivots) for the yáng Meridians. For the yīn Meridians, taiyin opens, jueyin closes and shaoyin is the pivot, presented in that order. It seems to me that even with just this initial material from *Língshū* 5, one can conclude that there are two separate theoretical frameworks being mixed together.[133] I believe this might possibly be a

133 There is an alternate interpretation possible in which the two aspects of this chapter in *Língshū* might be reconciled. The ordering of open, close and pivot presented here is almost identical to the order of penetration of xié qì presented in the Shānghán lùn, and this order was also discussed in chapters in the *Sù Wèn*, added by Wáng Bīng (Táng dynasty, 618-906) to replace those which were lost. These chapters provide a basis for wǔ yùn liù qì (五運六氣) and zǐ wǔ liú zhù (子午流注) theories that are also versions of macrocosm/microcosm ideas pertaining to Chinese medicine elaborated in the later Sòng (960-1264, e.g. He Rou Yo's *Zǐ Wǔ Liú Zhù Zhēn Jīng*), Yuán (1264-1368, e.g. Dòu Hàn qīng's (竇漢卿) *Zhēn Jīng Zhǐ Nán* and Míng (1368-1643 e.g. Xú Fèng's (徐鳳) *Zhēn Jiǔ Dà Quán* and Gāo Wǔ's *Zhēn Jiǔ Jù Yīng*) dynasties. As far as I can determine, there is no evidence that these biorhythmic theories existed as such at the time the *Nèijīng* and *Nánjīng* were written. Certainly the use of Extraordinary Meridian Command Points in the Líng Guī Bā Fǎ and Fēi Mǎ Bā Fǎ approaches could not have existed prior to the Sòng dynasty, as these Command Points were elaborated much later. The open, close, pivot order of Meridians in *Língshū* is slightly different than the *Sù Wèn* order, differing only in whether jueyin precedes or follows shaoyin, and there is disagreement as to which is the pivot of the yīn Meridians and which is their closing Meridian. Taking the *Sù Wèn* version, I have previously written a monograph linking this order to the Celestial order of the trigrams of the *Yìjīng*, an idea that certainly fits into a discussion of macrocosm/microcosm relationships. See Eckman, P. *The Book of Changes in Traditional Oriental Medicine*, The Traditional Acupuncture Foundation, Columbia, Maryland, 1988. With this in mind, it is possible that the two apparently distinct foci of *Língshū* 5 may have originated from the same school of macrocosm/microcosm thought. It should be noted however that the zǐ wǔ liú zhù explanation of how the Meridians and Points are opened differ completely from Simonov's interpretation, and lead to quite divergent acupuncture interventions. Simonov does not reference any materials dated later than the Hàn dynasty.

creation of editors and copyists who did not fully understand the material they rearranged. The opens, closes, pivots model just presented is not textually connected to a discussion of macrocosm/microcosm, as were the initial lines. Furthermore, once this bit of open, close, pivot material is done, chapter 5 returns to a macrocosm/microcosm discussion, never to return to open, close, and pivot material. There is one hypothesis I can suggest for why this chapter ended up mixing these two discussions together: Both my own and Unschuld's translation of the chapter title are valid, and this reflects an aspect of the Chinese language. At times the same phrase can have two entirely different meanings, and because of that fact, mistakes such as this potential one are almost inevitable. In any case, my intention is to attend to the macrocosm/microcosm thread, so let's look at the next passage in the *Língshū* text.

Here is the famous statement that is paraphrased in chapter 1 of the *Nánjīng*, and which is at the root of Simonov's research: "As for the 12 conduits, the qi in 1 day and 1 night pass through 50 circuits (zhōu 周)…"[134] The *Língshū* passage continues, "… and supply the 5 zàng with jīng. One grasps mài kǒu (脉口) and counts their arrivals." There are various terms here that are often considered to be interchangeable, mài kǒu (脉口), cùn kǒu (寸口), and qì kǒu (氣口), merely reflecting the author's historical epoch, but Simonov makes a strong case for the original meaning of cùn kǒu in the *Nánjīng* as being quite different from its later usage. What is clear from this *Língshū* passage however is that the subject under discussion is radial pulse diagnosis. One of Liao P'ing's objections to accepting the *Nánjīng*'s teachings as accurate was that the *Nánjīng* emphasized radial pulse diagnosis over the use of pulses throughout the body for diagnosis, but here we see this early passage in the *Língshū* definitely emphasizing radial pulse diagnosis, although the details of how one performs and interprets radial pulse diagnosis were not explicitly described. So, what are these 50 diurnal circuits, what is it in the radial pulse that reflects them, and what implications do they provide for guiding acupuncture treatment in order to coordinate microcosm with macrocosm, or the individual with their environment? The *Língshū* goes on to elaborate on the 50 cycles, but doesn't provide concrete advice on how to apply its ideas in practice.

134 The *Nánjīng* version states, "A person, in the course of one day and one night, …the pulses (mài) proceed through 50 passages."

Chapter 15 of *Língshū* is titled, per Unschuld, "The 50-fold Circulation". Actually, a literal translation is simply, "50 Yíng" (五十營). Yíng (qì) should be a familiar term, often translated as nourishing energy, which is said to circulate within the Meridians as opposed to wèi (qì), often translated as protective energy, and which is said to circulate outside the Meridians. Thus this chapter's discussion is about the qì that circulates within the Meridians having 50 parts. It is a relatively short chapter, but it specifically mentions Heaven (天) and the sun (日) as the reference points for when the water clock (outflow clepsydra) descends through 100 marks (kè 刻) during the course of one day and night. The meaning of kè is actually one quarter of an hour, so one day and night would take 25 hours, rather than our modern 24 hour standard.[135] This discrepancy accounts for some of the difficulty in translating classical Chinese statements about time into contemporary measurements. It is much easier to comprehend the Chinese text about 50 diurnal circuits if the measurements are based on a 25 rather than a 24 hour day. In the very first paragraph of this chapter it says, "This way daytime and nighttime can be distinguished." In the explanation Simonov proposes, this emphasis on distinguishing between what happens during daytime and nighttime turns out to play a central role, one that is elaborated in the *Nánjīng*. Although surprisingly, "hour" is never mentioned in the text under discussion, it is interesting that the Chinese word for hour (shí 時) can be equally correctly translated into English as time, season, or period. To decide which meaning is appropriate, the context is necessary.

In the next passage, *Língshū* 15 says, "The passageways of the qì through the human body cross each other". Ordinarily we might expect the passageways of qì to refer to the Meridians, but in this case a different meaning

---

135 The Chinese have made use of two different systems of measuring time. We commonly see reference to "12 double hours" in which the various Meridians are said to be at peak activity. This reflects a lunar calendar, as there are 12 months (moons) matched to the 12 Meridians. *Língshū* 15 however is based on a solar calendar, as indicated by the mention of the sun in the text, but also the importance of the separate daytime and nighttime segments of the day, which have nothing to do with the moon. As mentioned, the word "hour" does not occur in the text, only circuits (zhōu 周) and marks (kè 刻), also translated as "cuts" or "quarters". To keep the Chinese time systems reasonably comparable, these 50 circuits might be thought of as "25 solar hours". As a consequence, each of these "circuits" would last 28.8 of our contemporary minutes. Simonov has designated the term "period" for these mysterious circuits, each period being composed of 2 kè. The results of this system of time measurement are consistent with the 50 periods covering the 100 kè of each day during which the water clock completely empties.

would seem to be implied, as the Principal Meridians don't generally cross each other. We should remember that this statement is in the context of the 50 periods. While no definite interpretation of that bit of information is possible, I would point out that Simonov's theory presents an interpretation in which each yīn period of 28.8 minutes is followed by a yáng period of equal duration, and vice-versa. That situation seems to comport with the passage in question. The text does present a gloss on the idea of the passageways crossing each other. It says, (That means) "the numbers of their passages are identical." Jumping ahead for just a moment, the reader can look at Simonov's slide rule (Fig. 24) at the beginning of this chapter and see that the little dark rectangles (yīn) are always followed by a clear rectangle (yáng), and that counting all the dark and clear rectangles, the number of each is identical. The format of this slide rule comports with the idea of yīn and yáng crossing each other, while maintaining an equal number of passages adding up to 50. I think visual aids such as that one make these abstract concepts easier to comprehend. *Língshū* 15 concludes with a passage stating that, "In the course of 13,500 breathings, the passage of (yíng) qì covers 50 circulations in the human body, and the water (in the clepsydra) descends by 100 marks." Thus each of these "circulations is equivalent to 2 marks on the clepsydra. It is noted that the sun has now passed through all 28 constellations, so any Heavenly influences have been fully transmitted. What remains is for us to figure out how to use this information in a practical way to better coordinate ourselves with the cosmos around us.

There are two additional chapters in the *Língshū* that bear on the topic under discussion. *Língshū* 18 is titled, "Yíng and Wèi, Generation and Meeting". This chapter states that the yíng are in the Meridians while the wèi (衛) are outside the Meridians, but both are forms of the essence (jīng 精) qì. Although the text does not say so explicitly, my impression is that the essence qì of the body (what I think TCM refers to as the zhèng qì), being in constant circulation, acquires different names depending on where it is located and what it is doing. When it is in the Meridians it is yíng qì and when it is outside the Meridians it is wèi qì. The text here implies this notion by stating, "Yīn and yáng qì penetrate each others realms." Thus there must be a constant balance between these two forms of qì. If yíng qì moves out of a Meridian, it becomes wèi qì. In other words, an increase in wèi qì is accompanied by a decrease in yíng qì, and vice-versa. According to the text here, "Wèi qì (also) passes through the yīn 25 times and it passes through the yáng 25 times". These notions are employed in deciding the appropriate acupuncture treatment to harmonize the organism with the environment by addressing both yīn and yáng, and yíng and wèi. It is a mistake to assume that these two pairs of terms are synonymous, as this text chapter makes clear. Practical use of these ideas will be elaborated when it is time to discuss treatment strategies. In any case, chapter 18 states that both yíng and wèi circulate without stopping, and after 50 (periods) a grand meeting occurs at midnight [136] in the hand taiyin Meridian, where the cycle starts all over again, like a ring without end. The text further states that the yíng qì moves in the yáng conduits 25 times and in the yīn conduits 25 times. Also, the wèi qì passes through the yīn (territory?) 25 times and it passes through the yáng (territory?) another 25 times, divided by day and night. Since the wèi qì circulates outside the conduits, we must think of yīn and yáng somewhat differently when it relates to wèi qì circulation. The premise in CCA however is that all diagnoses must be reflected in the pulses, so let us wait until the presentation of Simonov's interpretation of the *Nánjīng* to address that issue, since terms such as yíng, wèi, yīn and yáng are all correlated to specific pulse criteria by Simonov. For now, I will simply share an analogy that I explain to my patients when I perform treatments (via contact needling) based on yíng and wèi qì. These terms refer to movements similar to those of the waves in the ocean. Each wave may be heading to shore or retreating from the shore, but the water in the ocean is the same regardless of its direction of movement. The waves moving towards the shore are like the wèi qì in that they depart from their source in the ocean (analagous to the zàng fŭ and their associated Jīng Luò), displaying a yáng direction of movement. The waves returning towards the ocean can be thought of as similar to the yíng qì which return to the Meridians and their associated Organs, a yīn direction of movement. Ocean waves and qì waves are similar natural phenomena, and treatment to synchronize yíng and wèi qì reinforces the natural functions of the human organism.

136 Midnight is the time of the first Earthly Branch, zǐ. Although the circulation of qì is like a ring without beginning or end, it makes sense to describe this cycle starting from midnight when the yīn is at its maximum and the yáng is just beginning to grow.

I believe that this example illustrates the obscure passages in the classics about yíng and wèi qì in a manner that patients can understand without getting confused by technical terminology. The specific treatments to be described must be done at definite times, easily chosen by using Simonov's slide rule. The goal of such treatments is to coordinate the movements of yíng and wèi qì (microcosm) with the movement of celestial qì from the sun (macrocosm).

Finally, *Língshū* chapter 76, titled "The Movements of Wèi Qì," reiterates that wèi qì also completes 50 circulations during one day and night, during which time the water in the clepsydra drops 100 cuts. "One follows the different lengths of a day as the norm and pierces accordingly. When the time sections are carefully observed, the course of a disease can be predicted. If one ignores the time sections and acts contrary, none of the hundred diseases will be curable." Important notions here include the ideas that specific time units must be attended to for carrying out the harmonization of microcosm to macrocosm, and that these time segments vary as each day is composed of changing parameters. The specific temporal details of each day must be attended to in this kind of treatment planning.

I think many practitioners of Chinese medicine have scratched their heads over these passages in the *Língshū*. As there are no practical treatment guidelines, I think that most practitioners simply ignore these passages, but perhaps that is one of the main reasons the *Nánjīng* was written, and has been considered to be a commentary on difficult to understand statements in the *Nèijīng*. *Nánjīng* chapter 1 reprises the preceding subject matter from the *Língshū*, "Cùn kǒu constitutes the great meeting point of the Meridians, where the hand taiyin pulse can be felt.... In 1 day and 1 night, a person has

13,500 breathings[137], and the movements of the pulse proceed through 50 periods (zhōu 周) circulating through the body while the water (in the clepsydra) drops 100 marks (kè 刻). Róng (荣)[138] and wèi move through the yáng 25 times and through the yīn 25 times. This constitutes one cycle. Meeting again at cùn kǒu after 50 periods is the beginning and end of the (circulation through) five zàng and six fǔ. Thus the pattern is obtained from cùn kǒu."[139]

My impression is that many practitioners and their teachers assume that the 25 circuits through the yáng must refer either to the Principal Meridians, the yáng Meridians specifically, or to the yáng time of day, when the sun is shining. Likewise the 25 circuits through the yīn would take place either in the zàng fǔ, the yīn Meridians specifically, or the yīn time of night when the sun is not visible. To my mind there are problems with each of these possibilities. Qì must continuously circulate everywhere in the body, at all times, in order to maintain life. So it must circulate both inside and outside all the Meridians at every moment, regardless of yīn or yáng considerations. It cannot circulate only in the yáng Meridians during the day and only in the yīn Meridians during the night. For the same reason it cannot circulate only in either róng or wèi states simply based on whether it is day or nighttime. The only solution that makes sense to me is that qì is continuously moving in (as róng) and out (as wèi) of the Meridians in a cyclical fashion, with yīn phases alternating with yáng phases, while at the same time alternately moving in

---

137 This is the same statement found in the *Língshū*. 13,500 breaths per day (equaling 9.375 breaths per minute) is much slower than we observe in people today. Medical commentaries, e.g. Johns Hopkins University, state that the normal breath rate is 12 to 16 breaths per minute. In medical school I was taught that 20 breaths per minute was the norm. My belief is that the ancient Chinese were excellent observers of such natural phenomena, and barring the unlikely possibility that humans have changed significantly in their breathing rate in the past few thousand years, we must consider that the reference to breathings (xī 息) must have to do with something other than pulmonary activity. The lungs are never mentioned in these passages, but the pulse (mài 脈) repeatedly is, as are róng and wèi, which in turn are forms of qì (breath!). Since the subject matter is clearly about the pulse, I think it is worth considering the possibility that the reference might be to "pulse breaths," meaning the time required for the qì to complete one unit of contraction and expansion during which it moves the distance of 6 cùn, as stated in the first chapter of the text (人一呼脈行三寸，一吸脈行三寸，呼吸定息，脈行六寸). In the same way that the chest rises and falls (expands and contracts) with pulmonary breathing and has intervals between and after the two movements, so the blood vessels rise and fall (expand and contract) with each pulse beat and have a rest between beats. This is the image I have in mind when thinking of a pulse breath. Using this interpretation of the text, each pulse breath lasts exactly 6.4 seconds (the mathematical precision of this figure seems highly significant to me in the quest to figure out the original author's intent). Simonov has used this duration clinically in timing the optimal contact between acupuncture needle and acupoint needled – more on this in a bit. As I've said, the text never defines the term breathing (息), so it must remain a matter of scholarly and clinical conjecture for the present.

138 Róng is used in the *Nánjīng* instead of yíng.

139 Author's translation.

and out of the Meridians. Looking at the slide rule again, one can see that there are 25 dark rectangles and 25 clear rectangles, alternating with each other in such a way as to comprise 50 "periods" exactly as stipulated in both classical texts under discussion. For now, let me just say that this slide rule does not say where qì is located in the body at any specific time, since it is always everywhere in the body. What the slide rule might show is what type of qì is available for the body to absorb, so as to coordinate with the macrocosmic environment.

Simonov's interpretation of the *Nánjīng* is certainly controversial, and while I agree with many of his assertions, I disagree with others. In order to proceed with the topic being discussed (macrocosm/microcosm), we need to look carefully at the *Nánjīng* text. Difficulty 1 begins with a question about the inch opening (cùn kǒu). The text states that cùn kǒu is examined to determine the pattern of death or life, good or evil auspices.[140] I think this statement must mean that cùn kǒu reflects the status of the zàng fǔ and their Meridians. This would however contradict Simonov's contention that the cùn kǒu pulse simply reflects a timing phenomenon induced by the macrocosm, and that it doesn't provide information about the Organ or Meridian states, which latter information he believes is reflected more proximally along the radial artery. There is only one way I can think of to reconcile these two different viewpoints (macrocosm/microcosm interaction vs. Organ/Meridian functional state). Let me describe Simonov's theory first, along with the experiment I carried out in Beijing to test it, with its favorable results, before discussing a possible reconciliation of the conflicting interpretations of the classical text.

According to Simonov, the cùn kǒu pulse is located exactly at the wrist crease, where acupoint LU 9 is located. Many texts and practitioners agree with him, that the cùn kǒu pulse is located at LU 9, while many other texts and practitioners find the cùn kǒu pulse elsewhere. Simonov claims to have discovered that by comparing the LU 9 pulse simultaneously on both left and right wrists, one will find that there is a cycle occurring where one side is more superficial and strong for 28.8 minutes, followed by a switch to the

140 十二經皆有動脈,獨取寸口,以決五藏六府死生吉凶之法,何謂也?

other side being more superficial and strong for an equal amount of time. He attributes this finding to an anastomosis between the radial and ulnar arteries, branching off from the radial artery at LU 9, being affected by some mysterious mechanism (mysterious only for the moment). The flow in this anastomosis reverses direction every 28.8 minutes, according to Simonov, accounting for this claimed pulse phenomenon. The switch in laterality always coincides with solar noon and solar midnight, in addition to other times dependent on the 28.8 minute periods of each polarity. Obviously the *Nánjīng* doesn't literally make such a statement, but let us consider its implications. The left side is usually attributed to yáng and the right side to yīn, so yīn and yáng are putatively alternating in this scenario. If this 28.8 minute cycle is confirmed, then 50 such cycles would take 24 hours, or exactly 1 day, with half of the cycles reflecting yáng and the other half reflecting yīn. Furthermore, Simonov reported that in any given geographical location, these cycles were synchronized, with everybody switching polarity at LU 9 at the same moment, every 28.8 minutes. Additionally, Simonov observed that males and females always showed this stronger superficial pulse on opposite wrists from each other, when examined at the same time and place. The reader must see that this mathematical description happens to perfectly match the statements in both the *Língshū* and the *Nánjīng* about the 50 passages of yíng and wèi qì in the course of each day, so a search for verification of a 28.8 minute cycle of superficiality alternating between left and right wrists becomes a matter of high priority, in my opinion at least. It is not always easy for me to determine which pulse is more superficial and strong, so I decided to use a large number of unprejudiced observers to see if statistically they found any corroboration of Simonov's claim. I had the good fortune to be a guest lecturer at the Beijing University of Chinese Medicine in 2019, where I had a whole evening to make a presentation to the entire freshman class. These beginning students did not have prior experience with pulse diagnosis (I asked them), nor were they familiar with most of the more subtle aspects of Chinese medical theory, such as the difference between the pulses of men and women, or their correlations with the left and right sides of the body. I considered this an ideal group of observers because they had no particular preconceptions about what they would find by following my instructions. Here is a report of that experiment:

On October 30th, 2019 I addressed the freshman class at the Beijing University of Chinese Medicine, on the topic of ancient technology and Chinese medicine, with 180 students in attendance. I asked them to work in pairs, and at exactly 7:50 pm, they were instructed to palpate the pulse at the radial wrist crease of their partner, and note which side was more superficial and strong, as well as the gender of their examinee. As they were all novices at pulse palpation, I did not expect 100% compliance, but 68 students did participate in the trial, a number that I felt was large enough to offer meaningful data.

57 examinees were female. 40 of them were diagnosed as having stronger and more superficial LU 9 pulses on the right and 17 were felt to have stronger and more superficial LU 9 pulses on the left (i.e. 70% on the right side). 11 examinees were male. 2 of them were felt to have stronger and more superficial LU 9 pulses on the right and 9 were felt to have stronger and more superficial LU 9 pulses on the left (i.e. 82% on the left side). None of the participants were given any explanation of what the meaning would be of their findings. According to Simonov's slide rule predictions (corrected for local solar noontime), females should have been experiencing a yīn period while males should have been experiencing a yáng period, and this is what these unbiased students actually found, based on their non-random results. I had come prepared with a PowerPoint slide to show the students, after the experiment was completed, what Simonov had hypothesized (a close up view of the slide rule, focused on 7:50 PM in Beijing). It showed how their results strongly conformed to his prediction. I was as surprised as were both the Professor who invited me (Zhou Xiaofei) and the students, at the degree of non-random findings, to say nothing of the fact that they perfectly reflected Simonov's contentions.

Let us think about the implications of this experiment. There are two possibilities that I can imagine: Either males naturally tend to have stronger, more superficial pulses on the left side and women naturally tend to have stronger, more superficial pulses on the right side; or else some external influence is affecting these pulses in a way that only depends on the time and place of the observation. There is an argument to be made for the first explanation, as some schools of thought in Chinese medicine have claimed such a difference between the pulses of males and females, although not specifically relating to the LU 9 location. I wish I had had enough time to repeat this experiment at a time when Simonov's theory would predict that males would show stronger, more superficial pulses on the right side and females stronger, more superficial pulses on the left side. Unfortunately I did not have the opportunity to do such an experiment, but it is one that would not be difficult for someone with the requisite large class of unbiased subjects to carry out. Simonov has been repeatedly confirming just such a finding for many years in his own practice, as have his students in theirs, so I think it is fair to provisionally grant the likelihood that his observation is accurate. That leaves the second explanation. Is there some external influence that affects all people in the same time and place in a predictable, regular pattern? The obvious answer is yes, the influence of the sun on the rotating earth, which we all know accounts for the phenomena of day and night (among many other cyclical phenomena), and upon which our common sense notion of "time" is based (i.e. 24 hours per day). And this brings us back to the topic of coordination between macrocosm and microcosm, as there can be no doubt that life on earth is absolutely dependent on the radiation from the sun. Let us not forget that *Nánjīng* chapter 1 specifically mentions the phrase, "one day and one night."[141]

Provisionally granting Simonov's interpretation, what might be explained about the macrocosm/microcosm passages in the classical texts? Looking at the slide rule, we find the 25 dark and 25 clear rectangles, each corresponding to yīn or yáng. The time is marked in hours, so the slide rule covers 12 hours on the top half and 12 hours on the bottom half. Since each day/night cycle takes 24 hours, these are conveniently separated into AM and PM sections exactly at solar noon and solar midnight, just as we enumerate time in modern terms. So far this hypothesis aligns with the dictum that both the róng and wèi circulate "through the yáng 25 times *(clear rectangles)* and through the yīn 25 times *(dark rectangles)*. This constitutes one cycle. Meeting again at cùn kǒu after 50 periods *(all the rectangles)* is the beginning and end of the (circulation through) five zàng and six fǔ. Thus the pattern is obtained

141 人一日一夜，凡一萬三千五百息

from cùn kǒu." Furthermore, "the movements of the pulse proceed through 50 periods (zhōu 周) circulating through the body while the water (in the clepsydra) drops 100 marks (kè 刻)." We might wonder why the water clock is constructed with 100 marks (kè) if there are only 50 periods of circulation (zhōu) corresponding to them. Simonov's answer is that each period (zhōu 周) is composed of 2 "cuts (kè)". Any period is either yīn or yáng, but its first and second halves are mirror images in terms of the movements of róng and wèi in the body, and need to be distinguished from each other. I will elaborate on this further along, but the mathematical correspondence between the classical texts and Simonov's model is striking. When this finding is supported by the observation of pulse fluctuations as described in the Beijing experiment, it becomes a fairly persuasive hypothesis. We can take the analysis one step further in asking why the technical term kè (刻) was chosen for the marks on the water clock. Its original meaning was one quarter. So we can ask, one quarter of what? Four cuts would last for two periods of 28.8 minures each, or slightly less than one of our modern hours. Using a solar clock, each day would have 25 hours, and one quarter (kè) of each solar hour would last 14.4 minutes, exactly as Simonov's model illustrates. Thus we can think of a cut (kè) as a quarter of an hour (14.4 of our modern minutes), a result that has practical implications for acupuncture, to be explained.

The preceding remarks support the hypothesis that cùn kǒu manifests the influence of the sun on the rotating earth in the pulse at LU 9, irrespective of the state of balance or imbalance of the Organs or Meridians in any given individual. As I remarked earlier, such an interpretation is incompatible with the idea than the cùn kǒu pulse can prognosticate a good or bad outcome, since it shows the same simultaneous fluctuations in all individuals. The only way I can reconcile this finding with the statement in *Nánjīng* chapter 1 about the prognostic use of cùn kǒu, is to propose that cùn kǒu has both a broad and a more narrow meaning in Chinese classical texts. The broad meaning is the distal part of the radial pulse at the wrist comprised of cùn, guān and chǐ, where the state of the Organ and Meridian activity is assessed, while the narrow meaning of cùn kǒu is the most distal part of the radial pulse at the wrist (LU 9), which reflects the macrocosmic cycles rather than the Organ or Meridian states. This duality of terms is far from unusual in Chinese medical writings. For example, qì (氣) can be translated any number of different ways (such as breath, influence or energy, for example), but when referring to the balance between qì and Blood (xuè 血), its meaning becomes more restricted. Similarly, the term shén (神) can have the broad meaning of all the psycho-spiritual aspects of an individual (including hún 魂, pò 魄, yì 意 and zhì 志), whereas the narrow meaning refers specifically to the psycho-spiritual aspect stored by the Heart. Although there is no way to definitively establish this dual use of the term cùn kǒu in the classical texts, such a hypothesis resolves the contradictions in the practical use of pulses to address either macrocosm/microcosm balancing or zàng fǔ/Meridian balancing. If cùn kǒu is interpreted literally, it means either the mouth or opening of cùn (inch), which is the most distal segment of the wrist pulses traditionally examined. Thus cùn kǒu could have originally indicated the pulse at the wrist crease, which is where qì enters the arm from its origin in the hand, using the imagery of well, spring, stream, river and sea Points, reflecting flow along the Lung Meridian. In that perspective, LU 9 is the "mouth" where qì enters the cùn segment. Other authors might then have interpreted cùn kǒu more broadly as a collective term for the whole region encompassing cùn, guān and chǐ, since that region is analogously where qì enters the arm from its origin in the hand, and where its reflection of Organ functions can be observed at cùn, guān and chǐ.[142] By positing these two different denotations for the term cùn kǒu, it is possible for Simonov's understanding to be a valid interpretation of the classical text, yet also to understand that in its broader sense, cùn kǒu reflects the Organ functions and can therefor serve as a prognosticator of health or illness, life or death.

Chapter 2 of the *Nánjīng* provides another example of the dual meanings of a technical term in separate passages of the same classical medical text. The character guān 關 first appears in this chapter, where it represents the dividing line between the portions of the radial pulse that resonate with yīn and yáng.[143] The text uses the term guān here as something like a mountain pass, occupying no territory, but merely separating the domains of yīn

142 Linguistically, this mode of interpretation is supported. One always speaks of cùn, guān and chǐ, never in the reverse order, so it is natural to think that the authors of the classics thought of qì as entering at cùn, not at chǐ.

143 從關至尺是尺內，陰之所治也；從關至魚際是寸內，陽之所治也。故分寸為尺，分尺為寸。故陰得尺內一寸，陽得寸內九分，尺寸終始一寸九分，故曰尺寸也。This text is copied from ctext.org. It is identical to Unschuld's published version. Simonov's version has 從 關 至 魚 際 (from guan towards yuji) 是 寸口內 (is inside from cun kou), instead of 是 寸內 (inside from cun). It would clarify matters if we knew the provenance of both versions.

and yáng on either side of it.[144] Personally, I cannot help but see an analogy between this "mountain pass" and the crest of the radial styloid process as being the dividing line between the territories dominated by yīn and yáng on either side. I think of this denotation of guān as referring to any border that acts as a transition between yīn and yáng. We might use the curving middle line of the tài jí symbol (太極圖) ☯ as an example of this boundary/transition between yīn and yáng that occupies no width of its own, similar to the dividing line called guān in *Nánjīng* 2. The more familiar use of the term guān is first found in *Nánjīng* 18, where it now refers to the pulse under the examiner's middle finger between the pulses at cùn, and chǐ. I think it is important to recognize the different denotations of guān in chapters 2 and 18, otherwise the text could be construed as contradicting itself as to the size and locations of cùn, guān and chǐ.

Returning to the topic of macrocosm/microcosm harmonization, chapter 3 of the *Nánjīng* introduces a new term that is again rarely attended to in the modern practice of acupuncture: chéng 乘, which means to take advantage of, or as Simonov translates it, victory. Actually this term was introduced in *Sù Wèn* chapter 9, in the discussion of two other terms: greatly excessive (太過 tài guò) and not reaching (不及 bù jí), concepts that were presented as part of the macrocosmic cycles in nature, but without any explanation of how these phenomena manifested in the pulse, making practical use of these ideas via acupuncture rather obscure.[145] Again, the *Nánjīng* can be seen to be presenting clarifying statements about material found in the *Nèijīng*. What chapter 3 is saying is that the following phenomena in the pulse are manifestations and confirmations of these particular technical terms, and what their implications are for evaluating the balance of yīn and yáng. To briefly summarize, yīn victory (chéng 乘) is indicated when the pulse between the guān position and the

144 Schuessler, on page 265, explains guān in old Chinese as meaning a barrier or frontier. Kroll, on page 143, supports this definition, adding its use as a pass that can be open or closed, noting that it also meant to close, which is how this character was employed in the *Nèijīng* in the chapters that discussed RYCK pulse diagnosis. Kroll adds that guān was specifically used to refer to critical points in the human body at which movements of qi and blood may be facilitated or blocked. Thus the original use of the term does not imply a segment, but rather a dividing line. It is only in the later chapter 18 of the *Nánjīng* that guān takes on a second meaning, as the pulse segment between cùn and chǐ.

145 *Sù Wèn* chapter 9, 帝曰:太過不及奈何? "Huang Di asked: Greatly excessive and not reaching, what do they mean?" The answer is given in terms of the timeliness of the seasons, their Five Element relationships, and the effects on their associated Organs. Clearly this is an issue of coordinating macrocosm with microcosm, but the particulars of diagnosis and treatment are left unanswered.

wrist crease exceeds 9 fēn in length[146]. This location is normally where yáng manifests, but if it becomes lengthened (tài guò), that means yīn has invaded yáng's domain. Likewise, yáng victory (chéng 乘) is indicated when the pulse from the guān position towards the elbow exceeds one cùn in length. This location is normally where yīn manifests, but if it becomes lengthened (tài guò), that means yáng has invaded yīn's domain. Now these defining characteristics of yīn and yáng victory can be understood rationally, but practically speaking, how readily can one distinguish 9 fēn, 1 cùn, or 1.1 cùn? Again, Simonov discovered a much easier way to perceive yīn and yáng victory via the pulse. If the examiner checks the pulse with the usual 3 fingers, but placed proximal to the guān position, and pulls the fingers towards the wrist crease and then towards the elbow, one of these maneuvers will show the pulse extending over a greater length under the examiner's fingers. If the pulse is longer when pulled towards the wrist, then yīn victory is indicated. If the pulse is longer when pulled towards the elbow, then yáng victory is indicated. It is not difficult to see the similarity between Simonov's findings and those described in the *Nánjīng*, however Simonov's method is a much easier skill to acquire. The findings are the same on both left and right wrists, however a remarkable result appears when comparing the pulses of males and females, examined at approximately the same time. When males have pulses that indicate yáng victory, females will be found to have pulses that indicate yīn victory, and vice-versa. Furthermore, Simonov found that all males have the same polarity victory at any given time, and females all have the opposite polarity victory at that given time. This discovery allowed Simonov to develop the slide rule previously pictured. Moreover, Simonov's findings explain that these states of yīn and yáng victory are reflections of the macrocosmic influence on human beings originally stated in the *Sù Wèn*, and that they are also a crucial aspect of understanding the classical texts' references to the 50 cycles of róng (yíng) and wèi. By examining the pulse, we can discover that the transitions from yīn to yáng victory and vice-versa always occur at solar noon and solar midnight. From midnight to noon, males exhibit yīn victory and females exhibit yáng victory in their pulses. From noon to midnight, males exhibit yáng victory and females exhibit yīn victory in their pulses.

146 One fēn (分) equals one tenth of a cùn (寸).

Now that we know how to identify these states, an obvious question is, what significance do they have for the practice of acupuncture? Simonov's answer is that yáng victory means "cosmic" qì is available to the patient as yáng qì, and that by correct choice of location, needle depth and timing of treatment, the practitioner can foster the increase of yáng in either the róng (yíng) or wèi parts of the patient's qì. Since róng (yíng) and wèi are opposite in their movement (into or out of the Meridians), by directing this cosmic yáng into a location that is in a state to receive wèi qì, it will have the effect of decreasing the yáng in the associated Meridian, which is always carrying only yíng qì. Thus cosmic yáng qì can be used to either increase or decrease yáng in any Meridian, depending on whether it is directed to the individual's yíng qì or wèi qì. At any given moment, the left and right branches of every Meridian are in opposite states of their qì behavior, one side exhibiting movement into the Meridian (yíng) and the opposite side exhibiting movement out of the Meridian (wèi). How to determine what is happening on a given side by checking the pulse will be described shortly. Analogously, yīn victory means that cosmic qì is available to the patient as yīn qì. This yīn qì can be directed to either the yíng or wèi areas of the patient to either increase or decrease the Meridian's yīn qì. My personal experience is that this cosmic qì can be used to reestablish proper functioning in any individual's Constitutional Meridian/Organ. If the Constitution of a patient is Excess of a yáng Organ/Meridian for example, then during yáng victory, needling the wèi side of the Constitutional Meridian will decrease its yáng qì, leading to more normal functioning. During yīn victory needling the Constitutional Meridian's róng (yíng) side will have the same beneficial effect. I think of this as using harmonization of macrocosm and microcosm as a way for the individual's Constitution to regain optimal functioning, and thereby benefit almost any type of malfunction affecting the individual. Obviously, this cosmic solar radiation is the same, whether it is impacting a male or female recipient. So yīn or yáng victory is created by the individual living organism, not by anything specific to the cosmic solar radiation itself. The type of needling I use for this harmonizing of microcosm and macrocosm treatment is non-invasive "contact" needling of a series of Points along the target Meridian. This approach was taught by Simonov, using about 6 seconds of contact at each Point.[147] Simonov's method was to choose Points on one

147 Recall Simonov's earlier reference to a "pulse breath" lasting 6.4 seconds.

hand Meridian and the opposite side's foot Meridian of the same named Great Meridian. For instance, Lung Meridian (Taiyin) on the left paired with Spleen Meridian (Taiyin) on the right. In most cases, this strategy involves a similar action on neighboring Elements, which naturally support each other. In the example just given, Lung Metal and Spleen Earth are mutually supportive. However in the case of Shaoyin and Taiyang, the Elements Fire and Water are in a control cycle (xiāng kè) relationship, so similar actions on both would seem to be counter-productive. For this reason, I typically restrict my contact needling to only one Meridian, and my best results have been when choosing the Constitutional Meridian, or its coupled Meridian of the same Element, for this type of treatment. During a yáng period (clear rectangle on the slide rule), the yáng Meridian of the Constitutional Element is available for this kind of treatment. During a yīn period (dark rectangle on the slide rule), the yīn Meridian of the Constitutional Element is available for this kind of treatment. Remember that contact needling is the appropriate technique indicated using instruments described in the passage on the nine needles mentioned in the *Língshū*. I typically reserve this type of macrocosm/microcosm balancing treatment for cases where the pulses are otherwise in good balance, i.e. as a form of preventative treatment, however Simonov uses it with good results to also treat individuals with imbalanced pulses. The scope of this technique is a matter for further research.

There is one missing piece in this overview of Simonov's methodology. It can only be used if one knows whether any target Meridian is expressing a yíng or wèi direction of qì flow at the moment of treatment. Simonov discovered that the pulse can also reveal this information, but let me just say that the 4 limbs follow a definite relationship in this regard. If the left hand Meridians are experiencing a yíng movement, then the right foot Meridians will experience the same movement, whereas the right hand Meridians and the left foot Meridians will experience a wèi movement. The opposite situation holds true if the right hand Meridians are experiencing a yíng movement. Thus we only need to find out the situation in one of the four limbs to determine the direction of movement in all the Meridians. To do this, Simonov recommends comparing the pulses on left and right wrists as a whole. If the pulse increases in strength on compressing the artery from its middle depth, then the Meridians of that arm are in a yíng phase. If the pulse gets stronger on lifting the fingers from the middle depth, then the Meridians in that arm are in a wèi phase. Readers should be able to confirm on themselves that the findings in their left and right arms are always opposite in this regard.

For a more detailed explanation of how to use these findings to harmonize macrocosm and microcosm in any specific case, the reader is encouraged to study Simonov's work, as it is not possible or even appropriate to present all of his methodology here. It is enough if I have established a reasonable basis for the role that acupuncture can play in addressing the coordination of macrocosm and microcosm, an approach that is fundamental to all of the Chinese classics concerning the use of acupuncture. I will include examples of this type of treatment in the case histories chapter, but the reader is cautioned that as with all methods of classical acupuncture, a thorough understanding is essential before applying any acupuncture techniques to actual patients.

One final thought about this topic is pertinent. For most people, the easiest example to understand of imbalance between macrocosm and microcosm is something we have almost all experienced, jet-lag. When we travel rapidly to a different time zone, we lose the coordination of our cyclical rhythms with those of the world around us. Common symptoms people complain about include insomnia, difficulty concentrating (brain fog) and fatigue. It's interesting to note that these symptoms are also very common in so-called chronic fatigue syndrome and other post-viral syndromes. Even now they are typical of people who have never fully recovered from Covid-19 infections. We call them long-haulers. In all of these post-infection cases there are obvious pathogens involved, but might it not be the case that the end result has been a provoked imbalance between macrocosm and microcosm? In ordinary jet-lag we don't usually think about pathogens, so with similar symptoms, might these other difficult, but common, illnesses result, at least in part, from a loss of coordination between macrocosm and microcosm as well (mechanism 3 in *Língshū* 9, which "requires acupuncture" for its resolution)? This is something I am currently investigating.

CHAPTER 8

The Luò Vessels: Connecting Pathology, Diagnosis and Treatment

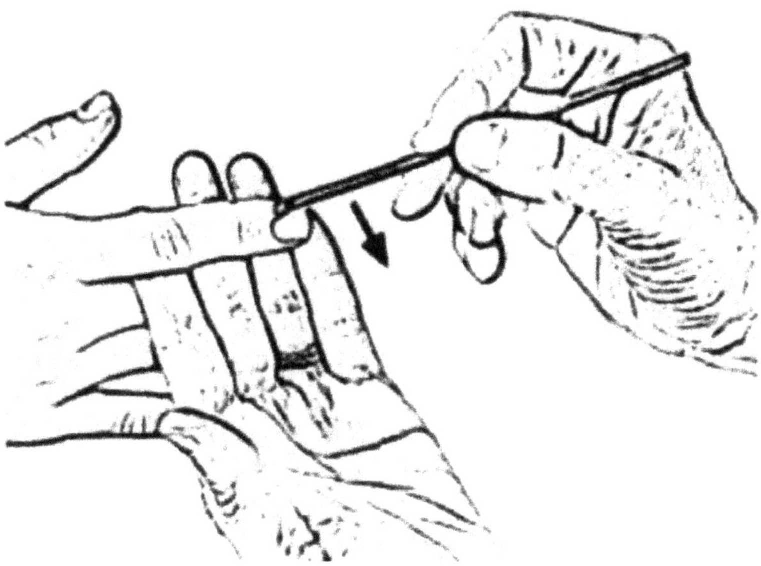

Figure 25: Akabane Testing [148]

Akabane Kobe was a 20th century Japanese acupuncturist who developed a test, using heat at the Jǐng-Well Points, to identify left/right imbalances of the Meridians (Fig. 25). His test became widespread, and was taught and used by both my Korean teacher Chae Lew and my English teacher J.R. Worsley. Lew

148 Source: Neal White.

used this test as a general measure of how well the Constitutional Meridian was balanced, but did not employ any special protocol for addressing such imbalances other than the 4 Needle Constitutional balancing technique he had learned from Kuon Dowon. Worsley, on the other hand, taught a different protocol, which was also different from the one proposed initially by Akabane. Akabane recommended treating the Back Shù Points to correct such imbalances, whereas Worsley's treatment involved needling the Luò Points, specifically tonifying the Luò Point on the deficient side. I have not as yet been able to discover if one of Akabane's students began using Luò Points, or if this was an innovation by Worsley or one of his teachers. Another possibility, and the one I consider most likely, is that this method derived from the writings of Niboyet[149], a colleague of Soulié de Morant, whose work is much less well-known in the English speaking world, but did have a significant impact on the early European lineage of acupuncturists, including Worsley. I will return to Niboyet's contributions shortly. As a number of Worsley's teachers had studied with Akabane directly, my suspicion initially was that probably one of these intermediaries was the originator of the use of Luò Points to treat Akabane imbalances.[150] I have not seen any published accounts of Akabane's ideas concerning the pathophysiological nature of the imbalances his test detected and his treatments addressed. This is a question that remains unanswered.

Although apparently Worsley interpreted Akabane imbalances as simply reflecting differences in the supply of qì between the left and right branches of the same Principal Meridian, he did not explain how tonifying the Luò Point on the deficient side could act to correct this imbalance. There is no trajectory that might serve such a function, unlike the situation with Entry/

149 Niboyet, J. E. H., *Nouveau Traité D'Acupuncture*, Maisonneuve, Paris, 1979, pp. 729, 962. His Asian teachers were never fully acknowledged, but among them was Chang Bin Lee, a renowned Korean teacher who was also associated with Worsley, supporting this possible line of transmission. Like Soulié de Morant, Niboyet's teachings reflect material from a much later period in Chinese history than the Han dynasty Classics. Putative sources include the *Great Compendium of Acupuncture and Moxibustion* (*Zhēn Jiǔ Dà Chéng* 針灸大成) and the *Gateway to Chinese Medicine* (*Yī Xué Rù Mén* 醫學入門). Neither of these important texts has been fully translated into English as far as I know. Niboyet began publishing his works on acupuncture as early as 1951, so this material could certainly have been transmitted to Worsley, whose Akabane teachings began no earlier than the late 1960s, and most likely started in the 1970s.

150 Eckman, P. *In the Footsteps of the Yellow Emperor; Tracing the History of Traditional Acupuncture*, Cypress Book Co., San Francisco, 1996.

Exit Blocks where such an explanatory pathway is documented in the classical texts.[151] When speaking of the Principal Meridians (also translated as Vessels), the most common Chinese term is Jīng Mài (經脈). However, when speaking of the Meridian system as a whole, another Chinese term is used, Jīng Luò (經絡), and so the Luò Vessels have been considered an essential component of the Meridian system for thousands of years, apparently more fundamental than the less commonly discussed Divergent Meridians (Jīng Bié 經別), for example. I should emphasize that in the Chinese classical texts there is no discussion, or even implication, that the Luò Vessels (also translated as Connecting or Junctional Meridians) cross the body's midline, and by this route have any function in the coordination between left and right sides of the body. However, since many phenomena related to acupuncture do not have a linear or semi-linear anatomical basis, but are rather delocalized resonant (radiant) phenomena, such anatomical arguments do not exclude the possibility of Worsley's methodology being reasonable as well as effective. My experience, both personal and in reports from numerous Worsley trained colleagues, is that correcting Akabane imbalances by tonifying the deficient Luò Point is highly successful in both correcting the heating test results and in alleviating multiple diverse signs and symptoms of the individuals receiving this treatment. I will go even further in stressing the importance of diagnosing and correcting Akabane imbalances. In many cases, such imbalances can last for decades or more, and produce serious abnormalities of both physical and non-physical functioning, that in turn can color every aspect of an individual's life. Because no widely accepted

151 Posted on Scholars of Chinese Medicine Facebook group by Par Scott on 3/10/21, "So, I have been trying to better understand theories about the luo mai and how they are treated, which lead me to Su Wen 63, which seems to be the basis for some statements about them. Among these is the idea of the luo becoming blocked and spreading into the 大絡. We have a 大絡 of the spleen, but I am assuming this is not precisely that. I assume some folks would perhaps equate this with the spleen 大絡 in that there is also some focus on the extremities in the same part of the text. We also have a notion that things may move from the left to the right, and above to below, engendering misleading or tricky illnesses. I can understand the notion of a localized area flooding over into an adjacent place, but how do folks square the idea of the pathogen being transmitted across the body? Is it a resonance thing? I've generally thought of the early strata of CM theory to be more or less materialistic, and for lack of a better word, hydraulic in nature, which to my mind dictates some sort of directed path for this migration. It seems like the mechanism of normal progression of the pathogen, through the main channels to the viscera is precluded here, so we know the viscera are not affected in the "normal" way, so the root of the opposite same organ channel would not be affected directly, and other than some sort of sympathetic relationship I don't see how the evils jump over?" My posted answer was, "I think it's a mistake to assume that the early strata of CM was purely based on hydraulic imagery. Even the *Nèijīng* has multiple passages that make no sense without ganying reasoning. Why then should we exclude such ideas in reference to the luò systems? Also, the issue is not always about the movement of pathogens, but the integrity of the homeostatic balancing mechanisms of ying/wei, yin/yang, wuxing and zangfu. These mechanisms are not necessarily based on a hydraulic model."

method of diagnosing these imbalances was taught prior to the 20th century, they were not routinely addressed. Complicating this problem is the issue of subjectivity in carrying out the Akabane test with incense sticks passed over the Jǐng-Well Points. A diagnosis of Akabane imbalances using pulse examination has the potential to markedly improve the accuracy and ease of locating such imbalances, and I will report the way I have discovered of using such a method further on in this chapter. In order to make sense of this topic, we might start with examining the classical understanding of the nature and roles of the Luò Vessels (Fig.26).

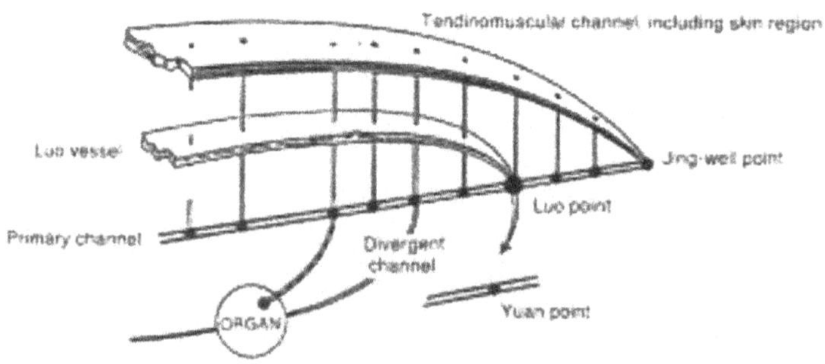

Fig. 26: Meridian and Organ Connections [152]

Língshū Chapter 10 describes the trajectories of all the Luò Vessels. In this chapter there is no clear statement of any rule governing the trajectories or connections of these Vessels, but some typical patterns find repeated mention, and this seems to be the basis for some of the common ideas about the Luò Vessels. They are all described as divergences from the Principal Vessels (Jīng Mài) that originate from their named Luò Points, and are often visible because they frequently are filled with blood. The pathways generally are described as "separating from" or "pouring into" the coupled Primary Vessel of the same Element, but of opposite polarity. Another branch of the Luò Vessel is usually described as running parallel to its Principal Vessel, but also continuing onwards to connect with structures higher up in the body, such as the heart, tongue and eye in the case of the Luò Vessel of the Hand Shaoyin (Heart). This pattern of description would appear to be the basis

152 Source: Mikhail Teppone, American Journal of Acupuncture, vol. 26, p.129, 1998.

for the idea that there are both transverse Luò Vessels connecting paired Vessels such as LU and LI, and also longitudinal Luò Vessels connecting Organs to the sensory portals such as the Heart to the eyes and tongue. The terminology of transverse or longitudinal aspects of the Luò Vessels does not however appear in the classics. A common feature of both these descriptions of the Luò Vessels is the idea of connecting inside to outside and vice-versa, reminiscent of the biǎo lǐ relationship to be discussed in this text's Chapter 10, in connection with the climatic factors (potential pathogens). From this basic notion it is possible to picture the role these Luò Vessels might play, transporting xié qì from the Exterior to the Interior in the development of an illness, and vice-versa, from the Interior to the Exterior in the recovery from an illness, the latter being a potential mechanism for the expulsion of pathogens. There is also a popular notion that the Luò Vessels connect the Luò Point of one Meridian to the Yuán Point of its coupled Meridian of the same Element, and as a consequence the treatment of these two Points together was advocated as a clinical strategy in later historical epochs, but this idea does not appear in the foundational classical Chinese medical texts. Before concluding this general introduction, I should also mention the material in *Sù Wèn* Chapter 63, which discusses miù cì, or contralateral needling, a topic that can be important in the context of Akabane imbalances and the use of Luò Points to treat them. This chapter also describes the order of penetration of the body by xié qì, and how to treat such an occurrence. First the skin is attacked, then the tiny network vessels, then the main Luò Vessels, then the Principal Vessels, before finally reaching the Organs (zàng fǔ). In such a situation, the treatment recommended in the case of invasion by pathogens is first via the Principal Vessels[153] (details however, not given in *Sù Wèn* 63), although if the Principal Vessels are successfully resistant to this xié qì, then it can overflow into the Luò Vessels, and then these could now be the lodging place of a lingering pathogen. Furthermore, when xié qì follows this Luò Vessel path, it is presumed to flow from right side to left side and vice-versa, so it must be treated by contralateral needling per *Sù Wèn* Chapter 63. I think it is quite likely that this mention of contralateral needling in connection with xié qì in the Luò Vessels was ultimately the inspiration for experimenting with treating Luò Points to correct Akabane imbalances.

153 This is the strategy normally used in CCA, as previously discussed.

I have speculated a bit about the possible role of pathogens (xié qì) in creating Akabane imbalances, but I will keep my remarks on this topic brief, as the thrust of CCA treatment is to restore the normal qì (zhèng qì), and let it take care of dealing with any pathogenic qì if such is present. If a pathogen does invade the body, it seems reasonable to assume that it could penetrate a Meridian on one particular side. In such a case, that side would develop an excess state due to the battle between pathogen and defensive qì[154]. From that moment, there will be an imbalance between the left and right branches of the attacked Meridian. Keeping the prior remarks about the connection between Luò Vessels and opposite side needling (miù cì) in mind, the idea of treating the Luò Points to correct Akabane imbalances does not seem surprising. More research is needed to determine if there is an involvement of xié qì in Akabane imbalances, but I will preview the answer by disclosing that Chapter 10 of this text provides evidence which convinces me that Akabane imbalances are not especially associated with the presence of xié qì, a problem that manifests with quite a different set of pulse findings.

Putting aside the issue of xié qì then, a more pertinent question regarding the CCA approach to diagnosis and treatment would be to ask "is there any change in the pulse when Akabane imbalances are present?" When I was a student in England, I remember Worsley diagnosing my classmates as having specific Akabane imbalances after examining their pulses, but he never taught such methods as far as I know. I can think of two ways to use the pulses to possibly detect Akabane imbalances, but I believe one of them is accurate, while the other detects a different kind of disequilibrium. Both of my thoughts are grounded in the success of Luò Points in treating both

154 It is common in TCM literature to ascribe the excess due to pathological invasion to a combination of defensive and pathogenic energies entangled in a struggle. CCA, on the other hand, is based on the proposition that the pulses reflect only what the organism is doing, not with any direct signals from pathogens showing in the pulse. From this perspective, the excess pulse after pathogenic invasion would be seen as representing a hyperactive state in the Meridian on that side, which is the natural response to being attacked. The Meridian on the opposite side would not be hyperactive, and if a connection between the two sides exists, then it could easily become hypoactive (deficient) as a result of sharing its qì with the branch on the attacked side. Thus there are two different physical/metaphysical explanations for the creation of these pulse changes, and I believe it is best to consider both possibilities when evaluating treatment strategies, as is discussed in this chapter. Again, the answer proposed by herbalists may differ from that proposed by acupuncturists, as the targets of treatment might be the two different mechanisms mentioned in this footnote.

Akabane imbalances and imbalances between coupled Organs/Meridians of the same Element.

The first possibility that I investigated was based on the finding that sometimes the Subdosha pulses on left and right sides were exact opposites, a situation that always implicates the same Element. I will return to this pulse finding in Chapter 10, as I have determined that it does not correlate with the presence of abnormal Akabane test results. The second pulse finding I've investigated does appear to actually represent Akabane imbalances, and has a rational explanation within the whole approach of CCA. It involves the *Màijīng* deviation pattern of the radial artery pulses that I have explained earlier. The same deviation pattern should be present in both the yīn and yáng finger positions on both left and right wrists. There are only six possible deviations associated with the six Great Meridians. I have recommended that this deviation pattern also be confirmed in the radial pulse at positions 4, 5 and 6 further proximally up the arm, so as to exclude the influence of Extraordinary Meridian imbalances. These Extraordinary Meridian imbalances only show up as abnormalities of deviation at positions 1, 2 and 3 in my experience. The new discovery pertinent to Akabane imbalances is the finding that sometimes the deviation pattern at positions 1, 2 and 3 doesn't fit either a six Great Meridian form or an Extraordinary Meridian form on only one wrist, left or right, while the deviation pattern is still normally present on the other wrist and at positions 4, 5 and 6. In such a case, it appears that one of the pulses has shifted away from its Constitutional position. This could manifest on one of the pulses in either the yīn or yáng finger placements, and occur on either the left or right wrist. Thus each deviated pulse represents the state of a specific Organ/Meridian. When such a displacement is found, it appears to correspond to an Akabane imbalance of its Meridian, as many of my clinical cases have confirmed. Whichever way the pulse deviates, to the patient's left or right (judged with the patient's palm facing upwards), is the side whose Luò Point should be tonified in order to correct both the heating test and Constitutional deviation pattern in the pulse. Unlike the Subdosha findings just mentioned, this deviation finding does not occur more frequently on the Constitutional Element. Such a finding is in agreement with what most practitioners of the Worsley school have reported in my experience. Now the astute reader might ask, why tonify the Luò Point on the side to which the pulse deviates? Wouldn't that be the excess side? Actually, I believe the deviation does point to the excess side, but there's one more step to the analysis. The correct side is indicated by looking at Fig. 22 again. When the body is in the Chinese Anatomical Position, the deviation now points to the opposite side than when the pulse is taken with arms at one's sides, facing palm up! In this normal pulse taking position, the deviation points to the deficient side. This interpretation is also consonant with the remarks I made above about the possible involvement of xié qì initiating Akabane imbalances. The attacked side becomes excess, while the treatment side is on the opposite (deficient) limb, following the instructions in *Sù Wèn* Chapter 63. My experience is that these "deviated" (Akabane imbalanced) pulses are quite common, however it is rare to find more than one in a given pulse exam. As this finding reveals a block to the normal flow of qì, it should be treated before any other balancing of Five Elements or Organs/Meridians. I do not believe either of the methods I've discussed could have been Worsley's way of diagnosing Akabane imbalances from the pulse because he didn't pay attention to deviation patterns associated with the "CF" or to Subdosha pulses, nor did he separately examine the Organs using what I have called the yīn and yáng finger positions. I am grateful to him however for suggesting the idea that Akabane imbalances are detectable in the pulse.

After composing the initial drafts of this chapter, I began to review Niboyet's massive French text cited earlier, *Nouveau Traité d'Acupuncture* (1,247 pages). Therein one can find many ideas that have been corroborated as to their efficacy after incorporating them in CCA, including remarks about left/right imbalances between the two branches of the same Principal Meridian. Niboyet claimed that if a left/right imbalance of that sort was present, the deficient side would show a "broken" pulse wave on its lateral edge, and that this situation could be treated by the Luò Point. This idea is very close to my own experience, but rather than perceiving a broken edge to the pulse wave, I find an actual displacement, medially or laterally, of one segment of the whole pulse wave. However, the edge under discussion is where I am most clearly able to read the pulse displacement. In hindsight, Niboyet's teachings would nicely account for Worsley's use of Luò Points to correct Akabane imbalances.

CHAPTER 9

Pathological Theory's Connection to Treatment Goals; From Nèijīng *to* Nánjīng *to Korean Constitutional Acupuncture*

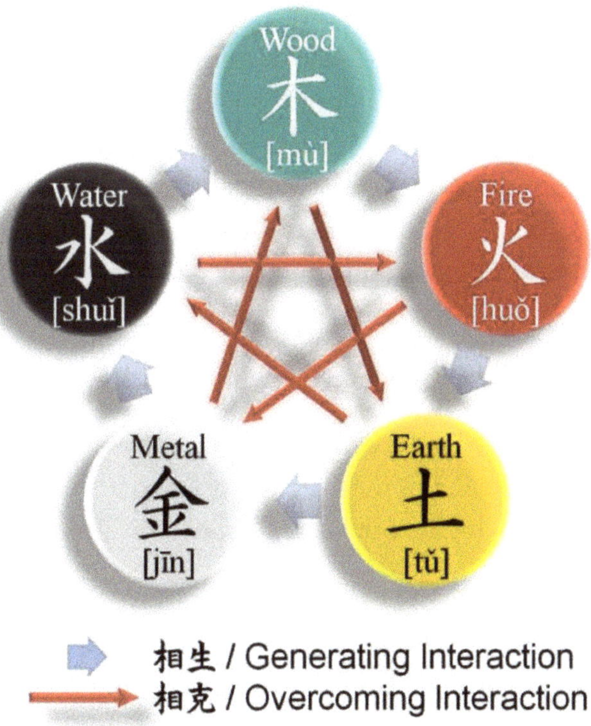

Figure 27: The Five Element Creative and Control Cycles [155]

---

155 Source: commons.wikimedia.org

Both the *Nèijīng* and the *Nánjīng* discuss the need to supplement (bǔ 補) what is deficient (xū 虛) and to disperse (xiè 泄) what is excess (shí 實), which strategy in CCA is often accomplished by use of the Five Element Creative and Control Cycles (Fig. 27). In my studies I have not come across a consistent diagnostic criterion for making the distinction between deficiency and excess, as these terms may apply to pulses, Organ functions, Meridian states, symptoms and even pathogens. What I have found, as usual by focusing on the pulse, is that an excess state corresponds to a pulse that is relatively hard in the middle of its appropriate positional depth (per *Nánjīng* 5). A deficient state corresponds to a pulse that can be felt at the upper edge and lower edge of its appropriate positional depth, but that is absent (or hollow) in the middle between upper and lower edges as shown in Fig. 28.[156] The overall depth can be superficial, middle or deep, but should be evaluated in this regard at its Element's proper depth per *Nánjīng* 5. This reminds me of the imagery of the *Yìjīng*, where a solid line is yáng (corresponding to excess), while a broken line is yīn (corresponding to deficiency). What then would a normal pulse feel like? I experience normal pulses as having a full thickness at their proper depth, which can be thought of as the upper and lower edges being palpable along with the presence of a pulse of equal or slightly greater intensity in the middle between its upper and lower edges. At this depth, the pulse should also be wider than at any other depth reached by increasing or decreasing one's finger pressure. Such criteria imply that excess and deficiency cannot be accurately identified by the size of the pulse. Excess pulses may be quite small, while deficient pulses might be quite large, but the reverse is equally possible: A large pulse that is distinctly harder in its center would indicate excess, while a small pulse that had an empty feeling in its middle would indicate deficiency. A large pulse that had an empty feeling in its middle would indicate deficiency, while a small pulse that had a hard

156 Careful attention to depth is crucial here. E.g. when checking the Heart pulse (left cùn, yin finger position) a strong pulse might be felt deeply at the Water depth, quite strong, big, and hard in its center. If you check the Fire depth however, the pulse will show a smaller, weak pattern, often hard to feel, but hollow in the center. Such a pulse reflects Heart deficiency (not excess!) with invasion of Water (from either Bladder or Kidney). The opposite situation is also encountered at times: The pulse at the Fire depth might be small. But quite hard in its center, while at the Water depth the pulse might be bigger than at any other depth, but hollow in its center. This latter situation represents Heart excess with deficiency of the Water Element (either Bladder or Kidney). These examples represent a true Element in Element diagnosis, in this case representing a xiāng kè relationship (control cycle). The classics often consider xiāng kè relationships as either problematic or prognostic of fatal outcomes compared to relationships classified as xiāng shēng (creative cycle). I have not found this teaching to be a reliable prognostic indicator.

feeling in its middle would indicate excess. While these criteria don't make pulse diagnosis simple, I believe they make it a lot easier than learning to distinguish 28 qualities (which can also present in combination with each other) and also memorize their correspondences with excess and deficiency. Perhaps in practicing herbal medicine the latter type of pulse quality diagnosis may be more important.

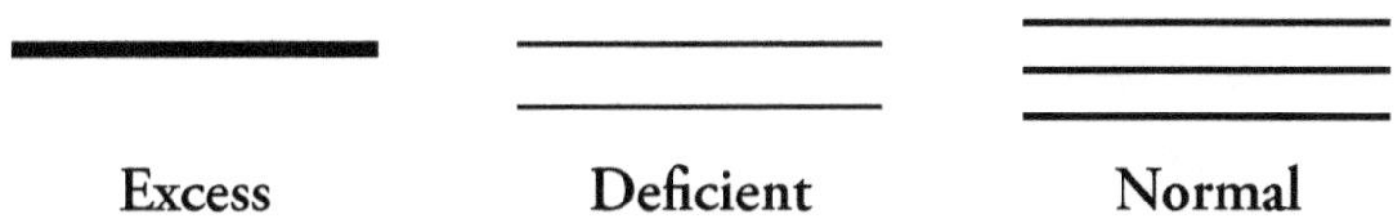

Figure 28: Images of shí 實, xū 虛, and zhèng 正 Pulses [157]

One useful application of this description of excess and deficient pulses occurs in diagnosing Entry/Exit blocks from one Meridian to the Meridian that sequentially follows it in what has been termed the superficial circulation of the qì. When the preceding Meridian shows an excess pulse and the following Meridian shows a deficient pulse, then a block of this type is likely. Entry/Exit blocks can be resolved by needling the Exit Point of the preceding Meridian and then the Entry Point of the following Meridian, both with tonification technique, a maneuver taught by Worsley among others. As noted, these blocks can occur both between Meridians of differing Elements, and also between Meridians of the same Element. Only the former situation is generally taught in the Five Element schools. I believe this is because if the block is felt between the two Meridians of the same Element in that tradition, it would ordinarily be treated by needling the Luò Point of the deficient Meridian. I have found that diagnosing splits between the two Meridians of the same Element using only the yáng finger position is not reliably accurate, whereas comparing both yáng and yīn finger placement pulses for the same Element is a better indicator of this type of imbalance, and could be either an Entry/Exit block or a Luò Vessel problem. I have not identified a pulse signal that would distinguish between these two possibilities, but the change in pulses should confirm which treatment restores a normal state. If neither

157 Source: Peter Eckman.

approach is successful, the RYCK treatment guidelines from the *Líng Shū* should provide an effective alternative.

Nánjīng 50 presents a very interesting model for interpreting the interrelationships between different Organs in the creation of pathology, dividing such cases into five groups. If an Organ becomes imbalanced from within itself, that is called being affected by a regular evil (zhèng xié 正邪). If the imbalance is transmitted from the Organ's parent (in Five Element terms), that is called being affected by a deficiency evil (xū xié 虛邪). If the transmission is from the Organ's child, that is called being affected by an excess evil (shí xié 實邪). If the transmission is from the Organ's grandparent, that is called being affected by a thief evil (賊邪 zéi xié). And if the transmission is from the Organ's grandchild, that is called being affected by a slight evil (wēi xié 微邪).[158] The examples in *Nánjīng* 50 of evils (attacking the Heart, for instance), given as illustrations of this doctrine, are wind, heat, improper diet or activity, cold and dampness. These are basically exogenous pathogenic factors, mostly of the climatic type, and yet the associations given here differ a bit from those listed in the *Nèijīng*. The order is clearly that of the Creative Cycle of the Five Elements starting with Wood, but here dampness is correlated with the Water Element (instead of Earth), and cold is correlated with the Metal Element (instead of Water). Dryness gets omitted, and the Earth correlation is to diet and activity (instead of dampness). Unschuld cites 26 commentaries on this chapter, but only one of them refers to these discrepancies: Ting Te-yung (1062 CE) noted that the reason given for these three changes in correlations is inadequate, and that they need to be explained in detail. I think this is actually an excellent example for illustrating the concept of "lenses" that I proposed earlier. The *Nánjīng* is using the Five Element lens here (as throughout most of its chapters) to correlate exogenous pathogens with Organs, while the *Nèijīng* uses the Six Great Meridians lens to make these same correlations (because the Meridians are connected to the Organs). What each lens reveals is true in its own domain, but can differ from what is seen through a different lens, examining a different domain. We can ask why these lenses provide apparently conflicting information. My best guess is that any theory of correlations, such as that between exogenous pathogens and

158 *Nánjīng* 50, author's translation and reordering: 從後來者為虛邪，從前來者為實邪，從所不勝來者為賊邪，從所勝來者為微邪，自病者為正邪

Organs, is only an approximation of reality. To assert that cold, for example, only can affect one of either the Bladder or Lung would be to force observations of nature into a rigid structure that does not exist in nature itself. These theories should be viewed simply as guides to help us understand the mechanisms most likely to be responsible for our clinical observations, and thus provide possibilities for appropriate therapeutic intervention. It is up to us as practitioners however, to find the lens that best reveals the actual pathogenic mechanism in any given clinical case.

While *Nánjīng* 50 explains pathology by reference to exogenous pathogens, the later chapters present quite a different orientation. Naturally, pathogens like wind, cold, dampness and heat need to be removed from the body, with the implication that acupuncture treatment should target the Organs and Meridians where the pathogens have lodged, with the goal of expelling them (although simply transforming them into non-pathological states is not explicitly excluded). I believe that the expulsion strategy is the principal basis of treatment as presented in the *Nèijīng*. However, in the chapters of the *Nánjīng* that discuss acupuncture treatment, the goal changes to one of "leveling" the qì (píng 平) in the affected Organs and Meridians, regardless of the pathogen that created the illness. Could leveling the qì lead to restoring each Organ's ability to transform pathological qì into physiological qì? These treatment chapters begin with number 69, which reprises the model of chapter 50, i.e. deficiency reflects transmission from the "parent", and so the strategy is to supplement the parent's Meridian. Likewise, excess reflects transmission from the "child", so the strategy is to disperse the child's Meridian.[159] Some commentators have interpreted chapter 69 differently, claiming that parent and child refer to distal Shū Points on the affected Meridian, but I can't agree, because the text clearly says that treating the affected Meridian itself is the proper response to illnesses that did not originate from pathogens affecting other Meridians. These other situations involving transmission must therefor be treated at the transmitting Meridian rather than simply at the affected Meridian. I certainly am not claiming that treating points on the affected Meridian, even when there has been transmission from other

159 The common usage in Chinese and English writings is to refer to "mother" and "son" rather than "parent" and "child". I prefer the latter approach, as it allows for using gender descriptions to differentiate effects both from and to Yīn and Yáng Meridians and Organs (zàng fǔ).

Meridians, is wrong (the interpretation stated by some commentators mentioned above), but rather that this chapter is not advocating that type of treatment. Actually, the combination of points on both the transmitting Meridian and the affected Meridian became recognized in later historical epochs as a very powerful treatment strategy, and reached its acme with Sa Am's 15th century addition of the combination of points on the controller and affected Meridians to form a four needle combination for treating either deficiency or excess in any Meridian.[160] What I would like to emphasize here is that if the strategy in treating deficiency of a Meridian is to supplement its parent Meridian, then the old paradigm must have changed, because the parent Meridian was harboring an "evil", and supplementing such a Meridian would in effect strengthen the evil qì along with the normal qì, and the *Nèijīng* warns us that evil qì represents a form of excess (even if we call this excess a "deficiency evil"), which therefore should not be supplemented. The example given in the *Nánjīng* is of an illness of the Heart that resulted from the Liver (its parent) being struck by wind, and then transmitting that to the Heart (the Liver's child, which now has become excessive). We may call the wind evil a deficiency evil, but it would not be sensible to supplement this wind qì, therefore only a paradigm shift in understanding the nature of pathology can resolve this contradiction.

Perhaps the most controversial chapter in the *Nánjīng* is number 75, which essentially answers the question of how to treat (for example) an excess in the Liver accompanied by a deficiency in the Lung. The first sentence states the governing principle: the Five Elements should level (píng 平) each other.[161] The concrete treatment recommended is to disperse the Fire and supplement the Water, however as the passage makes clear, Wood in this case means the Liver and Metal in this case means the Lung, therefore we are justified in concluding that the protocol recommended is to disperse the Heart and supplement the Kidney. It seems obvious to me that this passage has definitely changed the treatment goal from expelling pathogens (xié qì)

160 Sa Am also proposed a less well-known 4 needle combination for treating either excessive heat or excessive cold in any of the Meridians, and that has been a frequently applied clinical strategy by some Korean practitioners, such as Byoung Soon Kim, as explicated in his text. The diagnosis Kim uses for heat and cold is essentially made on the basis of pulse rate (beats per minute), and I have not found this to be entirely reliable.

161 金木水火土，當更相平.

to harmonizing the Organs and Meridians associated with the Five Elements (zhèng qì). As explained earlier, this idea reflects the interpretation of excess and deficiency as hyperfunctioning and hypofunctioning. The character píng 平 is emphasized, with the meaning of making level or equal. I believe the *Nánjīng* introduced a new standard of health (and a new paradigm) with the idea that each of the Organs should be equal in its functioning, including for example, supply of qì and Blood. At first glance this seems to be following the logic of yīnyáng wǔxíng theories, but actually it is not a necessary consequence of those theories. CCA is based on the recognition that individuals are conceived with different supplies of qì and Blood to their Organs (manifested as strongly functioning vs. weakly functioning Organs and their associated Meridians), and as a result, the goal of treatment to restore health is not really píng, but rather a return to one's original nature (i.e. Constitution), which differs for each individual, (the thesis examined in the first chapter of this treatise). From this point of view it becomes necessary to reexamine the nature of excess and deficiency. An excess becomes "functioning above the bounds of an Organ's Constitutional state" (which may have been merely strong initially), while a deficiency becomes "functioning below the bounds of an Organ's Constitutional state" (which may have been merely weak initially). CCA is thus based on a second paradigm shift, which I believe was first made explicit with Lee Jema's development of Sasang Constitutional Medicine (SCM) in the late 1800s, as discussed in *The Compleat Acupuncturist.*

Returning again to *Nánjīng* chapter 50, we may find the seed of a profoundly novel hypothesis about the way pathology develops, that only flowered with the introduction of Korean Constitutional Acupuncture (KCA) theory in 1965. This new paradigm involves incorporating some ideas derived from Western medicine within the overall understanding of traditional Oriental medicine. This new approach invariably seems as implausible to practitioners of Western medicine as it does to those of Eastern medicine. In short, the 5 possible situations depicted in *Nánjīng* 50 were correlated by Kuon Dowon, the originator of Korean Constitutional Acupuncture (KCA), with frequently occurring pathological mechanisms in Western medicine such as inflammation, infection, degeneration, autonomic dysfunction and others. The reason this idea is so counter-intuitive to almost everyone not trained in KCA is that in Kuon's conceptualization, the various Organs and Meridians do not have only fixed functions independent of the individual's Constitution, but also have relational functions based on their positions with respect to the Constitutional Organ/Meridian. E.g. someone with a Liver Constitution would need treatment of the Stomach Meridian to deal with any kind of infection, while someone else with a Kidney Constitution would require treatment of the Small Intestine for identical infections. The 20 Constitutions theory I have proposed in CCA would expand this proposition in such a way that while these 2 cases of infection would still be treated as described (by Kuon), someone with a Gall Bladder Constitution (not described by Kuon) would require treatment of the Large Intestine to treat the identical kinds of infection, and so forth for all the 20 Constitutions. Essentially, CCA is in many ways an expansion of KCA. I call this aspect of Kuon's system of acupuncture "constellational", because this same relational approach is applicable to inflammation, degeneration, etc.[162] *Nánjīng* 50 does not adopt constellational theory, and yet it is hard to conceive of Kuon developing such ideas without the intermediary paradigm shift displayed in *Nánjīng* 50. I might also add that *Nánjīng* 75 comes very close to constellational thinking, in that it clearly advocates treatment of Organs that are imbalanced, not by their own Meridians, but rather by Meridians that relate to them by the laws of Five Element theory. While this may be thought of as a "proto-constellational" idea, it is still part of the paradigm where "leveling" is the goal, rather than "returning to original nature", as in CCA. Some specific treatment approaches used in KCA will be discussed in a later chapter as will the topic of treating the pathological mechanism in cancer, which as far as I have seen involves the only systematic acupunctural methodology that treats this illness itself, rather than its outward symptoms or the side effects of Western medical approaches.

Since Kuon Dowon did not ever publish anything other than the simplest aspects of his methodology, I think this would be a good place to explain my personal understanding, based mainly on observation at his clinic in the 1980s. His methods evolved over time, as did his terminology, but what I will present is the material from that time period, however translated into

162 See the case description of zàng inflammation treatment I performed on my wife in chapter 5 as a good example of this "constellational" approach to pathology.

my own terminology, so as to render his teachings more user-friendly. One characteristic of constellational thinking is a strong emphasis on symmetry. In this regard it is apparent that symmetry is also a basic characteristic of yīn yáng theory. Just as all things belonging to the yīn category are opposite to their complementary members of the yáng category, the order of enumeration of formulas to treat afflictions of people whose Constitution is based on a zàng Organ (yīn) is opposite to the order for those whose Constitution is based on a fŭ Organ (yáng). Let us call the Constitutional Organ the "Root," and its 4 needle treatment the Root formula. This formula always belongs to one of the Five Elements. For treating inflammation of the zàng Organs in someone of a zàng Organ Constitution, the yīn grandchild (i.e. granddaughter) Meridian is chosen. It is called the zàng inflammation auxiliary formula. To treat inflammation of the fŭ Organs in the same individual, the yīn grandmother Meridian is chosen instead, and is called the fŭ inflammation auxiliary formula. For example, if the Constitution is Liver excess, the zàng inflammation auxiliary formula is a version of the 4 needle technique to supplement the Lung, while the fŭ inflammation auxiliary formula is a version of the 4 needle technique to supplement the Spleen. But if the Constitution was Gall Bladder excess instead (yáng Wood excess replacing yīn Wood excess), these auxiliary formulas are the exact opposite: the zàng inflammation auxiliary formula is a version of the 4 needle technique to supplement the Spleen and the fŭ inflammation auxiliary formula is a version of the 4 needle technique to supplement the Lung. The way I conceptualize this type of symmetry is to think of the auxiliary formulas for zàng Constitutions as following each other in a clockwise direction of enumeration (according to the shēng cycle), while those same auxiliary formulas in an individual having a fŭ Constitution follow each other in a counterclockwise enumeration order. I refer to this type of symmetry as "mirror-image". This dogma will be made clearer in a separate chapter (see Figs. 40 and 41), and also by the case histories to be presented.

CHAPTER 10

Pathogenic Invasion in the Sù Wèn: *the Liù Xié (六邪)*

Figure 29: Windmill-Holland (All Kinds of Weather) [163]

Almost all, if not all the Chinese classical medical texts discuss the role pathogenic factors (xié qì), characterized as penetrators of the body in the guise of climatic forces, play in the initiation, spread and recovery from illness (Fig. 29). Kuon Dowon, the initiator of KCA, treated all types of disease however, by strategies to simply balance the functioning of the Meridians and Organs in such a way as to return them towards their natural states. Expelling pathogenic factors was never part of his rationale in

163 Source: The late artist Leonid Afremov. Palette knife oil painting owned by the author, and used here courtesy of the painter, whose work can be found at www.afremov.com

treatment formulation, at least judging from his public statements. The other well-known teacher of a Constitutional approach to acupuncture, J.R. Worsley, took a slightly different approach. As discussed in chapter 3, following the example he learned from Lavier, he emphasized the importance of expelling pathogenic factors ("Aggressive Energy") from the body prior to balancing the Five Elements via their Meridians and Officials, in order to avoid spreading the pathogenic factors to places they had not yet affected. The major difference between Worsley and Lavier in this regard, is that the latter advocated treatment of the presenting symptoms as the first step in acupuncture treatment, so omission of his "Aggressive Energy" draining protocol was not seen as an impediment to any form of pattern (syndrome) treatment, such as those typical of TCM. Lavier viewed Five Element balancing as essentially preventative, rather than curative, or even symptom targeting treatment, best reserved for healthy individuals. I interpret this clinical teaching of Lavier's as supporting the idea that pathogens are not driven deeper by acupuncture treatments that address imbalanced functions of the organism, such as is the approach used in CCA. There is no doubt that dealing with xié qì forms a major focus of the *Nèijīng* teachings, so I have investigated pulse abnormalities that might indicate the presence of xié qì, and perhaps even disclose its nature.[164]

164 This may be another issue in which herbal and acupunctural lineages and theories differ. The following quotation exemplifies how an influential herbalist stated the case: "When a body that has deficiency contracts an evil, you must eliminate the evil. As soon as you eliminate the evil, you can supplement the right. Otherwise, if you supplement the right when there is a lingering evil, the supplementation will go into the evil. When the evil obtains supplementation, there is more excess and the right becomes more deficient." Wú Jūtōng (Qīng dynasty), translated and posted by Lorraine Wilcox on the Facebook Scholars of Chinese Medicine Group, 8/30/19. "Wu Jutong is regarded as one of the master figures in the school of seasonal warm diseases. In his representative work, Wen Bing Tiao Bian (Treatise on Differentiation and Treatment of Seasonal Warm Diseases), he was not only an expert in adopting and adjusting classical prescriptions, but also good at treating diseases by using warm - hot drugs. On the other hand, Treatise on Differentiation and Treatment of Seasonal Warm Diseases makes up some deficiencies and thus perfects Shang Han Lun (Treatise on Cold Pathogenic Diseases). So, it is a fact that Wu Jutong inherited and developed Zhongjing's doctrine," from an abstract by H. Huang of Fujian College of Traditional Chinese Medicine in Zhonghua Yi Shi Za Zhi. 2002 Jan;32(1):36-8. I am not alone in questioning whether or not acupuncture and herbology are based on identical theoretical foundations, as even Chinese scholars are divided on this issue. Most recently, William Ceurvels posted on the same Facebook group (4/2/22) regarding the views of different scholars concerning the fundamental texts of acupuncture and herbal medicine, "Hu Xishu, who along with Feng Shilun have both advanced this idea that SHL is completely unrelated to the Neijing and we should remove *Nèijīng* theory from our understanding of the SHL and JGYL." The question remains, does supplementation in acupuncture mean the same thing as in herbal medicine? Is it adding something to the organism or rebalancing what is already within the organism? Can acupuncture supplement xié qì? Even if acupuncture could not supplement xié qì, it might conceivably still be able to spread it within the organism, and so it is an important topic in any case. So far my clinical experience is that CCA treatments have shown no indication that they drive pathogens deeper or cause any detrimental effects when patients are correctly diagnosed and treated. Herbal medicine, on the other hand, does introduce new biologically active substances into the body, and these could easily have direct effects on pathogens, i.e. not mediated by the body's own systems (jīng mài and zàng fǔ). If that is possible, then perhaps supplementing herbs could inadvertently strengthen pathogens rather than the person's normal functions if prescribed incorrectly.

When it comes to pulse diagnosis, pathogenic factors have usually been identified by specific pulse qualities (rather than quantities or positions/ locations) such as wiry, slippery, choppy, etc.[165] The standard list of 28 qualities reflects the systematization of Lǐ Shí zhēn (李時珍 1518-1593)[166], but it is my impression (given his 1578 massive herbal publication *Běn Cǎo Gāng Mù* (本草綱目, *Compendium of Materia Medica*), and his biography, that Lǐ was primarily an herbalist, not an acupuncturist, and this analysis by pulse quality has not been something that I have fully been able to incorporate in CCA without it ending up as a "forced marriage". Certainly the *Nánjīng* approach to pulses was different, as exemplified in its chapter 5, on the resonances of five degrees of pressure with the Five Elements at the radial artery positions cùn, guān and chǐ. These are foundational aspects of CCA.

Many treatises had been written about pulse qualities prior to Lǐ Shí zhēn's work, but the most influential ones have been lost. Descriptions of how the various pulse qualities feel to an examiner have been reported, along with several pictorial representations. These latter could be most useful, if they were accurate, as it is unclear what to make of analogies such as "feels like pearls on a string" or "like a knife scraping bamboo," commonly cited pulse descriptions. A text called *Túzhù Màijué Biànzhēn*[167] by the Míng author Zhāng Shìxián (張世賢) was presumably based on a number of prior texts named *Màijué*, which can be translated as "Pulse Rhymes," since they often

---

165 See Table 1. I have underlined the pathogenic factors to be discussed, including wind, cold, damp and heat, and show their association with named pulse qualities. Of the standard 6 climatic factors only dryness and fire are missing. Tentatively, I think it is possible to extend this list by associating a hesitant (choppy) pulse with dryness, and a "pointy" pulse with Fire pathogens. One of my Korean mentors, Chae Lew, used to describe infectious conditions as exhibiting a pulse that feels like something sharp hitting your finger. The table shown is from the following Internet web link: https://theory.yinyanghouse.com/theory/chinese/pulse_diagnosis#diagnosis accessed 11/15/2018.

166 "Numerous pulse qualities are documented in ancient Chinese medical texts. For example, the *Nèijīng* describes over 30 types, and the *Màijīng* documents 24 types. The most common 28 TCM pulse qualities were (sic) used in clinical practice come from the references *Bīn Hú Mài Xué* (*Lakeside Master's Pulse Studies* 瀕湖脈學) by Lǐ Shí zhēn and *Zhěn Jiā Zhèng Yǎn* (*Correct Pulse Diagnosis* 診家正眼) by Li Zhongzi, 1588-1655. These qualities are floating, sunken, slow, rapid, surging, fine, vacuous, replete, long, short, slippery, rough, string-like, tight, soggy, moderate, faint, weak, dissipated, hollow, drumskin, firm, hidden, stirred, intermittent, bound, skipping, and racing. These pulse qualities are largely described qualitatively and are not clearly defined." Cited in Tang AC, Chung JW, Wong TK. Validation of a novel traditional chinese medicine pulse diagnostic model using an artificial neural network. *Evid Based Complement Alternat Med.* 2011;2012:685094. This article cites 2 references that show poor inter-rater and intra-rater reliability by Chinese medicine physicians when it comes to identifying pulse qualities.

167 *Pulse Key Annotated Diagram for Authentic Differentiation* (圖註脈訣辨真)

are poetic in structure, according to Hanson and Pomata.[168] Túzhù màijué literally means "annotated diagrams providing keys to identifying the pulses," so one would expect that texts such as this one would make pulse quality identification both easy and of high inter-rater reliability. Unfortunately, this outcome has not been demonstrated, a conclusion based on my research of the available literature. Several of these *Túzhù Màijué* texts even claimed Western Jin dynasty (265-316 CE) authorship by Wáng Shūhé (王叔和), but I have seen no evidence to establish such an early date for any illustrated works. Fig. 30 shows one example of such an illustration, from a 2006 publication that is said to be an adaptation from the *Túzhù Nánjīng Màijué*.[169]

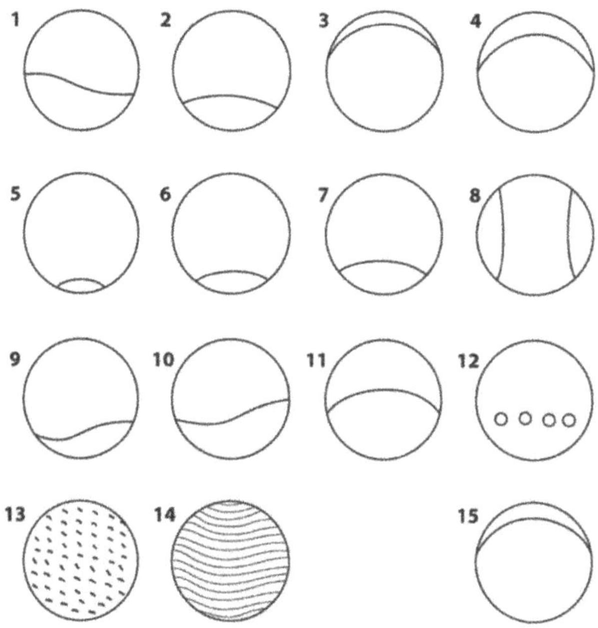

Figure 30: Pulse images from the Túzhù Nánjīng Màijué

168 For example: *Jiàozhèng Túzhù Màijué* 校正圖註脈訣 / By Wáng Shūhé (西晉王叔和撰); commentary by Zhāng Shìxián 四明張世賢註. 2 threadbound vols. 11.7 x 17.7 cm. Shànghǎi: Rixin Shuzhuang, 1905. An excellent account of the Western translations of *Túzhù Màijué Biànzhēn* can be found in Chapter 1 of *Translation at Work; Chinese Medicine in the First Global Age* by Hanson, M. and Pomata, G. pub. Brill: Leiden, 2020.

169 JBC, Vol. 32, No. 1 2006. www.jbiocommunication.org The caption claims a 2nd century CE dating, and gives the following names and Five Element correlations for the pulses: 1. Floating (Metal), 2. Bow-string (Wood), 3. Weak (Water), 4. Slow (Earth), 5. Hollow (Fire), 6. Firm (Wood), 7. Deep (Water), 8. Hidden (Wood), 9. Smooth (Water), 10. Extensive (Fire), 11. Gentle (Earth), 12. Damp (Water), 13. Filled in (Fire), 14. Rough (Metal), 15. Weak (Metal). If one compares this figure to the following figure, it is clear that the Elemental associations are incorrect, often being mirror image associations to those where the Chinese characters specify the Elements.

Figure 31, by comparison, is quite similar, indicating that both illustrations shared the same original source, but the interpretations of these two versions are totally different, obviously the result of an editorial mistake as explained in Footnotes 169 and 170. Unfortunately, we don't have a reliable date for the original version of this diagram, (let alone for exactly where to place ones fingers when taking the radial pulses). Personally, I find these illustrations of pulse qualities an argument against, rather than for, any consistent identification of such qualities. I doubt that anyone could distinguish numbers 2 from 7, 3 or 4 from 15, or 9 from 10 in Fig. 30, and these are just the most glaring examples of the ambiguity of how different pulse qualities feel.

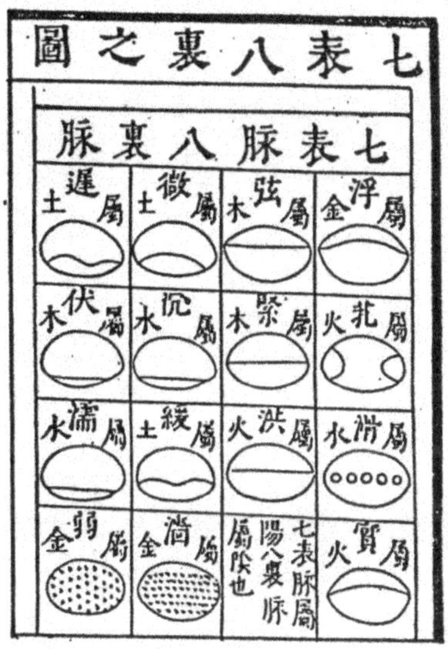

Figure 31: Pulse images from Sāncái Túhuì *(1607) by Wáng Qí* [170]

170 The text *Sāncái Túhuì* (三才圖會) from 1607 by Wáng Qí (王圻) shows a very similar illustration of pulse qualities to the preceding Figure. This illustration was certainly not originated by Wáng Qí, as it is identical to Zhāng Shìxián's 1510 *Túzhù Màijué Biànzhēn*'s version. The earliest publication of this figure is undetermined, but illustrations of the Chinese pulses can be dated as early as the 13th century, per Hanson and Pomata. This image was published in 1953 by Professor P. Huard, and republished by Soulié de Morant in 1983. On the right half it shows 7 yáng Exterior (biǎo) and on the left half it shows 8 yīn Interior (lǐ) pulse forms with their Elemental categories and names. From left to right, starting with the top row these are retarded (chí 遲) Earth, minute (wēi 微) Earth, stringy (xián 絃) Wood, superficial (fú 浮) Metal, hidden (fú 伏) Wood, deep (chén 沉) Water, tight (jǐn 緊) Wood, hollow (kōu 芤) Fire, weak (xū 濡) Water, relaxed (huǎn 緩) Earth, flooding (hóng 洪) Fire, slippery (huá) Water, faint (ruò 弱) Metal, rough (sè 澀) Metal and solid (shí 實) Fire. I have chosen the most evocative English terms that connote differences between these various qualities, and also chosen the Chinese characters that are closest to those in the poor quality original character images. It is interesting to note that the Five Element character of each pulse quality was emphasized without any mention of a specific Organ. Yīnyáng and wǔxíng theories seem to play a more fundamental role than either zàngfǔ or xié qì theories in this Míng dynasty text, regarding pulse qualities.

Lacking meaningful images of the pulse qualities, we are left with verbal descriptions as our most useful guide, as presented in Table 1. For now, note in the right hand column the underlined pathogens: wind, cold, heat and damp. These are four of the six classical exogenous pathogenic factors that I will return to further on.

TABLE 1

Pulse Descriptions, Qualities and Clinical Significance
(Adapted from: https://theory.yinyanghouse.com/theory/chinese/pulse_diagnosis#diagnosis)

DESIGNATION	SENSATION	INDICATIONS
By Depth:		
Floating (superficial)	easily felt at the superficial level not as significant as you feel deeper	**external condition/pathogen** + empty = yin and/or blood def + rapid = wind heat + tight/slow = wind cold
Sinking (deep)	felt only at the deep level	**Interior condition/obstruction** + rapid = internal heat + slow = internal cold + slippery = internal damp/phlegm + empty = qi or yang def
By Frequency:		
Slow	less than 4 beats per breath (< 60 bpm)	**cold condition a/or pathogenic factor** + floating = exterior wind cold + sinking/empty = yang def
Rapid	more than 5 beats per breath (> 90 bpm)	**hot condition a/or pathogenic factor** + floating = external wind heat + sinking = internal heat + full = excess heat + empty = empty heat
By Quality/Shape:		
Hesitant (choppy)	rough and uneven	**blood and/or jing stagnation**

Slippery	smooth with a viscous sensation	**excess dampness,** retention of food, pregnancy + rapid = damp heat + slow = cold damp obstruction
Tight	tension with side to side movements (thicker than a wiry pulse)	**excess cold** interior or exterior, commonly associated with pain
Wiry	tension with no side to side movements (thinner than a tight pulse)	**LV/GB disharmony**
By Width:		
Big (excess, overflowing)	broad but with distinct edges	**excess heat,** commonly in ST or Intestines
Thin (thready, fine)	fine but with distinct edges	**blood and/or qi deficiency**
By Strength:		
Empty (deficient)	wide but not strong, disappears with slight pressure, forceless	**blood and/or qi deficiency**
Full (excess)	wide and strong, felt with strength at all levels	**excess condition,** often excess heat with rebellious Qi
By Length:		
Short	not felt in all 3 positions	**qi deficiency**
Long	felt beyond the 3 positions	**excess, heat,** generally considered normal in absence of other qualities
By Rhythm:		
Hurried (abrupt)	rapid with irregularly missed beats	**heat agitating qi & blood**
Intermittent	regularly skipped beats	**Heart disharmony,** exhaustion of zang qi
Knotted	slow with irregularly missed beats	**cold obstruction,** HT qi or yang deficiency, general def of Qi, Blood and/or Jing

This difficulty in integrating pulse qualities, and perhaps specific types of pathogenic factors, into CCA exemplifies an important question to which I have already alluded: Do acupuncture and herbal medicine share all the same underlying theoretical premises, or are there significant differences?[171] I will only propose that there were many streams of thought and practice that contributed to the *Nèijīng*, and one can find contradictory statements in various chapters, an example of which I have already illustrated, presumably from these different streams of contributors. One can easily imagine that some practitioners were "specialists" in acupuncture, while others were "specialists" in herbal medicine. For example, the *Nánjīng* is primarily an acupuncturist's text (in addition to the claim that the *Líng Shū* was originally a separate treatise called the *Needle Classic*), while the *Shénnóng Běncǎo Jīng* and *Shānghán Lùn* are primarily herbalist's texts. From this point of view it may be easier to understand why the traditional pulse qualities, of major importance in herbal medicine, have not been dealt with as yet in CCA, which is definitely an acupuncture-based system of healthcare. This discussion urges me to mention one other crucial question that has been viewed differently by various authors; that is regarding the nature of Excess 實 shí and Deficiency 虛 xū (aka repletion and depletion or solidity and hollowness). These terms in modern TCM are usually interpreted via the saying, "Excess refers to the pathogenic factors (xié qì), while Deficiency refers to the anti-pathogenic resistance factors (zhèng qì)."[172] There have been opinions expressed that TCM primarily reflects the thinking of Chinese herbalists who held a higher status in much of Chinese history than did the acupuncturists. However, keep in mind that there were classical texts primarily devoted to acupuncture and its theoretical basis which viewed Excess and Deficiency not in terms of the relationship of xié qì to

171 Liao P'ing (1851-1914) is quoted in Unschuld's *Nánjīng* as follows, on p. 519, "In distinguishing between (illnesses caused by) wind, cold, moisture, and heat, the *Shang-han* (*lun* of Chang) Chung-ching considers the symptoms as most decisive, while it regards the movement in the vessels … as secondary. It is a false doctrine of the *Nan-ching* to determine illnesses on the basis of the (movements in the) vessels alone." This quotation illustrates the historical idea that the doctrines of the *Shānghán Lùn* (herbalism's primary clinical text) were considered by some as fundamentally correct, while the doctrines of the *Nánjīng* (acupuncture's comparable clinical text) were fundamentally incorrect, especially the importance of pulse manifestations, upon which CCA is fundamentally based. In fact, CCA is founded on the idea that treatment based on symptoms alone as primary diagnostic indicators (the Condition) leaves out the crucial importance of the Constitution in acupuncture practice.

172 Zhāng Shìxián 張世賢 (1510 CE), previously cited as the author of *Túzhù Màijué Biànzhēn*, is quoted by Unschuld as commenting on *Nánjīng* 48 thusly, "虛 xū stands for empty; that is the proper influences are lost. 實 shí stands for vigorous and full; that is, evil influences are present in abundance."

zhèng qì, but rather to the intensity of functional activity of the Organs or Officials such that an Excess implied hyperfunctioning and a Deficiency implied hypofunctioning, as I have already explained. In this theoretical interpretation, qì was not really lost of gained in understanding the mechanism of illness, but rather a state of imbalance was created in which one Organ's hyperfunctioning was matched by another Organ's hypofunctioning. Such a view provides the basis for employing the method of transferring qì by the shēng and kè cycles, as taught by Worsley and others, and for employing the Four Needle technique of Sa Am, that is the fundamental treatment strategy in KCA. Both of these disciplines are acupuncture-only Constitutional approaches, unconnected to herbal theories or techniques of diagnosis and treatment. CCA was originally formulated in accordance with this latter interpretation of Excess and Deficiency, explaining them as alterations in the functional activity of various Organs away from their original Constitutional state, regardless of which one of the three classes of etiological categories[173] provoked this change. Some Organs become hyperactive, others hypoactive. Treatment is aimed at restoring the original state determined by the Constitution, and from this state the organism is ideally situated to deal with ridding itself of any pathological factors, if such are found to be present.

Over the years (decades) I have studied with many teachers of acupuncture and Oriental medicine, and every once in a while I hear something that sticks in my mind, like an "earworm". Such was the case with a concept enunciated by the late Maurice Mussat, a leading figure of the French Acupuncture Association (AFA), who was a frequent lecturer in the USA as part of the Medical Acupuncture training course organized by Joseph Helms, and sponsored by UCLA. What I was struck by was Mussat's reference to something he called the "Operators" of the 6 Great Meridians (liùjīng) of yīn and yáng. Each Great Meridian has both a hand and a foot branch, and the Operator was said to be the branch of each Great Meridian to treat when its corresponding level in the 3 yīn and 3 yáng Great Meridian diagnosis showed an abnormality. In fact, diagnosing which of these 6 levels needed treatment was a hallmark of the French approach, and at least one other French

173 Classically these are exogenous, endogenous and neither (miscellaneous).

medical acupuncture teacher, Yves Requena, treated this idea very much as a Constitutional level of diagnosis, which he referred to as the individual's "terrain". Returning to the topic at hand, Mussat claimed that the Operator concept was something he found in the *Nèijīng*, but even when pressed he was not able to cite chapter and verse. After gestating within me for years and after numerous fruitless searches in the text, it somehow dawned on me that there might be a connection of the Operator concept to the different types of xié qì (邪氣) with which the 6 Great Meridians have a resonance. In pursuing this idea, I focused on the later chapters in the text of the *Sù Wèn*, starting with chapter 66, as these chapters deal extensively with the relationship of pathogenic factors to the 6 Great Meridians. The chapters I discuss, including 66 and those that followed it, are believed to have been inserted into our extant copy of the *Sù Wèn* by Wáng Bīng, based on his belief that the original chapters were deemed to be lost. They were supposedly replaced by material from other early treatises, now also lost. During my research, I was not able to establish that the concept of "Operator Meridians" had a verifiable relationship to any specific pulse qualities for the longest time, but neither was I ready to reject it out of hand. The remainder of this chapter reports my observations while investigating this topic.

Sù Wèn 66 is quite a significant chapter for a number of reasons. Firstly, this chapter starts out by enumerating only 5 climates to match the 5 Elements: i.e. cold, heat, dryness, dampness and wind. Fire was not initially included as a climatic factor (qì), but later in the same chapter it is added as a sixth climatic factor. Why was that done? Secondly, we have the question (unasked): How does the system of 6 (i.e. the Great Meridians, 3 yīn and 3 yáng) relate to the system of 5 (i.e. the 5 Elements)? - a crucial issue in the practice of acupuncture according to classical theory. It would appear that one major intention of this chapter was to present an explanation of how the resonances of 5 and 6 are related. Let me start an analysis with the text that begins chapter 66:

> "Heaven has the Five Elements (wǔ xíng)...(which) generate cold, heat, dryness, dampness and wind.... Man has five zàng that transform the five qì thereby generating joy, anger, pensiveness, anxiety and fear."[174]

174 天有五行，御五位以生寒暑燥濕風. 人有五藏，化五氣，以生喜怒思憂恐.

This is an obvious correlation of climates with emotions, based on the Five Elements. It forms a clear parallel of exogenous pathogenic factors (climatic) with endogenous pathogenic factors (emotional). Later, however, in the same chapter, one of Huang Di's teachers, Guǐ Yú qū says,

> "cold, heat, dryness, dampness, wind and fire, these are the yīn and yáng of Heaven. The three yīn and the three yáng, they act on their behalf. Wood, Fire, Earth, Metal, Water and (Minister) Fire, these are the yīn and yáng of the Earth. Generation, growth, transformation, gathering and storage, they correspond to them below."[175]

What we have is fire added as a sixth climatic factor and Ministerial Fire added as a sixth Elemental aspect to match it. On first glance this passage ends up implying that Heaven resonates with the 6 climates, while Earth resonates with the 5 Elements, but notice that there is an extra (second) Fire listed with the Elements, even though there are only 5 stages of development allocated to them. I think the implication is that in the descent of the qì of Heaven toward Earth, one of the 6 qì becomes hidden, and it is obviously Ministerial Fire (which in the human body has a function but not a form!).[176] I think we all intuitively recognize the difference between the 5 obvious types of climates and fire, which we do not ordinarily think of as a climate.[177] This doubling of the Fire Element into a normal part of the sequence of the 5 Elements plus a hidden part of Fire that is outside the sequence (xiāng shēng) of the 5 Elements perfectly comports with why Ministerial Fire is treated outside the set of rules governing the other

---

175 寒暑燥濕風火，天之陰陽也，三陰三陽上奉之。木火土金水火，地之陰陽也，生長化收藏下應之. The Five Elements (wǔ xíng) are presented here in the order of the Creative cycle (xiāng shēng), starting from Wood to Fire to Earth to Metal to Water. The Creative Cycle is a recurring model in these chapters of the Sù Wèn added by Wáng Bīng.

176 Zhū Dānxī (1280-1358): "In the great polarities, movement generates yang, stillness generates yin: yang moves to transmute, yin settles to unite. From them, water, fire, wood, metal, and earth are generated, each with its own nature; yet, there are two types of fire: there is the sovereign fire, which is the human fire; and there is the ministerial fire, which is the heavenly fire.... As for its name, the physical form and qi are generated and it is categorized within the five phases, thus it is called the sovereign (fire). As for its role, it is generated in the empty void, to safeguard its role, to receive the (heavenly) decree, thus when it is in movement, it becomes visible thus it is called the ministerial (fire)." Translation by Allen Tsaur, who posted it on the Scholars of Chinese Medicine Facebook group on 6/7/19.

177 Interestingly, the ordinary climatic factors have influences (qì) but not form (xíng 形). We cannot see heat, cold, wind, dryness nor dampness, only their effects on material substances. However, we can see fire in both the flames from burning objects and the sun in the sky. The 10 Organs associated with the ordinary climatic factors have form, but the Ministerial Fire Organs do not!

Elements and their associated Officials (zàng fǔ) in CCA, including the thesis that there are no Ministerial Fire Constitutional types.[178] The doctrine of Five Elements is thus kept by introducing this double nature of Fire, while at the same time clearly associating the Ministerial Fire functions and their associated Meridians with the Fire Element, a matter that has been critiqued by other commentators who associate them (for good reasons) with the Water Element.

A third important aspect of chapter 66 is that it appears to be the original locus of the enumeration of these 6 climatic factors (liù qì 六氣). When these 6 climates appear at the proper times and in proper degree, they help maintain a healthy state in the people exposed to them, but when they are untimely or not of proper degree, they become injurious to health and are referred to as the 6 evils (liù xié 六邪, or liù yīn 六淫). In other words, we now have a theory of pathogenic factors that can disrupt the proper functioning of their resonant partners, the 6 Great Meridians, but I want to emphasize that in the way this material is presented, there is no indication that the way climatic factors might disrupt the normal functioning of the qì dynamic is essentially different than the way the emotional factors might, since they are initially presented as strictly parallel concepts. I point this out because it appears to me that both exogenous and endogenous factors create identical types of pulse readings, but I'm getting ahead of myself. I do believe, however, that most Oriental medicine texts refer to some exogenous and endogenous pathogens by identical terms such as wind, damp and cold. In the case where these are endogenous, it is not really a problem to imagine that they are created by mental/emotional issues rather than by climatic influences.

Sù Wèn 67 continues the discussion of how the systems of 5 and 6 (Heaven and Earth)[179] can be integrated in a medical context. The reasoning is a bit complex, but it turns out to be logically coherent. In order to understand

---

178 "The Japanese self-healing system of Jin Shin Jyutsu, which shares many of the basic Five Element concepts, disaggregates V and VI (*PE and TH*) from Fire and views them as a different and deeper level within the system." Posted by John Wheeler on the Five Element Acupuncturists Facebook site 5/12/19.

179 The association of numbers with resonant terms depends on context. While Heaven sends down 6 climatic factors, it is associated with the 10 Heavenly Stems (5 doubled by yīn and yáng). While the Earth manifests the 5 Elements (wǔ xíng), it is associated with the 12 Earthly Branches (6 doubled by yīn and yáng).

it, one needs to start with the traditional dating system of China, called the 10 Heavenly Stems and 12 Earthly Branches, which is said to date back to a minister of the Yellow Emperor, Da Rao Shi, c. 2700 BCE. Legend aside, this system can be found on oracle bones dated to c. 1350 BCE during the Shāng dynasty. While this system was originally used for marking the succession of days, by the Hàn dynasty and the time of the *Nèijīng* it was already being used to additionally mark the succession of years. There are 10 Stems, which can easily fit into correspondence with the Five Elements, having a yīn and a yáng Stem for each Element. Since there are 12 Branches, the correspondence with the Five Elements is not so simple, however most references give the following associations, noting that the numerical cyclical order of the Branches never changes, or the system would not serve to identify any sequential phenomena.

The 12 earthly branches are 1) zǐ: yáng Water, 2) chǒu: yīn Earth, 3) yín: yáng Wood, 4) maǒ: yīn Wood, 5) chén: yáng Earth, 6) sì: yīn Fire, 7) wǔ: yáng Fire, 8) weì: yīn Earth, 9) shēn: yáng Metal, 10) yǒu: yīn Metal, 11) xū: yáng Earth and 12) haì: yīn Water. The logic of the Elemental associations indicates that they follow the shēng cycle progression starting and ending with Water, except that instead of Earth following Fire, it is moved to a central position in each set of three Branches.[180] These different positions of Earth, either centrally or on the periphery following Fire are both accepted as traditional variants of the wǔ xíng relationships.

Chapter 67 however gives us something new in the associations of the Branches. It states:

> What is above in zǐ and wǔ, shaoyin rules it.
> What is above in chǒu and weì, taiyin rules it.
> What is above in yín and shēn, shaoyang rules it.
> What is above in maǒ and yǒu, yangming rules it.
> What is above in chén and xū, taiyang rules it.
> What is above in sì and haì, jueyin rules it.

180 Water, Wood, Fire, Metal, with Earth between each of them.

In order to make sense of this information, it is first necessary to figure out the logic behind the ordering of the pairs of Branches. This turns out to be quite simple: The 12 Branches are divided into the first six and the second six. Each pair now follows the order: 1 and 7, 2 and 8, 3 and 9, 4 and 10, 5 and 11, 6 and 12. The original sequence of Branches was shown to reflect the creative cycle (xiāng shēng) of the Five Elements (wǔ xíng). Can we perhaps find a similar pattern here? Shaoyin includes both Heart Fire and Kidney Water. Taiyin includes both Lung Metal and Spleen Earth. The only creative cycle relationship here is Heart Fire being followed by Spleen Earth. This reasoning, when extended to the subsequent pairs and their associated Meridians leads to an almost perfect match with the creative cycle.[181] The only exception is that Ministerial Fire is out of sequence (it follows Earth here instead of preceding it), but I have already suggested that Ministerial Fire does not follow the ordinary rules governing wǔ xíng and zàng fǔ relationships. Thus I think my interpretation, that the ordering of these coupled pairs of Branches reflects the creative cycle of the Five Elements once again, is justified, however only one member of each pair comprising a Great Meridian is responsible for establishing this ordering. In short, my interpretation is that the text is implying that shaoyin is governed by the Heart, taiyin is governed by the Spleen, shaoyang is governed by Sanjiao (Triple Heater), yangming is governed by the Large Intestine, taiyang is governed by the Urinary Bladder and jueyin is governed by the Liver. The word the text uses, 主 zhǔ, I have translated as governs or rules, but there is not much difference if we say operates instead, and I think this is probably the genesis of the Operator terminology of Mussat, having been translated from Chinese to French and finally to English. Let me repeat, my interpretation of the relationship of pathogenic factors to specific Meridians is not stated explicitly in the text here, but it is implicit if one searches for a rationale for the ordering presented in the text. Curiously, these exact choices of which half of each Great Meridian to select for treating a given pathogenic factor will result in a shēng cycle order that happens to reflect the same choice as the half of each Great Meridian that Mussat called the Operators.

181 Fire, Earth, Minister Fire, Metal, Water, Wood.

To simplify the previous passage, we may make the following associations:

> Shaoyin is governed (operated) by Imperial Fire (Heart)
> Taiyin is governed (operated) by Earth (Spleen)
> Shaoyang is governed (operated) by Ministerial Fire (Triple Heater)
> Yangming is governed (operated) by Metal (Large Intestine)
> Taiyang is governed (operated) by Water (Urinary Bladder)
> Jueyin is governed (operated) by Wood (Liver)

Or, Fire creates Earth, (which creates Ministerial Fire), which creates Metal, which creates Water, which creates Wood. These 6 Organs and their associated Meridians are the Operators of the 6 Great Meridians (3 yīn and 3 yáng), establishing a rationale for the relationship between the theories based on 5 with those based on 6. That is, we now have the creative cycle as the basis of the succession of the 6 Operators, but we have not yet established any connection with pathogenic factors, whether exogenous or endogenous, nor with purported pulse qualities. That final piece of the puzzle is partially discussed in the following chapter, *Sù Wèn* 68.

Sù Wèn 68 is quite an extensive chapter, but I will focus on the stated connections of the climatic factors with the 6 Great Meridians. Here is my translation of the significant passage from this chapter:

> "Shaoyang above (taking Heaven's position for this time period), fire qì governs, in the middle (from the position of Earth, we) see jueyin. Yangming above, dryness qì governs, in the middle (we) see taiyin. Taiyang above, cold qì governs, in the middle (we) see shaoyin. Jueyin above, wind qì governs, in the middle (we) see shaoyang. Shaoyin above, hot qì governs, in the middle (we) see taiyang. Taiyin above, damp qì governs, in the middle (we) see yangming."[182]

182 少陽之上，火氣治之，中見厥陰。陽明之上，燥 氣治之，中見太陰。太陽之上，寒氣治之，中見少陰。厥陰之上，風氣治之，中見少陽。少陰之上，熱 氣治之，中見太陽。太陰之上，濕 氣治之，中見陽明. The commentaries on this passage give a number of differing interpretations of what is meant by this term zhōng 中, which I have translated as middle. Most have to do with the climatic factors that change halfway through the time period governed by each of the 6 Great Meridians following each other in the course of a year. This interpretation does not seem to me to match what can be observed in nature, so I have offered an alternative one.

It is not entirely clear what the text means by the middle (zhōng 中) in these instances, but we can see from the structure that this is a case of coupling the Meridians of the zàng and fǔ Organs of each of the 5 Elements. Thus shaoyang is always coupled with jueyin, yangming is always coupled with taiyin, taiyang is always coupled with shaoyin, etc. The text uses the terminology of root (本 běn) and tip (標 biāo) to describe the principal meridian above and its coupled meridian in the middle (center-zhōng 中). These relationships are exactly the same as the more common description of these coupled meridians as having a surface (表 biǎo) and inside (里 lǐ) connection. The two words biao (標 biāo and表 biǎo) are close homophones, but are written with different characters. It is tempting to see this passage as an early iteration of this cardinal teaching about Meridian relationships. However, for our purpose at the moment, the information I wish to emphasize is that the following connections exist between specific Meridians and their associated climates, again taking into account the "Operator" designations from the preceding *Sù Wèn* chapter:

> shaoyang – fire – Triple Heater is the Operator. It is coupled with jueyin (Pericardium).
> yangming – dryness – Large Intestine is the Operator. It is coupled with taiyin (Lung)
> taiyang – cold – Urinary Bladder is the Operator. It is coupled with shaoyin (Kidney)
> jueyin – wind – Liver is the Operator. It is coupled with shaoyang (Gall Bladder)
> shaoyin – heat – Heart is the Operator. It is coupled with taiyang (Small Intestine)
> taiyin – dampness – Spleen is the Operator. It is coupled with yangming (Stomach) [183]

In passing, it is interesting to note that the coupled Great Meridians, referred to as zhōng (中) in *Sù Wèn* 68, are associated with opposite pathological factors to those of the Operator Meridian. Dryness is the opposite

183 In a previous monograph mentioned earlier in this work, I have demonstrated that this order of enumeration represents one manner of expression of the Celestial (or Fú Xī) circular order of the *Yìjīng* trigrams, and is presented this way in *Sù Wèn* 68.

of dampness, cold is the opposite of heat, and wind is the opposite of fire. This last pairing is not as intuitively obvious, but is illustrated by blowing out a lit match (wind opposing fire).

Having established the classical citations that underpin the 6 Operators, including their Organs, Meridians and associated pathogenic factors, it is now finally possible to ask if there are any clinical findings (in the pulse) that would allow practitioners of CCA to diagnose and treat imbalances caused by these 6 climatic pathogens or their psycho-emotional resonant partners. One would think that having presented a systematic association of the 6 pathogenic factors with the 6 Great Meridians, that there would follow a description of how to identify these pathogenic incursions. The *Sù Wèn* chapters I've discussed and the ones that follow them[184], do list the signs and symptoms of imbalances induced by these climatic pathogenic factors, however I don't recall seeing a single reference there to pulse qualities associated with either the 6 Great Meridians or specific pathogens. That fact strikes me as astonishing, and seems to indicate that this exegesis of seven chapters did not place much emphasis on the relationship between the pathogenic factors, their correlated Meridians and any associated pulse qualities. In addition, my initial personal impression was that the Operator designations did not seem to directly correlate with any observed (xié qì) pulse signature that can reliably be associated with either the Constitution or the Condition of the patient. Well, negative findings are as important as positive findings in any scientific pursuit. These chapters however, added by Wáng Bīng, do not actually discuss acupuncture, when it comes to treatment advice, but rather almost exclusively refer to the taste and thermal nature (qì) of remedies, whether they be foods or herbal products. I might also point out here that in herbal medicine, the vast majority of attention is paid to the state of the zàng, as opposed to the fǔ Organs (notably in the zàng fǔ biàn zhèng methodology of TCM). This is quite different from the situation in acupuncture, where the Points chosen for treatment are as likely (or more

184 *Sù Wèn* Chapters 66-71 and 74 were added to the surviving text by Wáng Bīng, and form an integrated thesis about the topic under discussion.

likely) to be on the fǔ as on the zàng Meridians.[185] Think of the most frequently used acupuncture Points: Who could ignore the importance of ST 36 or LI 4? This is just one more indication of the historical difference in the theoretical postulates of acupuncture and herbal medicine. In CCA, there is no preferential focus on either the zàng or fǔ Organs regarding either the Constitution or the Condition.

Despite the negative results just described, I have followed the trail of the mysterious Operators, hoping to find some explanation of how to use this French analysis of the *Sù Wèn* clinically, especially as regards the pulse diagnosis of pathogenic factors. Ironically, after having written a draft of this chapter, I found a posting by a contemporary French acupuncturist, Jean-sylvain Prot, who did illustrate the connection of these *Sù Wèn* chapters to the pulses, and I have summarized his teachings in Fig. 32. From this figure we can see that the Operator associations are still being attended to, but not in terms of pulse qualities. Rather they are obviously being used as keys to wǔyùn liùqì (五運六氣) theory, which after all was the central focus of Wáng Bīng's added chapters. This theory, and its later application to acupuncture, was never intended to correlate pulse qualities with pathogenic factors, but rather is an elaborate system of Macrocosm/Microcosm harmonization, analogous to the one presented in Chapter 7 of this treatise.[186] It is a highly abstract system that relies on calculations to determine open acupuncture points depending on the year, month, day and hour. I prefer Simonov's method because it can be empirically verified in the patient's pulse.

185 Consider the famous 12 stars of Mǎ Dānyáng (馬丹陽), initially proposed in 1329, and finalized subsequently by Xú Fèng (徐鳳) in the Sòng dynasty (c.1439), which were said to be able to treat virtually any health problem. Three fourths are on yáng Meridians and only one fourth are on yīn Meridians. The 12 stars are ST 36, ST 44, LI 11, LI 4, UB 40, UB 57, UB 60, GB 30, GB 34, LV 3, HE 5 and LU 7.

186 Rhonda Chang posted the following remarks on her Facebook group, Yinyang Wuxing Yi on 6/6/21: "主 zhǔ means a master.... 主 is an important character to remember particularly in connection with 五運六氣 wǔ yùn lìu qì, which are the theoretical principles of Yi. For example, within 五運 there are 主運 zhǔ yùn (*Master*) and 客運 kè yùn (*Guest*) which are to do with the position of the planets in relation to the earth (italics added by author)." This is a macrocosm/microcosm application, unrelated to any reference to pulse qualities.

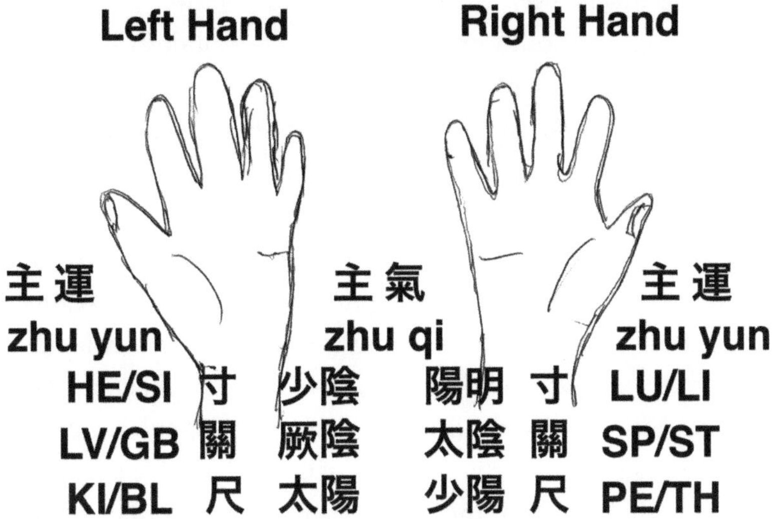

Figure 32 Wǔ yùn liù qì theory [187]

Never one to give up the search, I believe I have more recently finally found a connection between the Operators and CCA pulse patterns, but I doubt that it is anything like what has previously been described! The explanation, like much of this chapter, is rather complex, but in the end it will be shown to perfectly match the chapters in the *Sù Wèn* I have been discussing. I will start with the unusual finding that sometimes the Subdosha pulses are exactly opposite on the two wrists. This is always a case of the same Element showing signs on both sides. Every one of the 20 Constitutions has a unique pattern of normal Subdosha pulses. In a fully healthy individual, a necessary but not

187 Source: Peter Eckman. This drawing is adapted from the Facebook post of Jean-sylvain Prot of 12/7/20. It illustrates the ongoing use of the "Operator" designations in contemporary French acupuncture. The Operator designations are given in Chinese characters in the middle of the figure so as to fit them in the illustration. From top down (cùn, guān and chǐ) on the left hand they are Heart (Imperial Fire) as the operator of shaoyin, Liver (Wood) as the operator of jueyin and Bladder (Water) as the operator of taiyang. On the right hand they are Large Intestine (Metal) as the operator of yangming, Spleen (Earth) as the operator of taiyin and Triple Heater (Ministerial Fire) as the operator of shaoyang. Zhǔ 主 is the term being translated as operator and yùn運 here means movement, referring to the Five Elements or Phases. This type of illustration is a way to summarize the relationships discussed in this chapter, but it is not something found in the *Nèijīng*, as that text did not even use the terminology of cùn, guān and chǐ, and thus could not have assigned pulse qualities to operators manifesting their activity at these radial artery positions. Prot's teachings in this regard are otherwise consistent with those in the classical texts, as emphasized by Chang in the preceding Footnote.

sufficient condition for its good state of health is that this Subdosha pattern is the same on both their left and right wrists. The most common finding in people with significant health problems is that the Subdoshas on the left and right sides reflect different Elements. Such individuals need to be treated in a way that balances the Organs and Meridians of the Five Elements, so as to return their level of functioning to that ordained by their Constitution. After this state of balance is achieved, one may still find however, that although the left and right side Subdoshas reflect the same Element, they may not do so in the same way. For example, in someone with a LU Excess Constitution, either both left and right Subdoshas might show Vata/Kapha (spikes on the radial edge of the index and ring fingers), or one side might show Vata/Kapha (spikes on the radial edge of the index and ring fingers) and the opposite side might show Pitta (a spike on the middle finger on the radial edge). These two outcomes are both patterns of the Metal Element, and both of these patterns of Subdosha findings have appeared in patients I have treated. When the Subdosha patterns are the same on both sides there is no further analysis needed, as this reflects a healthy state, but what are the possible reasons for the occurrence of opposite Subdoshas? Before answering this question, let me review the findings I am calling opposite Subdoshas.

The following pairs of Subdosha patterns on opposite wrists illustrate this dynamic: Wood shows both Kapha (Wood type) and Pitta/Vata (Wood type), Earth shows both Kapha (Earth type) and Pitta/Vata (Earth type), Metal shows both Pitta and Vata/Kapha and Water shows both Vata (Water type) and Pitta/Kapha (Water type), while Fire shows both Vata (Fire type) and Pitta/Kapha (Fire type). The only difficulty is with Fire, which always shows spikes at 2 of the 4 Subdosha locations on each examining finger. These spikes are in opposite places on the left and right sides when this kind of pathology is present however. E.g. if the <u>left</u> side shows spikes at positions 1 and 3 on the index finger, then on the <u>right</u> side the examiner will feel spikes on the index finger at positions 2 and 4.

There are several theoretical interpretations possible for opposite Subdoshas, but after eliminating the ones that seem less likely, as I will demonstrate, we will be left with this hypothesis: When the Subdoshas are exact opposites, in my experience it appears to indicate that the individual is harboring some type of retained xié qì pathogen. What then are the theoretical models that might possibly account for the presence of opposite Subdoshas? I can think of three possibilities to be considered: Akabane imbalances, imbalances between the two Organs of the same Element, and pathogenic invasion. I have already explained the pulse manifestation when there is an Akabane imbalance. That pulse deviation abnormality does not seem to ever occur in conjunction with the presence of opposite Subdoshas, so I think we can rule out Akabane imbalances as the explanation. As for imbalances between the two Organs of the same Element, the situation is a bit more complex. If the two Organs are experiencing opposite states of activity (one excess and the other deficient), then there should be a RYCK imbalanced ratio present, but this has never been the case in my experience. Alternatively, both Organs and Meridians of the same Element may be out of balance with each other, but sharing the same situation of either excess or deficiency, just to different degrees. For example, in the case mentioned above, one side may show Vata/Kapha indicating LU excess, while the other side might show Pitta, indicating LI excess. If that were the explanation, then I would expect that the first side would show CK>RY in a ratio of 3/1, while the second side should show RY>CK also with a ratio of 3/1. These are big differences in pulse ratio, and so should be easy to notice, but as I just said, I have never encountered any abnormal RYCK size findings when opposite Subdoshas are detected, so I think we can also dismiss that explanation. We are left with pathogens as the remaining possibility. This perverse energy can be supposed to have attacked a Meridian on one side of the body creating a state of hyperfunctioning, as the body sends its wèi qì to deal with the pathogen. Since the two sides of each Meridian are connected (as demonstrated by Akabane diagnosis and treatment, discussed in Chapter 8), the opposite side's same Meridian will have diminished wèi qì, and thus be hypofunctioning. Such a supposition can at least explain the occurrence of opposite Subdoshas on opposite sides of the body. One thing I have observed, unexpectedly, is that this situation of opposite Subdoshas on the two sides only seems to occur with regard to the Constitutional Element. Such a pathogenic invasion might occur on whichever side of the body was initially malfunctioning to some degree, might attack any Meridian or Organ, and which then might call on the Constitutional Meridian/Organ for assistance. This interpretation is entirely consistent with classical references, which posit that a properly functioning organism is relatively immune to penetration by xié qì. So how might we determine which Meridian was attacked by xié qì, and is possibly harboring and/or spreading a stuck pathogen? Fortunately, there is one more pulse sign that provides the key to understanding the presence of opposite Subdoshas and their connection to pathogens.

As I've stated, I have only found such pulse presentations in the Subdoshas of the Constitutional Element. In every case where I've noticed this Subdosha pattern, I've found that one, and only one, other Meridian's pulse will have moved to the depth associated with the Constitutional Element, while the pulses of both Meridians of the Constitutional Element have stayed at their normal depth. I was amazed when I discovered that the Meridian whose pulse moved to the Constitutional depth always belonged to one of the Operator Meridians identified in the *Sù Wèn* chapters discussed above. Furthermore, every time I found this pulse pattern, there was something in the case history that clearly associated the pathogen connected with that Meridian/Operator with a significant disruption in that person's life, even if it occurred 50 years ago or more. Bolstering the *Sù Wèn*'s teaching, the pathogen involved could be either an external climatic one (an EPF) or an internal emotional one (IPF)[188]. Lastly, dispersing the Luò Point of the Operator Meridian in such cases has been remarkably effective in eliminating residual symptoms that all the previous strategies I tried were unable to accomplish. I chose to disperse the Luò Point because its nature is to mediate exchanges between interior and exterior plus left and right sides, thus providing a pathway for expelling xié qì, if the organism is strong enough to do so. I am still researching the question of which side's Luò Point to disperse in order to discover if there is a consistent rule, although my preliminary impression is that it is always the side displaying the correct Constitutional Subdoshas that needs dispersing. The method I use is to test both sides with a silver colored press pellet on the Korean Hand Acupuncture's corresponding point. So far, I have found that only one side will show a correction of the pulse in response to the silver pellet, and so far it has been the side that shows the expected Subdosha associated with the Constitutional Organ. Only the Luò Point of the Operator Meridian appears to show this corrective response to the silver pellets. Very

188 EPF is short for external pathogenic factor, while IPF is short for internal pathogenic factor.

superficial needling of this Luò Point, similar to Worsley's technique for treating "Aggressive Energy" has been effective in normalizing the pulse, as well as addressing the patient's symptoms, however deeper needling also has the same effect, and may induce a better expulsion of pathogens. I have not seen the type of erythema typical of Aggressive Energy however, so the types of xié qì these techniques address might be quite different from each other.

Let me summarize these findings with regard to the diagnosis and treatment of retained pathogenic factors. My theory is that when an individual is in a weakened or vulnerable state, any pathogen might succeed in penetrating its resonant Meridian (the Operator) on the more vulnerable side, presumably via its Luò Point, which connects outside to inside. This attacked Organ/Meridian would naturally call on its Constitutional Organ/Meridian for help in fighting the pathogen, which explains why the attacked Organ/Meridian shifts its depth to that of the Constitutional Element, but also why the Constitutional Element is the one that shows opposite Subdoshas, since it is sending its Defensive Energy preferentially to one side of the body, the attacked side. This highly theoretical explanation may not be convincing in itself, but it serves to account for my clinical observations. The reader is encouraged to pay attention to the case histories for a better understanding of this topic, as these cases hold the most interesting connections I've found between an individual's life experiences and the retained pathogens that have colored their lives ever afterwards. In any case, as mentioned, dispersing the Luò Point of the Operator Meridian has produced both symptomatic improvement and a normalization of the previously imbalanced pulses. However, the practitioner is cautioned not to necessarily stop treatment there, even though there has been such a successful outcome. Remember that the penetration by xié qì was hypothetically only possible due to a deficiency of zhèng qì in the originally attacked Organ/Meridian. Such a deficiency can now be looked for via the appearance of a hollow pulse pertaining to the treated Organ/Meridian. If found, this deficiency could now be treated in order to restore normal functioning of the attacked Organ/Meridian. I recall having read a discussion of this exact topic in an early book by Lavier, in which he claimed that after removing any pathogenic factors from an Organ/Meridian, it was necessary to tonify the Horary Point of that Meridian in order to strengthen its resistance to xié qì.[189] Lavier did not treat the Luò Point in order to disperse xié qì, so our methods, though similar, are not identical. My inclination is to try using the Yuán (Source) Point for tonifying the depleted Meridian, as the combination of Yuán and Luò Points, needled in opposite polarity, has been used in acupuncture historically, as an application of Guest/Host theory. I do not have enough data yet to form a conclusion as to whether or not this approach is a universally reliable one, but it has seemed helpful in several cases so far. Another possible follow-up treatment that has worked clinically has been to rebalance the yíng and wèi movements as reflected in the pulse by Simonov's contact needling method, previously described in Chapter 7.

189 Lavier, J. p. 163.

CHAPTER 11

Korean Sasang Medicine Prefigured in the Màijīng

Figure 33: Li Jema, founder of Sasang Constitutional Medicine [190]

I have discussed Sasang Medicine in my prior books, but most acupuncture practitioners have never heard of it, so a brief review is appropriate. This Korean system of herbal medicine was introduced by Li Jema towards the end of the 19th century. Li's idea was that every individual had an inborn Constitution that was one of four types: Soyangin, Taeyangin, Soeumin or Taeumin. It is clear that this idea is a development from the *Yí Jīng*, or *Classic of Changes*, which divided yīn and yáng into the four images (sì xiàng). According to Li, these four groups were defined by their Organ strengths in the following ways: Soyangin has a strongly functioning Spleen and a weakly functioning Kidney. Taeyangin has a strongly functioning Lung and a weakly functioning Liver. Soeumin has a strongly functioning Kidney and a weakly functioning Spleen. Taeumin has a strongly functioning Liver and a weakly functioning Lung. In the 1960s Kuon Dowon adapted these ideas into Korean Constitutional Acupuncture, but he realized that Li's definition of the four types was not complete, as there were individuals whose

190 Source: jameschoiacupuncture.com

Constitution was based on the functioning of the Hollow Organs (fǔ). Kuon proposed that there are really eight types of Constitution, while CCA now posits that there are in fact 20 types of Constitution, but all these approaches are based on the idea of the four images as foundational. In that case, what is the meaning of the four images according to CCA theory? Li's idea about strongly and weakly functioning Organs is kept in CCA, but these original four Organs are not necessarily the strongest or weakest of all the Organs; It is only their paired relationships, as proposed by Li, that are maintained while other Organs may also take the strongest or weakest positions. My own observation is that there seems to be a reliable association between the two yáng types (Soyangin and Taeyangin) and extroversion, as well as between the two yīn types (Soeumin and Taeumin) and introversion. Clinically, this can be a very helpful sign when diagnosing Constitutions. Often the Element and even the Organ at the root of a person's Constitution might be easy to determine, but whether this Organ tends toward hyperfunctioning or hypofunctioning can be a challenge to distinguish. In effect, the Sasang type is really about dividing people into yīn or yáng types. That distinction is obviously quite important in Oriental medical diagnosis and treatment. While the *Yí Jīng* and *Nèijīng* contain the basic ideas upon which Sasang medicine was founded, there is an interesting connection between the *Pulse Classic* (*Màijīng*) and Sasang theory.

The *Màijīng* (Book 10) introduced a second set of correspondences between the cùn, guān and chǐ positions and the Meridians that are reflected in the pulses located there, as I have described in *The Compleat Acupuncturist*. I am not aware of any acupuncture clinicians who use this unusual mapping other than my students, but perhaps I am simply ignorant of their work. In the *Màijīng*, no explanation is given for this new set of associations, but as they have been repeatedly verified in my practice as clinically accurate guides to Constitutional diagnosis, I have searched for a basis in the fundamental theories of yīnyáng, wǔxíng and zàngfǔ. I believe this map provides a basis for Li Jema's later formulation of Sasang Constitutional Medicine (Fig. 33). The standard associations of the radial artery pulse positions were given in the *Nánjīng*, and essentially adopted by most practitioners of classical acupuncture. They can easily be understood as reflecting the Creative Cycle order of the Five Elements, starting from the left chǐ position, which corresponds to the Water Element (Bladder and Kidneys). The Water Element is the first Element in the "Cosmogonic Order" mentioned in the *Book of Documents*[191], one of the Confucian classics that long preceded the *Nèijīng*, so it is fitting that the enumeration of correlations should begin starting from the chǐ position within the cùnkǒu section of the radial pulse, as it is the first place where information is transmitted via pulse waves emanating from the Heart[192]. The next more distal (the direction of the pulse wave) position is the left guān, which corresponds to the Wood Element (Gall Bladder and Liver). Finally at the most distal left cùn position, the correspondence is to the (Imperial) Fire Element (Small Intestine and Heart). This first half of the Creative Cycle concerns the first three Elements of the Cosmogonic Order, Water, Fire and Wood, although not in the exact same order. There are two Elements left, but three positions on the right wrist at its Cùnkǒu location. Once again we encounter the mystery of how to reconcile the lens of 5 with the lens of 6. This problem is resolved by including the Organs that have no form, namely Triple Heater and Pericardium as the proximal continuation of the Creative Cycle from Imperial Fire to Ministerial Fire at the right chǐ position. The doctrine of only Five Elements is preserved by this double appearance of Fire. Next to it distally (again following the direction of the pulse wave) is the right guān position correlated with the Earth Element (Stomach and Spleen). Finally, the most distal position on the right is cùn, correlated with the Metal Element (Large Intestine and Lung). The Creative Cycle is now complete and can start over again with Metal on the right wrist giving birth to Water on the left wrist. The shape of this movement through the pulses follows that of the infinity symbol, so it is important to remember that continuous cycles are not only represented by circles, but also by the "figure eight" shape. Another way to look at this is via the seasonal associations. The left side reflects the growth of yáng from Winter to Spring to Summer (Water to Wood to Fire), which explains why the left side is correlated with yáng. When we move from Ministerial Fire to Earth to Metal (late Summer to Autumn), it is yīn that is growing, so the right side is correlated with yīn, providing the second half of

191 Hóngfàn section (洪範) of the *Shūjīng* or *Shàngshū* (尚書).

192 A similar argument could be made for starting the enumeration on the right side with Ministerial Fire. Since Ministerial Fire is without form, it might logically precede the energies of the formed Organs. The argument comes down to whether the jīng of the Kidney or the shén of the Heart is established first. This is a chicken/egg situation with no real answer.

the Cosmogonic Order's Elements. Although it is common to think primarily of the Five Element associations of cùn, guān and chǐ on the two wrists, it is equally true that the associations of yīn and yáng are also being displayed. This link between yīnyáng and wǔxíng holds the mystery of resolving the different phenomena revealed by the lenses of 5 and 6.

The *Màijīng* now presents us with an additional map (not a replacement) of the associations of cùn, guān and chǐ with the 6 Great Meridians (or levels) of yīn and yáng, as shown in Fig. 34. We see that Taiyang is paired with Jueyin, as they represent either radial (yáng) or ulnar (yīn) deviations of the radial artery at the cùn position on both left and right wrists. Let us examine the Taiyang deviation first. Taiyang is comprised of the Meridians of the Small Intestine and Bladder. Each of these can tend towards excess or deficiency on a Constitutional level. By referring to my prior publication, we see that both Small Intestine deficiency and Bladder excess are grouped in the Taeyangin classification of Sasang based Constitutional Conditional Acupuncture (CCA). Small Intestine excess and Bladder deficiency are grouped in the Taeumin classification of CCA. Now let us examine the Jueyin deviation at the cùn position. Here we have the opposite deviation (yīn instead of yáng), but the CCA possibilities are exactly the same. The Constitutional types are either Liver deficiency, which is Taeyangin or Liver excess, which is Taeumin in both Sasang Constitutional Medicine (SCM) and CCA. Thus any deviation at the cùn position belongs to either Taeyangin or Taeumin. There is a reason I am using the Korean spelling here. Taeyangin and Taeumin refer to old yáng and old yīn types, which are contrasted with Soyangin and Soeumin, which I translate as young yáng and young yīn types. This Korean terminology additionally avoids confusion with the Chinese use of these same names as part of the Six Levels of yīn and yáng doctrine, while specifically indicating the use of these terms as labels for the Four Images (sì xiàng).

When we look at the *Màijīng* deviations at the chǐ position we find Shaoyang and Shaoyin. Shaoyang (yáng deviation) on the Constitutional level is represented by the Gall Bladder, which can be deficient (Soyangin in CCA) or excess (Soeumin in CCA). Shaoyin (yīn deviation at chǐ) represents Heart and Kidney. Heart excess and Kidney deficiency tendencies are classified as Soyangin in CCA while Heart deficiency and Kidney excess tendencies are classified as Soeumin in CCA, Constitutionally. Again we find young yáng or young yīn as the only possible Constitutions that lead to deviations at the chǐ position. What can we learn from this consistent distribution of typologies? The proximal pulses at the chǐ position correlate with the young yáng and young yīn Constitutions, while the distal pulses at the cùn position correlate with the old yáng and old yīn Constitutions. Once again, starting from the positions closest to the Heart (chǐ) we first find indicators of the young yīn and yáng Constitutions, while more distally (at cùn) we find indicators of the old yīn and yáng Constitutions. This is exactly following the movements of yīn and yáng in the tài jí symbol derived from the *Yìjīng*. Young yáng transforms into old yáng and young yīn transforms into old yīn. In addition old yáng transforms into young yīn and old yīn transforms into young yáng, so in describing this aspect of the radial pulse at cùnkǒu, we see the infinity symbol again, for old yáng at cùn transforms to young yīn at chǐ, and old yīn at cùn transforms into young yáng at chǐ. Symmetry is once again seen to play an important role in understanding these relationships that are reflections of natural law.

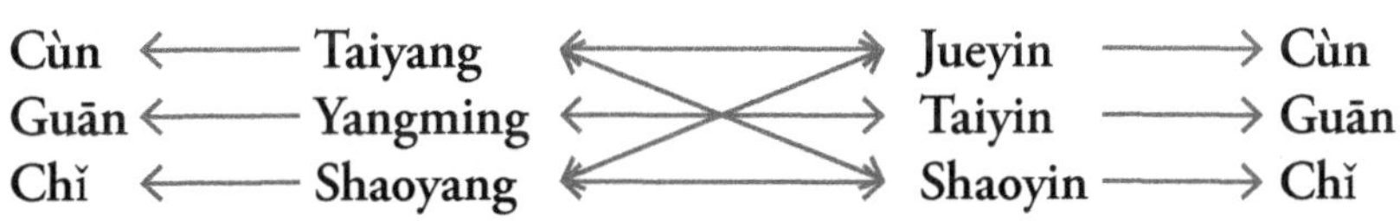

Figure 34: The Màijīng *Deviation Map* [193]

There is obviously a missing part of this model: what happened to Yangming and Taiyin? These Meridians deviate in the guān location in individuals who have either of these Constitutions, that is excess or deficiency of Metal or Earth in their correlated zàng or fǔ Organs. It turns out that half of these Constitutions belong to the young yīn or yáng Constitutional types (Soyangin or Soeumin in SCM and CCA), while the other half belong to the old yīn or yáng Constitutional types (Taeyangin or Taeumin in SCM and CCA). I think of this as a representation of how young yáng is able to transform to old yáng, and how young yīn is able to transform

193 Source: Peter Eckman

into old yīn. Between young (just starting out) and old (ready to change into its opposite) is the mature expression of fullness. Thus Yangming and Taiyin are said to be the Meridians most filled with qì and blood, and are correctly positioned between the young and old extremes of life. The *Yìjīng* shows the progression of one to two to four to eight, but does not include an understanding of six in this progression (although it has a profound understanding of six in the hexagram expressions!). This middle position of Yangming and Taiyin allows us to integrate the lens of six with the lenses of one, two, three, four and five, based on the *Yìjīng*. The progression of lenses given in the *Yìjīng* (implicitly at least) includes tàijí (太極), yīnyáng (陰陽), sāncái (才三), sìxiàng (四象), wǔxíng (五行) and liùjīng (六經), the latter related to the six lines (yáo 爻) of each of its hexagrams (guà 卦). We can correlate this progression of lenses with the various pulse diagnostic findings used in CCA. Tàijí is the singular universal presence of all the body's pulses, indicating the vitality of life, (jīng shén 精神), or perhaps simply qì itself. Yīnyáng is expressed in RYCK pulse diagnosis. Sāncái is reflected in the three pulse positions, cùn, guān and chǐ. Sìxiàng is the basis of the four part Sasang Constitutional Medicine (SCM) version of pulse diagnosis elaborated by Puramo Chang. The wǔxíng are the obvious correlate of the Five Element style of pulse diagnosis as taught by Worsley and others, while the liùjīng find their reflection in the six deviations transmitted by the *Màijīng* in Book 10.

Having established coherence between these theories (lenses) implicit in the *Yìjīng*, we are still left with the question of how the author of the *Màijīng* came up with this alternate mapping of the Meridian associations at cùn, guān and chǐ. I believe the mystery can be resolved by a closer examination of chapter 6 of *Su Wen*, "Treatise on the Separation and Union of Yin and Yang". According to Liu Lihong (p. 123)[194], the text states that the two yáng (i.e. taiyang and shaoyang), when they unite, become yangming. Also, the two yīn (i.e. taiyin and shaoyin), when they exchange completely, become jueyin.[195] The following illustrations from my 1987

194 Liu, L. *Classical Chinese Medicine*, The Chinese University Press of Hong Kong, 2019

195 My reading of the text does not include such a statement, however I believe that Liu is inferring this intended meaning from a combination of the title of the chapter and the first appearance of the terms yangming and jueyin in the *Su Wen*. I believe this is not an unreasonable interpretation.

monograph on the *Yìjīng* (Figs. 35 and 36) graphically show how the *Su Wen* conceptualized this statement. Yangming is situated between shaoyang and taiyang, and can thus be seen as uniting them. Jueyin, on the other hand, follows the movement from taiyin to shaoyin, and thus can represent the state when these two have completely (jué) exchanged (yì) the potential of their qì. Although I cannot include the whole monograph, this exact map is expressed in perfect detail in *Su Wen* chapters 70 and 71. I've included this preliminary material on yangming and jueyin merely to situate the next section of material in *Su Wen* chapter 6. Chapter 6 is all about the three yáng and the three yīn, but its major point is that the three yáng are really one yáng (when in good health), and the three yīn are really one yīn (when in good health). Their separation into parts is quite suggestive of the pulse images in Fig. 17, illustrating RYCK diagnosis.

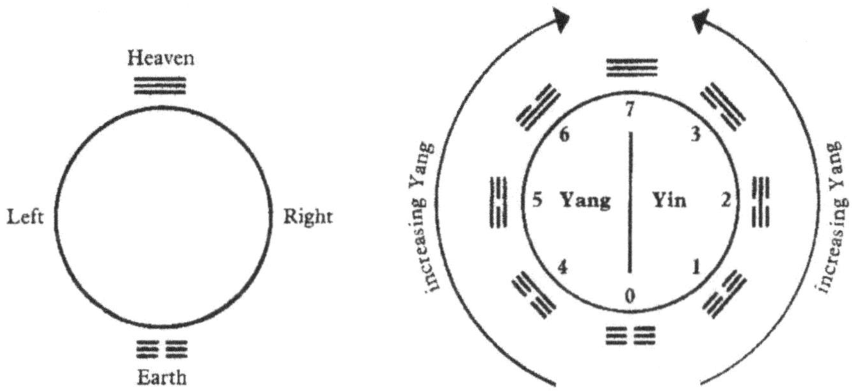

Fig. 35: Fu Xi's Circular Order of Trigrams; Levels of Yin & Yang [196]

196 Source: Peter Eckman.

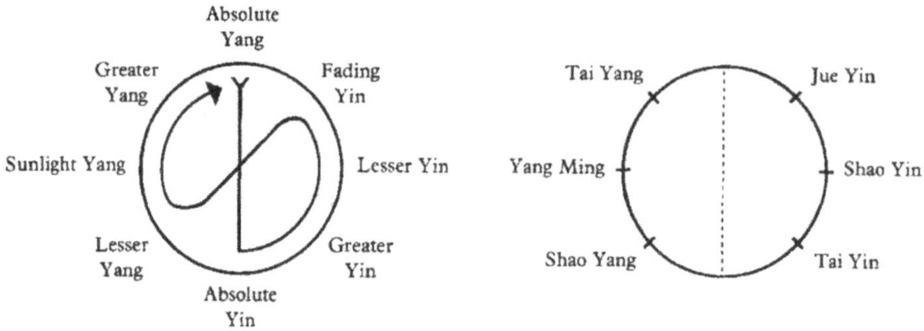

Fig. 36: The Meridian Correlates of the Fu Xi Circular Order of Trigrams [197]

Chapter 6 now introduces the concept of opening, closing and pivoting (kāi 開, hé 合 and shū 樞) for the Meridians of both yīn and yáng. For yáng, taiyang opens, yangming closes and shaoyang is their pivot. Likewise for yīn, taiyin opens, jueyin closes and shaoyin is their pivot. Unfortunately the text does not define what it means by these three terms, but Liu Lihong explains it in a way that makes sense to me (his chapter 3). We can think of two doors, a yáng door and a yīn door. In order to keep the balance of nature, when the yáng door opens, its complementary yīn door must close and vice-versa. A door may stay open or closed for an extended period of time, but it can also swing back and forth between open and closed, or in other words, it may pivot between these states. If we diagram this information, we come up with Fig. 34! In a yáng stage when the yáng door is opening (taiyang, radial deviation at cùn), its opposite yīn door at cùn, responds by closing (jueyin). However, in a yīn stage, when the yīn door is closing (jueyin, ulnar deviation at cùn), its opposite yáng door at cùn responds by opening (taiyang). Similarly, in a yáng stage if the yáng door is closing (yangming, radial deviation at guān), its opposite yin door at guān responds by opening (taiyin).

197 Source: Peter Eckman. As noted earlier, there is a slight difference in the order of the Six Levels in the *Sù Wèn*, *Líng Shū* and *Màijīng*. Simonov has used pulse changes to corroborate the *Sù Wèn* order for contact needling: Shaoyang, Yangming, Taiyang for the sequence of yáng Meridians and Taiyin, Shaoyin, Jueyin for the sequence of the yīn Meridians. Each Great Meridian is dominant for approximately 5 minutes during any "cut" which enables the practitioner to apply contact needling at exactly the optimal time. This order can be seen to reflect that of the cyclical increase of yáng. This text's Chapter 7 has discussed the essential material, but knowledge of the above classical sequence is necessary for proper treatment formulation. Once again the teachings I'm including end up being validated by pulse phenomena, not merely by scriptural doctrine.

In a yīn stage if the yīn door is opening (taiyin, ulnar deviation at guān), its opposite yáng door at guān responds by closing (yangming). Finally we have the pivots. In a yáng stage, if the door alternately opens and closes (shaoyang, radial deviation at chǐ), then its opposite yīn door at chǐ responds by alternating oppositely (shaoyin). In a yīn stage, if the yīn door alternately opens and closes (shaoyin, ulnar deviation at chǐ), then its opposite yáng door at chǐ responds by alternating oppositely (shaoyang). Both of these pivot Meridians are associated with chills and fevers, or heat and cold. Although this map was not explicitly developed in the *Su Wen*, it seems clear to me that the *Màijīng* merely made what was implicit in the *Su Wen* explicit, and this later shed light on the basis for the Korean system of Sasang medicine.

There is another part of Chinese medical theory that may help in understanding the *Màijīng* map. It lies in one aspect of traditional acupuncture theory that is often given less attention than it deserves: The idea that there are three circuits of qì flow, which have a degree of independence from each other. As far as I know, there is no agreed technical term in Chinese to designate these circuits, but they have been discussed in numerous books and articles for a long time, and like all phenomena comprised of three parts, they have often been interpreted in terms of the Three Powers (sāncái 才三), composed of Heaven, Earth and Human. I have already introduced this material, but it is worth reprising here in order to support the contentions made in this chapter.

The first circuit I will discuss comprises Shaoyang and Jueyin, which includes the Meridians of the Heart Master or Pericardium (PE), Triple Heater (TH), Gall Bladder (GB) and Liver (LV). There are several reasons why I'm enumerating this circuit first: It includes the two Ministerial Fire functions that have no form, and which I believe must come into existence prior to the formed Organs, as envoys of the Original qì. It also includes the Gall Bladder, which is classified as an "Extraordinary fǔ" Organ, thus relating it to the Extraordinary Meridians. I will explain later why I think the Extraordinary Meridians are established prior to the Principal Meridians in embryological development, and so I'm inclined to start any progression with something related to what is considered extraordinary (qí 奇). Being both an extraordinary fǔ and an ordinary fǔ, the Gall Bladder represents, for me, the transition from formless to formed (perhaps the transition from Anterior to Posterior Heaven). The Gall Bladder is said to have its peak activity at the hours corresponding to the first Earthly Branch (zǐ 子) from 11 PM to 1 AM, another reason to start an enumeration of the circuits of the 12 Meridians with the Gall Bladder. The Liver is then the second formed Organ, and this pair fits very nicely with the Wood Element being correlated to the first of the Five Transport (Shù) Points of the zàng Organ's Meridians and the appearance of new life in Spring. Additionally, these four Meridians, when imbalanced, have the smallest proportional difference in the RYCK diagnostic schema. Most interpreters say this ratio is two to one, but as I've previously indicated, I think it is closer to one and a half to one. I experience this phenomenon as the bigger pulse feeling like it has a small tag attached to it, often very slightly deeper, and not clearly divided into two separable impulses. It is almost an image of a transition occurring between the formless state and the formed state. Once the organism has established a communication system based on the Extraordinary Meridians (qíjīng bāmài), Principal Meridians (jīngmài), and Network Meridians (luòmài), it can develop more complex structures. One might ask why the Liver is involved in the first circuit, if that is primarily needed for establishing communication systems. My hunch is that just as the yáng jīngmài possess the ability to move (communicate) healthy qì, the yīn jīngmài have a similar role to play in moving (communicating) the Blood (xuè), and the Liver/Pericardium couple serves this yīn medium analogously to the way the Triple Heater/Gall Bladder couple deals with the qì, a more yáng medium. The key here is maintaining a communication system that depends on the free flow of qì and blood. In Fig. 34 we can see that Shaoyang and Jueyin are positioned diagonally opposite each other, forming a continuous circuit from PE to TH to GB to LV. This diagonal relationship, a known connection in traditional acupuncture theory, is one basis for devising the new *Màijīng* map's correlations.

By contrast, when the bigger one of the RYCK pulses has two clearly separable, but adjacent impulses, I find that the indicated Meridian imbalance is either in the Heart (HE), Small Intestine (SI), Urinary Bladder (UB) or Kidney (KI). For me, this is therefore the second circuit composed of Taiyang and Shaoyin Meridians. I see the dynamic here as representing the interplay of Essence and Spirit, rather than of qì and Blood. Other authors have discussed the concept of a "Heart/Kidney Axis", a term which I do not recall seeing in the classics. The Heart and Kidney are certainly responsible for housing the shén (神) and jīng (精) respectively, but how the Urinary Bladder and Small Intestine are involved is less obvious. Being yáng Organs, they could be envisioned as moving the functions of their paired yīn Organs from a state of storage into one of activity. Thus, the Small Intestine has a metaphorical function of dividing the pure from the impure, a task which I interpret as one of examining all inputs (physical, mental, emotional and spiritual) and looking to the shén in the Heart in order to know if any input is coherent or not with one's Original Nature. If coherent, the input can be absorbed, but if not coherent it must be rejected and passed along for elimination. I conceptualize this as "being true to one's Heart". The Urinary Bladder can then be seen as playing an analogous role with respect to the jīng stored by the Kidney, but here it is not a question of input that must be regulated. Rather it is the output of Essence that must be matched to the needs of Original Nature. What is involved here is the regulation of our lifestyle behaviors, especially work and other forms of exercise, sleep hygiene and sexual activity. All of these are important expressions of healthy behavior when they do not countermand the proper norms of our individual jīng, but become unhealthy if they do so. It is important to recognize here that because our Constitutions are different from those of other people, there can be no universal standards for how much exercise, sleep or sexual activity is optimal. Many passages in Chinese medicine texts would disagree, but if you look at the real world, you will find tremendous variation in people's needs for exercise, sleep and sex. Many famous people have lived well into old age in good health with less than six hours of sleep, or having done daily strenuous activity (farming for example) well past retirement age, and of course there are those people we might call libertines who sometimes flourish. Others might need ten hours of sleep each night, only passive stretching exercises and little to no sexual activity, yet otherwise lead productive lengthy lives. This variation comes from respecting the dictates of our Essence. The common examples of people who became sick and died young, who violated the conventional norms, would really be cases of people who violated the mandates of their own personal jīng. The most common reason that people visit doctors in contemporary society is for persistent back pain. Is it any wonder, when so many ask their Urinary Bladder Meridian to carry a load that is not coherent with their endowment of Essence? If we look at Fig. 34, once again we see the diagonal relationship between cùn, and chǐ from HE to SI to UB to KI.

The third circuit consists of Lung (LU), Large Intestine (LI), Stomach (ST) and Spleen (SP), representing Yangming and Taiyin Meridians. I find the RYCK diagnosis of imbalances in this circuit to be surprising at times. The basic pattern is that RY or CK occupy the space for three components in its pulse, although the middle impulse is usually absent. This circuit appears to deal with the storage of qì and blood, rather than with the maintenance of free flow and communication functions. *Líng Shū* 33 is helpful in understanding this circuit, especially in light of the previous discussion of the Meridians as analogous to streams or rivers of water running from the feet to the upper regions of the body. We are told in *Líng Shū* 33 that the "Stomach is the Sea of Water and Grains (shuǐ gǔ zhī hǎi)", and that its transport (shū) points are ST 30 and ST 36. Chapter 33 is specifically about the Four Seas, and right after the Sea of Water and Grains, the Chōng Mài (衝脈) is mentioned and is said to be the "Sea of the 12 Meridians (jīng 經)" but is also identified as the Sea of Blood. Its shū points are UB 11, ST 37 and ST 39. In his commentary on this topic, Lǐ Gǎo (李杲) later described the Stomach as not only the Sea of Water and Grains, but also as the Sea of qì and Blood, which it discharges into the five zàng and six fǔ. From these statements we can understand the close relationship between the Stomach and the Chōng Mài, and between the Principal Meridians and the zàngfǔ, as well as their relationship to the qì and Blood. In fact, the Acupoint Sea of Blood is on the Spleen Meridian, which coincides with the lowest branch of the Chōng Mài. The "Seas" are both repositories of qì and Blood flowing from the Meridian rivers, but can also help fill the Meridians when they suffer a deficiency[198]. Just as the Stomach takes in nourishment (gǔ qì 穀氣), the Lung takes in the breath (dà qì 大氣). These are the sources of the postnatal qì, which fills the Meridians in order to carry out the tasks of daily life. It should not be surprising that the Large Intestine participates in this function by absorbing a significant amount of water, so that it isn't lost in defecation. Thus of all the six Meridians, only Yangming (Stomach and Large Intestine) is said to have more qì and more blood. In Fig. 34 Yangming and Taiyin are opposite each

198 The symptomatology for excess and deficiency conditions of the Seas are given in *Líng Shū* 33, along with the implication that their treatment is via the named shū points, which can be supplemented or dispersed as needed.

other, but rather than presenting at an angle diagonally, they appear exactly in the middle, befitting the role previously mentioned of forming the transition from young yīn and yáng to old yīn and yáng.

It should be obvious that these three circuits are composed of two Elements each. The first circuit is composed of Ministerial Fire (xiàng huǒ 相火) and Wood. The second circuit is composed of Water & Imperial Fire (jūn hu 君火). The third circuit is composed of Metal & Earth. Each has a unique disproportionality when imbalanced, which shows in the RYCK pulse examination. The classics mention first, second and third yīn and yáng in a number of different contexts. Could this concept be related to these three circuits? When any of the RYCK imbalances are treated to the point of resolution, the RYCK ratio should return to even (1/1). This principle is stated in the classics. Is that because the fragmented components of yīn and yáng have been successfully reunited?

Before finishing this chapter, I'd like to point out one more teaching from *Sù Wèn* 22. Its organizing principle is the Five Element (wǔ xíng) theory, and how to explain its association with the different physical appearances of people (an early attempt to formulate a Constitutional theory?) The Elements and their associated Organs and Meridians are presented in a specific order, the Creative cycle: First is the Wood Element and the Liver. Its Meridians are "Foot Jueyin and Foot Shaoyang." (Incidentally, this is part of the first circuit discussed above, with yáng beginning to grow). Second is the Fire Element and the Heart. Its Meridians are "Hand Shaoyin and Hand Taiyang." (Second circuit, with yáng reaching its peak). Third is the Earth Element and the Spleen. Its Meridians are "Foot Taiyin and Foot Yangming." (Third circuit, with yīn and yáng achieving equilibrium, as yáng has declined past its peak). Fourth is the Metal Element and the Lung. Its Meridians are Hand Taiyin and Hand Yangming. (Still third circuit, but yáng has declined even further). Fifth is the Water Element and the Kidney. Its Meridians are the "Foot Shaoyin and Foot Taiyang". (From equilibrium, yáng retreats even further, returning to the Second circuit again). There is no mention of a sixth component, nor of the Ministerial Fire functions, nor their Meridians. In this proto-Constitutional theory presentation, Ministerial Fire is clearly excluded, which I take as supporting my conclusion that there are no Ministerial Fire

Constitutional types. I think we can take the mention of the zàng here as implying their associated fŭ, since their Meridians are also specifically named. This pairing of the Meridians of the zàng and fŭ Organs of the same Element is later incorporated into the treatment protocols for RYCK imbalances, in which both Meridians of the Element to be addressed are treated. Finally, this *Sù Wèn* chapter gives dietary advice that has a large degree of overlap with the later system of Sasang Constitutional Medicine (SCM), again emphasizing the kernel of a Constitutional theory being proposed[199]. And this Constitutional theory (SCM) is another link from *Sù Wèn* to the mysterious map in the *Màijīng*.

199 The Liver diet matches that for Taeumin, where Liver Excess is the prototype. The Lung diet is closest to that for Taeyangin, where Lung Excess is the prototype, but it recommends eating mutton, which is contrary to Sasang teaching. The Spleen diet matches that for Soyangin, where Spleen excess is the prototype. The Kidney diet matches that for Soeumin, where Kidney Excess is the prototype.

CHAPTER 12

The Dual Roles of the Eight Extraordinary Meridians/Vessels (EEVs)

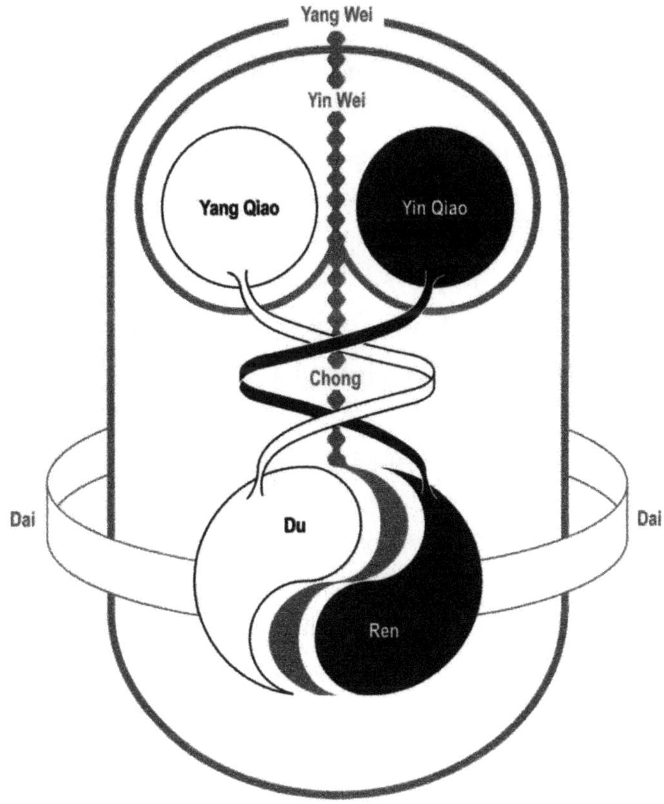

Figure 37: The Eight Extraordinary Meridians [200]

200 Source: http://www.extraordinarychinesemedicine.com. This figure appeared in *Extraordinary Chinese Medicine* by Thomas Richardson with the graphic help of Michael Corner, a publication of Singing Dragon.

Many authors have glossed the character qí (奇) as "odd" meaning the opposite of "even". While that is one of the possible translations of 奇, I think it is not a good choice in the phrase (qí jīng bā mài, 奇經八脈), because as I will propose, the pairing of the Extraordinary Meridians is one of their more salient characteristics, and is clinically quite useful (Fig. 37). The fact that they number eight, in itself is a hint that they are better characterized as even than odd numerically.[201] The diagnosis and treatment of Extraordinary Meridian problems was given a preliminary presentation in *The Compleat Acupuncturist*, and the guidelines presented there, while accurate, will be expanded here. A deeper understanding of how these Extraordinary Meridians relate to the Constitutional focus of CCA definitely warrants additional commentary. It is my belief that the Extraordinary Meridian "system" has yet to be explained with as much understanding as the Regular Meridian system, however both of these concepts achieve better clarity when they are approached as systems, rather than as being composed of separable isolated entities.

The first thing I would like to suggest is that these eight Meridians are part of a unified "Primitive Constitution", and act in a "constellational" manner, much as the Regular Meridians do. I have not come across this teaching anywhere, but it reflects my clinical experience, is supported by pulse findings, and additionally strikes me as an intuitively essential aspect of their role in the Oriental medical perspective on human physiology and pathology. As I will discuss, these Extraordinary Meridians are developed at a very early stage in embryonic life (before there are any Organs or Regular Meridians), and they take on the role of ordering the unfolding of Constitutional (Original Nature) information, which then is the basis for the subsequent development of the Regular Meridians and Organs.[202] This is a crucial point in the CCA approach to treating the Extraordinary Meridians, because if there is an imbalance present in the Extraordinary Meridian system, then it may

201 Kroll cites two possible antonyms for 奇: zhèng 正 meaning regular and ǒu 偶 meaning even in number. Certainly these Meridians are not odd in number. Also they are not the Regular Meridians we are familiar with. Extraordinary directly follows irregular in Kroll's list of Classical Chinese definitions, and as such makes sense as the better translation than odd, in this context (Page 354).

202 "Extraordinary Vessels are the first channels that we develop since foetal stages. They govern our prenatal developments and maturation until our regular channels are matured. They are connected with our 元氣 Yuan Qi/ Source Qi, and function as the axis, longitudes and latitudes of our sphere." – posted by Sze Kai Paau on the Facebook group Classical Chinese Medicine 9/20/20.

prevent treatment of the Regular Meridians from achieving a successful result, a belief enunciated by my teacher of Korean Hand Acupuncture Yoo Tae Woo. If one looks at the classics and their later commentaries, one can find descriptions of the signs and symptoms of both excesses and deficiencies in the Regular Meridians and Luò Vessels, but not the Extraordinary ones.[203] As far as I am aware, there has been no discussion of the option to either tonify or disperse an Extraordinary Meridian, as most texts simply assume that the Extraordinary Meridians serve as a safety reservoir to contain excess pathogens, and thus are always dispersed when the decision is made to treat them.[204] Certainly the pulse images of the EEVs transmitted historically reflect this idea, and as such are accurate to a degree. However, a more nuanced interpretation taking yīnyáng theory into account will explain why this approach may be incomplete, and might lead to less than optimal treatment outcomes.

Not the least extraordinary thing about the Eight Extraordinary Meridians is that this terminology is not found in the *Nèijīng*![205] The first appearance of this grouping in a medical text is in the 27th chapter of the *Nánjīng*. I have already suggested that the *Nánjīng* introduced a paradigm shift away from the viewpoint of the *Nèijīng*, and that CCA is based on a second subsequent paradigm shift. We can see aspects of both of these paradigm shifts in studying the Extraordinary Meridians. The first paradigm shift included the idea of Constitution (yuán qì) as the origin of each human life, thus creating a

203 See Mann, F. *The Meridians of Acupuncture*, William Heinemann, London, 1964. Mann presents concise descriptions of the excess and deficient signs and symptoms of the various Meridians.

204 Pialoux's approach (Pialoux, J. *Guide to Acupuncture and Moxibustion*, Fondation Cornelius Celsus, Switzerland, 2008), included tonifying the Opening and Closing Points of an Extraordinary Meridian selected for treatment, however he interpreted this as dispersing that Extraordinary Meridian's qì so that it flows back into the Regular Meridian via the tonified Opening Point. He also dispersed the Opening and Closing Points of the Extraordinary Meridian of opposite polarity, but didn't explain his reasoning. My protocols, explained in this chapter, are similar, but treat the polarities differently than Pialoux did. Earlier, Niboyet had clearly stated that the Opening Points of the EEVs were always treated by dispersing them – p. 881. This is exactly the same approach as Yoo Tae Woo used in Korean Hand Acupuncture, and was previously incorporated in CCA, but is now revised to include protocols to tonify EEVs when indicated.

205 In my reading, it appears that only the "First Ancestry" Extraordinary Meridians (Chōng Mài, Dū Mài, Rèn Mài and Dài Mài) are mentioned in the *Sù Wèn*. Additionally, the two Qiāo Mài (Second Ancestry Extraordinary Meridians) make their first *Nèijīng* appearance in the *Líng Shū*. The two Wéi Mài Extraordinary Meridians are not mentioned in the *Nèijīng* at all, but first appear in the *Nánjīng*. From this sequence, I understand that the knowledge base in Chinese Medicine has been acquired over the course of time, and that it is therefore contrary to common sense to limit one's ideas to a single source, such as the *Nèijīng*. CCA must be approached with this understanding in mind, as is true for propositions in all the natural sciences.

reference point for understanding how one's clinical Condition can reflect a loss of Original Nature. As suggested, the Extraordinary Meridians play an early role in the unfolding of Original Nature. In addition, the *Nánjīng* paradigm stressed the importance of leveling (píng) the functions of the twelve Organs and their Meridians as being the goal of treatment, rather than focusing on expelling pathogenic factors (xié qì), although it did state that the Extraordinary Meridians could serve as an "overflow" reservoir to contain pathogens that could not be expelled. CCA adopts these ideas from the *Nánjīng* but reinterprets the relationship among the twelve Organs and their Meridians to include a constellational interaction, which had not been fully developed in either of these classical texts. In a similar fashion, CCA posits a constellational way of understanding the Extraordinary Meridians that was not explicitly described in the medical classics, but does not contradict them either. My thinking in this regard is more reflective of Daoist teachings than what may be found in the medical texts. I am not alone in attending to the Daoist ideas in this regard.

Larre and Rochat have written that the Extraordinary Meridians are the "foundation of life from the very beginning to the very end" and that "they are part of the primitive constitution of the being."[206] This is quite a different point of view than one can find in the *Nèijīng* or *Nánjīng*, but one that is quite compatible with the viewpoint of CCA. What is this primitive constitution, and how can we understand the role it plays, not just in the origin of life, but also until life's very end?[207] Analogous to CCA's axiom that the Organs have an inherent unequal hierarchy of strongly functioning and weakly functioning members according to Five Element classification, I believe the Extraordinary Meridians are also likely to be present in an inherent individually specified hierarchy according to yīnyáng classification during embryological development, since the primitive Constitutions cannot generate unique individual

206 *The Eight Extraordinary Meridians*, Larre, C. and Rochat, E. Cambridge, Monkey Press, 1997, p. 1.

207 Gerald D. Barra posted the following on Facebook on 7/7/20: "In particular (the one we know today as) circulatory system seems to gain greater importance over the rest when we study not only the origins of the channels and their historical development, but also the development of medicine itself in China... after all who is the 'first' of our Zang? For evidently the (Emperor) Heart." Is there a relationship between the circulatory system and the Extraordinary Vessels as the earliest means of internal communication? Igor Simonov added the comment, "It's the first system to start developing in the fetus to give food to the development of all other systems not even the nervous system is so important." This idea makes me wonder if endocrine problems (blood mediated) might best be treated via the Extraordinary Meridians.

hierarchies (mature Constitutions) unless they have such a built-in hierarchical aspect themselves. Such an inborn hierarchy does not however govern which Extraordinary Meridian is showing a need for treatment in post-natal life. That is purely a Conditional matter. For example, Yoo taught that if RY is wider than CK, one can always treat this situation (a yáng state) by dispersing a yáng Extraordinary Meridian, but he also taught that an alternative treatment could be chosen to disperse a yáng Regular Meridian or tonify a yīn Regular Meridian to adjust this overly yáng state. Theoretically, one might also treat this state by tonifying a yīn Extraordinary Meridian. There are many questions generated by such a hypothesis. How can one determine if any given patient needs correction of an Extraordinary Meridian, or not? If several Extraordinary Meridians are showing pulse indications, how can one determine the pivotal one to treat? Which Points are needed and on which side of the body? As with all questions in CCA, it is crucial to first identify the individual's mature Constitution before trying to make sense of other pulse findings, including those reflecting the Extraordinary Meridians. Although still a preliminary clinical finding, it appears that the relationship between Constitutions and Conditional Extraordinary Meridian presentations is as follows: If the Constitution is based on an Excess Organ tendency, then Extraordinary Meridian imbalances might be best treated by dispersing one of them. However, if the Constitution is based on a Deficient Organ tendency, then Extraordinary Meridian imbalances might be best treated by tonifying one of them. This rule of thumb has been added just as this treatise was about to be published, so it needs further research concerning its universal applicability.

Almost any contemporary text on the Extraordinary Meridians will be found to present the Opening or Master Points and the Closing or Coupled Points of these Meridians, but it should be noted that this teaching was not present in the foundational classical acupuncture texts. The first mention of these four sets of coupled Points (SP 4 and PE 6, LU 7 and KI 6, SI 3 and UB 62, TH 5 and GB 41) dates to 1241 in a text by Dòu Hàn qīng (竇漢卿). His book *Zhēn Jīng Zhǐ Nán* or *Compass of Acupuncture* (literally *South Pointing Acupuncture Classic*) listed these Point combinations, but did not discuss their relationship to the Extraordinary Meridians. This latter association was introduced by Xú Fèng (徐鳳) in 1439 in his book *Zhēn Jiǔ Dà Quán* or

Encyclopedia of Acupuncture and Moxibustion. As far as I know, there has been no rationale proposed for why these particular Points were chosen as Command Points for the Extraordinary Meridians, especially considering the fact that some of the couples share the same Great Meridian (SI 3 with UB 62 and TH 5 with GB 41) while others have a different structure (LU 7 with KI 6 and SP 4 with PE 6), suggesting that these Point couples were chosen empirically, or for some other undisclosed reason. I will propose a hypothetical model that justifies this asymmetrically structured set of Points, but I'd like to state that Master Dòu was said to have been strongly influenced by Daoist colleagues, and the concept of a primitive Constitution governed by the Extraordinary Meridians is a reflection of Daoist teachings, so the idea of a primitive Constitution governed by the Extraordinary Meridians is likely to be a factor in the choice of these particular Points.[208] My model will be presented later in this chapter, but the rationale is as follows: The Extraordinary Meridians form a system, rather than merely being separate entities. This system is based on yīnyáng theory as will be demonstrated, but additionally these Meridians form the basis upon which the Regular Meridians and the mature Constitution will be formed. For that reason, it appears to be imperative that they must have the potential to influence the Five Elements and their associated Organs and Regular Meridians. With that in mind, there needs to be some connection to the Wood Element (putatively via GB 41), the Earth Element (putatively via SP 4) and the Metal Element (putatively via LU 7) in addition to the Fire and Water Elements (putatively via SI 3, UB 62 and KI 6)[209]. As mentioned several times, the Ministerial Fire functions are not represented in the mature (Organ based) Constitutions, but they are prominent in the primitive Constitutions (via TH 5 and PE 6). Indeed, the Ministerial Fire functions are the bridge between the primitive and mature Constitutions, as specified in the *Nánjīng*'s emphasis on their role in

208 "Most scholars feel that these theories (*about the Extraordinary Meridians*) are rooted in Daoist philosophy and contemplative practice." Matsumoto, p. 10

209 The Fire and Water Elements will be shown to play a dominant role in the organization of the Extraordinary Meridian system, and perhaps that is why more of their Points are included in its set of Command Points. Xú Fèng associated these Key Points with the Wén Wáng order of the trigrams, which are presented with Fire (Lí 離) on top and Water (Kǎn 坎) on the bottom, as the axis of this system. My own research, to be described, shows that in all cases of Extraordinary Meridian imbalance, the Subdoshas on one side (left or right wrist) reflect the Water Element, while on the opposite side they reflect the Fire Element. "These energies (Ministerial Fire) are the center of the body's energetic systems, the first materializations of yin and yang, symbolized as fire and water." ibid. p. 9.

supplying Original (Yuan) Qi to the Source Points of the Organ Meridians.[210] The asymmetric Point couples provide a basis for the evolution of the yīnyáng based primitive Constitution to differentiate into the Five Element based mature Constitution. The clinical details of Extraordinary Meridian diagnosis and treatment will have to wait until the Extraordinary Meridians are first presented individually in more detail.

There are certain fundamental requirements for the successful development of every living organism. Right from the beginning, each organism must have a "spatial grid" that specifies its three dimensions, so that the developing Organs and other body parts end up in the right places. Certainly we do not want a kidney growing out of our head, but we do want ears there (curiously kidney shaped structures). Thus we need to know where is our top, our bottom, our front, our back, our left, our right, our interior and our surface. I believe that the Eight Extraordinary Meridians provide this information (as one of their functions) in order to establish this primitive constitution. It is important to recognize, however, that in Chinese medicine the physical manifestations are always accompanied by the more metaphysical levels that express the symbolism of our Spirit, so that top, bottom, front, back, left, right, interior and surface all resonate with the non-material aspects of our lives, and we must not separate these indivisible realms, even if the presentation is couched in physical terms. Having said that, I should mention that one of the most common clinical settings in which I have found Extraordinary Meridian imbalances is following physical trauma such as occurs in motor vehicle accidents. In such cases it is not unusual for the body's alignment of left/right, top/bottom, front/back or interior/surface to be injured. I might suggest that we widen the concept of orientation, from the spatial realm to all aspects of life that require organization. With that in mind, I think we must also include a fourth dimension, time, in the functioning of this primitive Constitution. Our bodies (and beings in general) develop in a sequential manner, and something must guide the timing of this unfoldment. Thus it is not surprising that most biorhythmic approaches to acupuncture, such as the Líng Guī Bā Fǎ and Fēi Mǎ Bā Fǎ, use the Command Points of the Extraordinary Meridians, but I will not

210 "The triple warmer, master of the heart, pericardium and ming men are the energetic creations of no-form, the first materializations of energy that precede all other material and energetic processes in the body." ibid. p. 9

delve into that subject at this time, other than to note that four dimensions fits the Eight Extraordinary Meridian concept better than the standard three dimensions. With that in mind I should emphasize that it is crucial that we maintain our inherent organization in all aspects throughout our lives, if we are going to be able to function in a healthy manner. Several of my teachers have inspired me to develop the interpretation of the Extraordinary Meridians that I am now presenting, among them Larre and Rochat, Kiiko Matsumoto, Yoo Tae Woo, Jeffrey Yuen, (a transmitter of Daoist lineage teachings), and the writings of Jacques Pialoux (transmitter of the lineage of Soulié de Morant). The following ideas in this chapter represent my own sense of what role the Extraordinary Meridians play in physiology, pathology, diagnosis and treatment, but they should not be seen as a doctrinaire presentation of classical teachings. There are many texts that can be consulted for alternate viewpoints.

The title of this chapter uses the word dual advisedly. Actually, there are a number of ways in which the Extraordinary Meridians express this duality, some of which I have already mentioned, but I would like to emphasize this duality in the roles the EEVs[211] play in both embryonic and post-natal life. The embryonic role is purely physiological (organizing growth and development) while the post-natal role affords a mechanism for also addressing pathology. Let me start by classifying these Meridians into two (dual) groups. These two groups of Extraordinary Meridians were not given distinguishing category names in the *Nánjīng*, however this classification has been popularized by Daoist teachers such as Jeffrey Yuen. A lot of material on the Extraordinary Meridians derives from Daoist sources, as these Extraordinary Meridians, especially the Governing (Dū) and Conception (Rèn) Vessels, have long been used in Daoist meditation practices (e.g. to circulate qì along the "microcosmic orbit"). Yuen calls these two groups the First and Second Ancestries, indicating that they both refer to quite primitive aspects of human organization, in keeping with the quotation from Larre and Rochat. The First Ancestry, associated with Early Heaven (先天) and the Fú Xī trigram order of the *Yìjīng*, includes the Chōng Mài (衝脈), Dū Mài (督脈), Rèn Mài (任脈) and Dài Mài (帶脈). These early organizing mechanisms are the Meridians

211 EEV is an abbreviation of Eight Extraordinary Vessels, a term that I consider synonymous with Eight Extraordinary Meridians. I find the EEV terminology a useful shorthand for indicating this subject.

without Xì Cleft Points, perhaps a reflection of their extreme primitivity. The implication of "early and first" is that these Meridians develop before those of the Second Ancestry. The main trajectories of all the First Ancestry Meridians, as given in *Nánjīng* chapter 28, are confined to the trunk and head of the body, and their starting point is in the area of the "moving qì" between the Kidneys where our Original Nature is stored. It is only with the Second Ancestry Meridians that main trajectories on the limbs appear. The Second Ancestry includes the Yīn and Yáng Qiāo Mài (蹻脈) and the Yīn and Yáng Wéi Mài (維脈), which are associated with Later Heaven (后天) and the Wén Wáng order of trigrams. The reference to Ancestry supports the notion that the Extraordinary Meridians develop during embryogenesis, so we should consider what role they play from the very beginning of life.

For a new life to begin, what is the most important faculty to develop? I should think it would be the ability to distinguish what belongs within one's boundaries from what is outside oneself (self vs. other). The vehicle for transmitting what is deepest within oneself all the way to the surface would be a Vessel for distributing one's Original Nature or Constitutional Energy (yuán qì). There is no "Constitutional Meridian", but there is a Thoroughfare Vessel (one of the English translations of Chōng Mài), which is said to be the "Sea of the Twelve Meridians," which in turn could be thought of as its tributaries. Looking at the character Chōng 衝, we can see the image of something of great importance (zhòng 重) repeatedly (chóng 重) being circulated (xíng 行) from the center or core, shown between the two steps taken by the left and right feet (行) enclosing this important "thing". What are right and left here, but emblems of yīn and yáng? And what is this thing of great importance, but tài jí (太極), the great ridgepole, an emanation from the dào (道), prior to its differentiation into yīn and yáng. From this association it is understandable why Jeffrey Yuen refers to the Chōng Mài as carrying the "blueprint" for each individual life. Such a ridgepole would naturally resonate with the central axis of our body where we might expect to find a Thoroughfare Vessel. As with all matters in CCA, the pulse is a crucial diagnostic finding for identifying abnormalities in the functioning of Chōng Mài. According to Lǐ Shízhēn (李時珍), the Chōng Mài pulse is one where the beating at cùn, guān and chǐ is felt as a long, taut and straight set of impulses at the deepest level, appropriate for a ridgepole that supports any structure.[212] Later I will expand the methodology for identifying each of the Extraordinary Meridian pulses, as I believe the only reliable way is to consider the collection of Extraordinary Meridians is as a unified whole, so that these Meridians are not seen as unrelated to each other. However, it is best to keep in mind that the pulse images, as well as the symptomatology historically associated with the Extraordinary Meridians reflect situations where they are in an excess state (the overflow model). We might want to consider the question, what pulses and symptoms might be associated with deficiencies of each of the Extraordinary Meridians, if such Conditions were possibilities?

Before proceeding further, it is important to mention that four of the Extraordinary Meridians are of yīn nature (Chōng, Rèn, Yīn Qiāo, Yīn Wéi) and four are of yáng nature (Dài, Dū, Yáng Qiāo, Yáng Wéi), another example of their duality. Some authors favor the translation of 奇 as "odd" on the basis that they are not paired (by yīn and yáng?), but this is patently not the case. The reason this is an important distinction is that excess Conditions of the yīn group present with RYCK pulses where Cùnkǒu is wider than Rényíng (an excess yīn or deficient yáng indicator), while excess Conditions of the yáng group present with RYCK pulses where Rényíng is wider than Cùnkǒu (an excess yáng or deficient yīn indicator), at least according to Yoo.[213] In my experience so far, the yīn or yáng state reflected in the RYCK pulses in individuals with an EEV imbalance correlates with the yīn or yáng state of the mature Organ/Meridian/Official Constitution of the individual in every case. Thus in diagnosing EEV imbalances, one should attend to the RYCK pulse pattern in addition to the classical pulse forms of each of the EEVs. If one finds a pulse that fits Lǐ Shízhēn's description of a Chōng Mài (excess) pulse, one should also confirm that Cùnkǒu is wider than Rényíng before concluding that the classical approach to treatment of Chōng Mài

212 Descriptions of the Extraordinary Meridian pulses can be found in the following English references: Wang Shuhe. *Mai Jing (Pulse Classic)*. Translated by Yang Shouzhong. Blue Poppy Press, 1997, pp. 54-58, 354-356, Li Shizhen. *Lakeside Master's Study of the Pulse*. Translated by Bob Flaws. Blue Poppy Press, 1998, pp. 57-60, and Li Shizhen. *Pulse Diagnosis*. Translated by Hoc Ku Huynh and Gary Siefert. Paradigm Press, 1985, pp. 54-57.

213 This rule was proposed by, and is fundamental to, Yoo Tae Woo's use of the Extraordinary Meridians. Yoo only considered the excess state in his method, while I am proposing the possibility of deficient states as an expansion of his model, however my current belief is that Yoo's treatment choice of dispersing the Excessive EEV is generally the simplest and typically quite effective intervention.

(a yīn EEV) is indicated, which in CCA would be SP 4 dispersed and PE 6 tonified, both on the same side (where the Chōng Mài pulse image is felt. Of course it is important to recognize the clinical patterns that are mentioned in the classics for diagnosing Chōng Mài abnormalities, but CCA would add the observation that in some way, such an individual is also likely to be fundamentally disorganized in the sense of being without an appropriate blueprint that takes into account the environment, other people, socials norms, etc. I interpret this finding as indicating that this individual is not able to recognize that their original nature is not the only reality, since it is not being subject to any limits. Such people might lack a sense of how to behave with others, what they should (not can) do with their lives, or even a recognition of there being any purpose in life other than self aggrandizement. As an exercise, can the reader think of anyone who seems oblivious to social norms, ignores established boundaries, evinces no clear planning ability when challenges arise, and yet is so bottom heavy that no amount of pressure can alter their stance at all? That, in my opinion is most likely a case of excess Chōng Mài pathology. Such a state can occur for other reasons, but the Chōng Mài pulse pattern serves to confirm the differential diagnosis. For example, someone who is "possessed by internal demons" would be similarly inappropriately relating to their original nature, but as explained in *The Compleat Acupuncturist*, would have pulses all at the middle (Earth) depth, and a Rényíng bigger than Cùnkǒu presentation, both of which are opposite to the excess Chōng Mài pulse findings.

There are many ways practitioners have advocated for treating the Extraordinary Meridians, but I have found that a simple use of the Master (Opening) and Coupled (Closing) Points, when needled with the correct order and polarity, will often resolve the Extraordinary Meridian abnormalities. These rules however, need to be updated from those in my previous publications, which had been based on my study with Yoo Tae Woo. Yoo would demonstrate how he viewed the Extraordinary Meridians by drawing Regular (Principal) Meridian lines on a paper box, and then twisting the box so that its spatial orientation was seriously altered. He would ask, "How can you expect these Regular Meridians to function properly, now that they are so contorted?" His answer was that only the Extraordinary Meridians could reestablish the original shape (in three dimensions) of this "body". Yoo used a combination of RYCK pulse diagnosis and his own style of abdominal diagnosis[214] to decide which Extraordinary Meridian to treat in virtually any patient visit. Yoo's methodology is based on identifying where there is an excess state through locating abdominal points that are tender when pressed. My own observation is that this approach can be very helpful in corroborating a decision about treating an Extraordinary Meridian, but it does ignore the possible need to additionally or even perhaps instead, tonify one of the Extraordinary Meridians. I should add that others besides Yoo use Extraordinary Meridian treatments quite routinely. One of the faculty members of J.R. Worsley's college in England when I was a student there, Tony Evans, referred to the Extraordinary Meridians as "the method of the Kings," and he claimed to use them as his main method of acupuncture treatment. Again, I do not advocate such an indiscrimenate approach. If these Meridians provide the mechanism for controlling the primitive constitutional grid, there can be serious consequences if a mistake is made in treating an Extraordinary Meridian.[215] As we shall see, there are two groups of Extraordinary Meridians, and only one of them has Xì Cleft Points (郄穴). These points can be used to correct mistaken choices in treating either an Extraordinary or a Regular Meridian, but since four of the Extraordinary Meridians do not have xì cleft points, there is no systematic way to correct mistakes if these are treated inappropriately, other than by giving the more accurate Extraordinary Meridian treatment based on the pulses. Here's a preview to help readers decide when an Extraordinary Meridian treatment is called for: All cases that I have seen where there is the same RYCK ratio imbalance on both sides of the body imply the need to treat an EEV. I have not seen any exceptions to this rule, and I have not encountered EEV imbalances in anyone with only unilateral RYCK ratio imbalances.

---

214 Yoo's style of abdominal diagnosis, Three Constitutions Theory (Sam Il Che Jil, in Korean), is explicitly based on finding excess Conditions, not deficient ones, where palpatory pressure is uncomfortable rather than pleasant.

215 In a personal communication from my colleague Marcus Sommer, he wrote, "I had learned about pre- and post-heaven through the teachings of Michel Odoul in Paris, pupil of Nakazono and Lavier, in which he told us about his avoidance of using them (*the EEVs*). For one reason, because there is much to do before any therapist should consider working with them. And, in a more Faustian tradition, one has to ask oneself what exactly one is meddling with by using them, and if indeed he wishes to carry such responsibility. His warning got shouldered by another teacher, Jean Pelissier, pupil of Prof. Leung Kok Yuen. Furthermore the grand Jacques A. Lavier refrained from using them, according to his work 'bio-énergetique chinoise', as 'too much knowledge about them and their usage is lost'." As the reader will see, I disagree with this stricture, as long as the pulse criteria indicate an EEV imbalance is present.

Yoo Tae Woo's treatment protocol was to disperse the Opening Point and tonify the Coupled Point of the chosen Extraordinary Meridian, both on the same side of the body. For example, if RY is wider than CK, it means that a yáng state exists. In Yoo's teachings, in such a case, he always selected one of the yáng Extraordinary Meridians for treatment. I believe his choice may not necessarily be the optimal approach, but at least it does appear to be effective. A yáng state can represent either yáng excess or yīn deficiency. In the yáng excess example, Yoo's method makes perfect sense, as it disperses one of the yáng Extraordinary Meridians. But in the case of yīn deficiency, such a strategy might not necessarily be sufficient, for as I will explain, the optimal treatment needed might be to first tonify one of the yīn Extraordinary Meridians. As I said earlier, I have not encountered any scholarly presentations of deficient Extraordinary Meridians (or their pulses), so that topic needs further research, and I am not insisting that tonifying an EEV is ever a neccessity; I am only suggesting that the possibility exists and should be investigated. The 4 EEV treatment strategy that I will describe presently appears to support the efficacy of tonifying EEVs when chosen correctly. A similar situation exists when CK is wider than RY, indicating a yīn state. This can reflect yīn excess, and can be treated as Yoo recommended, but it could also reflect yáng deficiency, in which case the optimal treatment might be to tonify one of the yáng Extraordinary Meridians. While these suggested protocols make logical sense, the treatments can always be made stronger by adding Yoo's suggested protocol if one has first tonified an Extraordinary Meridian. The newer treatment rules are simple: To disperse an EEV, first needle the Opening Point in dispersion on the side with the abnormal pulse finding, and then needle the Coupled Point in tonification on the <u>same</u> side. This protocol is different than the ones which can be found in Classical Chinese texts, Pialoux's text, or any other teaching that I have encountered except that advocated in Korean Hand Acupuncture. However, it is both a very simple and a very effective protocol, and one that I will place on a firm intellectual footing throughout this chapter. To tonify an EEV, first needle the Opening Point in tonification on the chosen side and then needle its Coupled Point in dispersion on the <u>same</u> side. Because each polarized pair of Command Points can activate two different Extraordinary Meridians, one tonified and the other dispersed, it might be further helpful if at least one needle from the trajectory of the chosen EEV is added, in order to specify which EEV is the target of treatment. However, practically speaking, I have not found it to be necessary to add any trajectory Points if the Opening and Closing Points are correctly needled. As mentioned, treating Extraordinary Meridians is a very powerful intervention, and so I like to leave the needles alone for a few minutes after insertion, to ensure that the patient feels comfortable in themself. If so, the treatment can be strengthened by the addition of ion pumping cords, which further polarize the treatment in the optimal direction.[216] This approach should improve pulse parameters to a stable degree within less than ten minutes in my experience, and be very effective in producing a sense of wellbeing in the patient, even when performed with quite shallow needle insertions. If for some reason the pulse does not correct, and you are confident of your diagnosis, then there is the option to use a potentially more powerful treatment, based mainly on Pialoux's recommendations. I will explain it at the end of this chapter, as it involves general relationships among the Extraordinary Vessels yet to be described. However, my impression is that the simple combination of Opening and Closing Points, all on the same side, is a powerful Extraordinary Meridian treatment that can most often be successfully used by itself. In cases where the results are less than hoped for, the likelihood is that the diagnosis or point location was not accurate, rather than that the treatment was not strong enough. In any event, once an Extraordinary Meridian abnormality is resolved, it is usually possible to address the Regular Meridians on the same visit. In the past I did not generally treat the Regular Meridians on the same visit, so as to allow the primitive constitution time to solidify first, however this has not been an absolute rule. When time is a constraint, I have treated both in the same visit, as demonstrated on a cancer patient I treated in Beijing in 2018 who improved significantly after only a few days of treatment. She is included in the chapter of case histories (Case number 24). At present, I treat both Extraordinary Meridian and Regular Meridian imbalances on the same visit more often than not, with very favorable results. The EEV treatment should be done first, as it provides a better arrangement for the Regular Meridians to be able to correct their imbalances, much as removing a block would.

I expect that many readers will be scratching their heads at this point. The majority of teachings about treating the Extraordinary Meridians instruct the practitioner to treat the Opening Point on one side of the body and its

---

216 Ion-pumping cords were developed by the late Japanese acupuncturist Yoshio Manaka. They are essentially a wire that has a diode in it so as to allow electricity to flow in only one direction.

Coupled Point on the opposite side. Usually it is even more specific: treat the Opening Point in males on the left and females on the right. Of course these instructions are not accompanied by any pulse findings to explain such a protocol, but often these treatments are quite effective, so we must ask ourselves how such treatments can be understood in light of the standards I just proposed in using CCA. I will explain this seeming contradiction presently, but for the moment I can share my observation that many practitioners who use the EEV Command Points, insert the needles perpendicularly, so the body is left to decide whether tonification or dispersion is the intended result. In the CCA use of these Command Points, the choice of polarity of each needle insertion is an obligatory one, determined by the direction of insertion with or against the flow of the Point's Regular Meridian.

Let me resume discussion now of the other Extraordinary Meridians. Readers should recall that I emphasized the importance of symmetry in developing the understanding of nature embodied in CCA. That is because symmetry seems to be a commonly encountered principle in both Eastern and Western notions of natural law; science in other words. If we try to imagine what might be the symmetrical opposite of the Chōng Mài, we shall find an unusual answer. One's first impulse might be to choose the Extraordinary Meridian activated by the opposite polarity of the Master and Coupled Points of the Chōng Mài, which would be the Yīn Wéi Mài. In my opinion, that would not be an ideal answer. To start, the symmetrical opposite should be a yáng Extraordinary Meridian. The pulse pattern should also help guide us in answering this question. If the (excess) Chōng Mài pulse is deep and elongated, but presents in a straight line parallel to the flexor tendon, then we might expect its opposite to be both superficial and compressed, while still presenting the same straight line. This is in fact a description of the (excess) Dài Mài pulse. Its pulse is felt superficially at cùn, guān and chǐ, but the central guān section is either much stronger or wider. Usually it is both bigger and even more superficial. Now, if the Chōng Mài acts as an interior ridgepole, extending vertically as an axis, then the Dài Mài can be seen to act as an external horizontal ring at the middle zone of the body, binding all the Meridians passing it vertically. It is hard for me to envision any other image that would be opposite to that of a ridgepole than this central exterior ring, unusual as that might sound. We might think of it this way: In putting on our clothes, the most exterior article is usually our belt. And we can loosen or tighten our belt in order to change the location of the surface of our body, especially in the region between our Kidneys, where our Original Energy, embodying our Constitution, is stored. Whereas the Chōng Mài is responsible for spreading its blueprint from the center to the periphery, the Dài Mài is responsible for defining the surface limits of the body, and ensuring that the blueprint is not being replaced by information from the outside.[217] People with Dài Mài excess disruptions are not without purpose, but their goals in life may not be completely their own. Rather they might appear to have been inculcated by family, teachers or other social forces. My patients with Dài Mài pulses have often corroborated this idea, saying that they have struggled to free themselves from ideas instilled in their upbringing, which have felt oppressive (constrictive) later in life. The character Dài 帶 depicts a belt or girdle with objects hanging from it, but as Larre pointed out, it also implies the faculty of keeping things carefully, and in the correct place. I propose that what is carefully kept in place is one's original nature (essence/spirit), which belongs within an organism's boundary, and is not to be disturbed by outside influences.[218] In examining this first pair of Extraordinary Meridians, we can take the analysis one step further. If Chōng Mài is in an excess state, then it will be opposing the Dài Mài's attempt to maintain an appropriate boundary, thus favoring a weakness in Dài Mài. This same opposition holds true for all four pairs of Extraordinary Meridians. Likewise, if Chōng Mài is deficient, then Dài Mài will not be appropriately opposed and it would not be surprising if the individual felt constricted in the middle such that the lower body might end up feeling immobilized, as if "sitting in water" to cite a classical indication for treating Dài Mài. This reasoning supports the idea that Extraordinary Meridians can be in need of either dispersion or tonification as the situation demands. The combination of dispersing Chōng Mài and tonifying Dài Mài together on the same side of the body is an example of such a dual approach. This method is applicable to all pairs of symmetrical EEVs, as I will indicate further on in this chapter.

Having established one's core and one's exterior boundary (inside vs. outside), the next essential faculty needed is a means of orientation towards the external environment. Is it safe to move forward with our life or do we

217 "Dài Mài encloses the space of a living being within its natural boundaries, which gives us the beginning of a framework, a container." Larre, C. op. cit. p.17

218 Larre, C. ibid. p. 134

need to stand our ground, or even back away at any given moment? The front of our body contains the sensory portals that inform us if we are in danger, and the Rèn Mài tells us exactly which direction defines our front (which way to face). *Sù Wèn* Chapter 60 tells us that Rèn Mài passes through the face and penetrates the eyes. The character Rèn (任) depicts a person (人) bearing a burden (壬). Larre and Rochat suggest that this implies the ability to endure and continue.[219] What are we to endure and what do we continue? In many ways there is a synergy between Chōng Mài and Rèn Mài, both First Ancestry yīn Meridians. If the Chōng Mài "thrusts" us into life with our unique trajectory (destiny or mìng, 命) the Rèn Mài shows us the present path forward. Without strong resources (yīn) we could not endure even the slightest difficulty as we move forward in life. People with Rèn Mài abnormalities seem to get stuck, often embarking on projects, only to experience a feeling of overwhelm and an inability to accomplish their endeavors. Rèn Mài is most well known as the Sea of all the yīn Meridians. The excess Rèn Mài pulse is best understood in conjunction with the excess pulses of Chōng Mài and Dū Mài. All three are felt to be in a straight line parallel to the flexor tendon at cùn, guān and chǐ. For Chōng Mài, this line, where the pulses feel most coherent, is at the deep level. For Dū Mài it is superficial, while for Rèn Mài it is in the middle between the levels of the other two Extraordinary Meridians. The symbolism here is Heaven (Dū Mài, Sea of yáng Meridians), Earth (Rèn Mài, Sea of yīn Meridians), and Man (Chōng Mài, Sea of all twelve Meridians which interpenetrate each other). These three Extraordinary Meridians (Chōng Mài, Rèn Mài and Dū Mài) also share a common origin in the lower abdomen (the dān tián 丹田), but on reaching the surface of the body they split apart into yīn and yáng branches, with the ridgepole running up the center between them. In this context, the symbolism is Heaven (Dū Mài), Man (Chōng Mài) and Earth (Rèn Mài). As with the trigram arrangements in the *Yìjīng*, it is important to remember that there is no single correct order of correspondences. Many different orders can be manifested, depending on the context.

The Dū Mài is a symmetrical opposite to the Rèn Mài. Dū (督) means to supervise, direct, or govern. The character includes the primitive form of an eye 目,

219 Larre, C. ibid. pp. 86-87.

implying the role of overseeing something. We know that in the symbolism of the *Nèijīng*, the Emperor does not supervise. Rather he is surrounded and protected by the royal court, among whom there are other Officials to take care of such activities. This Governor Dū uses a character for a younger brother, not appropriate for the Emperor himself. What does Dū supervise? When the Chōng Mài branches into yīn and yáng, the resources (yīn) are collected in the front of the body, while the functional control of these resources (yáng) requires the masculine attributes of protection (backbone) and activity (younger brother's musculature). If the excess Rèn Mài presentation is of someone who cannot move forward in life because they have too much "stuff" congesting them, we might think of the excess Dū Mài presentation as someone who rushes forward without the means to sustain any projects, because of overblown management behaviors. They might react to this situation by adopting a "micro-managing" approach: Too many orders piling up, leading to a breakdown in processing. Although the disorders of Rèn Mài and Dū Mài may sometimes appear similar, they can be easily distinguished by diametrically opposite pulse findings. With excess Rèn Mài, Cùnkǒu is wider than Rényíng, and cùn, guān and chǐ beat at the middle depth. With excess Dū Mài, Rényíng is wider than Cùnkǒu, and cùn, guān and chǐ beat very superficially. I should point out that the pulse and symptom descriptions I'm giving reflect pathological states, whereas the functional descriptions reflect normal developmental processes. In a completely healthy individual, there should not be any Extraordinary Meridian pulses present. I should also reiterate that Rèn and Dū connect to each other forming one "Great Vessel". Thus while I mentioned the sensory portals as having an association with Rèn Mài on the face, they are actually located closer to the domain of Dū Mài, above the mouth. Thus both Vessels are necessary for us to move forwards appropriately, one providing the resources and the other managing the movement. Treatment of Rèn Mài and Dū Mài imbalances will be considered along with the other Extraordinary Meridians, since the approach follows the same general rules. In summary, there are two pairs of First Ancestry Extraordinary Meridians establishing: a) one's unique identity and physical manifestation (tǐ zhì 體質) in life, and b) the necessary orientation for manifesting one's destiny (mìng 命). Each pair has one yáng Meridian and one yīn Meridian, an important piece of information to keep in mind as this presentation continues.

The Second Ancestry Meridians strike me as more involved with fine-tuning our movements through life. We have already established the directions of interior/surface and front/back, but even when moving forwards (or backwards), do we do so in a straight line or do we deviate to the left or right? It is not unusual in my experience to see significant postural abnormalities when examining patients, and they do not necessarily reflect First Ancestry abnormalities. These individuals are often able to be quite active in life without transgressing social norms, but they never seem to accomplish their goals (hit the target). Instead, they act as if something is always pulling them off track. A common finding is that the head is not held over the midline of the body, but has shifted location towards either the left or right. In many of these patients I have found pulse indications pointing to either the Yīn or Yáng Wéi Mài. These Wéi Mài pulses at cùn, guān and chǐ are not aligned parallel to the flexor tendons, perfectly imaging both the individual's posture and the skewed path they habitually tend to follow in life. For excess Yáng Wéi Mài the cùn pulse (yáng) is displaced in the radial direction (yáng) while the chǐ pulse (yīn) is displaced in the ulnar direction (yīn). The opposite pulse picture appears in excess Yīn Wéi Mài abnormalities, with the chǐ pulse deviating radially and the cùn pulse deviating ulnarly. These pulse images reflect an inability to keep from deviating to the right or left, and as we know, right and left correlate with yīn and yáng. A common situation where these Wéi Mài pulses appear is after motor vehicle accidents where the body is violently pushed out of alignment, but the original alignment of right and left (the grid) can often be reestablished by treating one of these two Extraordinary Meridians. The character wéi (維) means to preserve, maintain, keep stable, hold in place or attach. Larre and Rochat refer to the Chinese expression sì wéi 四維 as indicating the four corners of the Earth.[220] These corners of the square Earth, with their diagonals, depict the pulse images of the Wéi Mài. It is these attachments that frequently are the locus of pathology in whiplash injuries. Larre and Rochat also explain that sì wéi 四維 can refer to the four limbs, and it is interesting to note that the Second Ancestry Meridians introduce trajectories on the limbs and also have limb pathologies as clinical indications. Furthermore, we hold things (in place) with our hands, and the Opening Points for both the Wéi Mài are located on the wrists, an example of symbolic resonance.

220 Larre, C. ibid. p. 207.

As for up and down, we have the last two Extraordinary Meridians, the Yīn and Yáng Qiāo Mài. Qiāo 蹻 means to raise one's foot, to stand erect, but also can refer to stilts. The common thread connecting these ideas is the polarity between top and bottom. We raise things up to the top, but we depend on a firm rooting down at the bottom. The radical is the foot, which gives us the ability to climb up by raising our feet. In English we might describe something as top-heavy, meaning it is unstable because it does not have its center of gravity low enough. The foot radical of Qiāo emphasizes the importance of the lower limbs where the Opening Points are located, in contrast to the Wéi Mài whose Opening Points are on the upper limbs. We see the same imagery in the pulses of the two Qiāo Mài. The excess Yáng Qiāo Mài pulse is one where cùn, guān and chǐ are aligned parallel to the flexor tendon, but the cùn position is either more superficial or bigger and stronger than the other two positions. The excess Yīn Qiāo Mài pulse is also aligned parallel to the flexor tendon, but here it is the chǐ position that has either a more superficial or bigger and stronger pulse. In what has been called the Chinese anatomical position (hands above the head), the cùn is at the top and the chǐ is at the bottom of the region where the pulses are examined, known as Cùnkǒu. Thus it seems evident that the Qiāo Mài are responsible for distribution along our vertical grid orientation. In patients whose structure suggests unbalanced development in either the head/chest (upper heater) or the pelvis/legs (lower heater), it is important to check for these pulse patterns.

The discussion of the Extraordinary Meridians so far has focused on one of their primary functions, from the moment of conception, and continuing throughout life. This is in keeping with CCA's orientation of restoring the Conditional state of people to their original Constitutional state. The Extraordinary Meridians are the earliest vehicles for distributing the Original Energy (yuán qì) that determines each individual's Constitution via the primal grid they form during embryonic life. This grid must be maintained throughout one's lifetime in order to stay healthy, but if it is disrupted, then acupuncture has the potential to restore it. Doing so is simply an example of transforming a Conditional state back to the Constitutional state. In a way, these Extraordinary Meridians provide a transitional step in the individual's developmental expression of the potential present in their yuán qì. From the Extraordinary Meridians, the next step is the emergence of the 12 Regular Meridians, a development arising from

Chōng Mài, and starting with the functions of Ministerial Fire (the Closing Point of Chōng Mài), which acts by initially establishing the three heaters (sān jiāo 三焦). Having no formed Organs themselves, the Ministerial Fire Officials distribute the yuán qì to the yuán Points of the 12 jīng mài, conduits that give form and function to the developing zàng fǔ. I must reiterate that this comprehensive interpretation of the role played by the Extraordinary Meridians was not discussed as such in the medical classics, but is my own interpretation, whose biggest influences have been wide study with different teachers and careful observation in my clinical practice.

There is something in common among Rèn Mài, Dū Mài, Yáng Wéi Mài, and Yīn Wéi Mài. The Opening Points for all four of these Vessels are on the Upper limbs. Similarly, the Chōng Mài, Dài Mài, Yáng Qiāo Mài and Yīn Qiāo Mài have, in common, their Opening Points on the lower limbs. I have never encountered a discussion of this way of parsing the Extraordinary Vessels, but it clearly suggests to me an alternate classification using Yáng and Yīn groupings. I have used these groupings of the EEVs in my pursuit of possible pulse findings that clearly corroborate the presence of an EEV imbalance, in order to lessen the chance of making a mistake in diagnosis. What I have discovered is that in all cases of EEV imbalances I have studied in my patients, in addition to the bilateral RYCK ratio imbalances previously noted, the pulse on one wrist will display the zào 躁 quality and have Subdoshas of the Fire Element, while the pulses on the opposite wrist will display the jìng 靜 quality and have Subdoshas of the Water Element. These pulse qualities were explained in Chapter 6. In addition, if the zào quality is found in this situation on one side, then it appears to indicate a need to treat an EEV that has an Opening Point on a hand Meridian on the opposite side. Likewise, the side that displays a jìng quality appears to indicate a need to treat an EEV that has an Opening Point on a foot Meridian on the opposite side. If the practitioner wishes to treat only one EEV, then the choice could be determined by the symptomatology, however, my present approach to EEV treatment includes a way to avoid having to pick between these alternatives by using four EEVs together, as will be described.

Another way of parsing the two groups of Extraordinary Meridians is to divide them as follows: Dū Mài, Rèn Mài, Yáng Qiāo Mài and Yīn Qiāo

Mài in one group and Chōng Mài, Dài Mài, Yáng Wéi Mài, and Yīn Wéi Mài in the other group.[221] The reason for this grouping has to do with the constellational aspect of the system of Extraordinary Meridians mentioned earlier, which becomes clinically important in understanding pulse patterns and choosing treatments of the Extraordinary Meridians. Let's consider the pulses of the excess EEVs (treated by dispersing that EEV) using the preceding analysis. Either it is a yīn EEV (Rèn Mài, Yīn Qiāo Mài, Chōng Mài or Yīn Wéi Mài) or a yáng EEV (Dài Mài, Dū Mài, Yáng Qiāo Mài, or Yáng Wéi Mài). Following yīnyáng theory, and attending to the discussion of the individual EEVs above, it seems logical to conclude that an excess of Chōng Mài will result in a deficiency of Dài Mài and vice-versa. The same holds true for the pairs Rèn Mài and Dū Mài, Yīn Qiāo Mài and Yáng Qiāo Mài, Yīn Wéi Mài and Yáng Wéi Mài. From this analysis we can understand that the pulse of any deficient EEV (if such a possibility exists) would present a similar image as the pulse of its partnered excess EEV just listed. Now, if one has a yáng deficient EEV, the treatment needed would logically be to tonify a yáng EEV, which I assume would result from reversing the polarity of the Command Points ordinarily used to disperse that EEV. For example, in someone with a deficient Chōng Mài, treatment might be to tonify SP 4 and then disperse PE 6, both on the same side. I have had good results using this treatment approach. Such a treatment would produce a similar result as dispersing the Dài Mài by dispersing GB 41 and then tonifying TH 5. My experience shows that either one, or both of these treatments together produce similar corrections of a perceived Dài Mài pulse, and the analogous approach holds for all the other Extraordinary Meridians if they are thought to be deficient (perhaps from the case history). At present I do not know how one can distinguish a deficient EEV pulse from an excess EEV pulse of its partnered Vessel, but there is a practical application of this theoretical analysis. Before illustrating this idea, let me give one more example. If a Yīn Wéi Mài pulse is felt in an individual, and one suspects that it is due to a deficiency, one might treat by tonifying the Yáng Wéi Mài (TH 5+ and GB 41-) on that side. After this treatment is in place, it can be expanded by adding needles to disperse the Yīn Wéi Mài (PE 6- and SP 4+) on the same side.

221 The first group has either an Opening or Closing Point on Fire or Water Meridians, while the second group has either an Opening or Closing Point on Wood or Earth Meridians. Fire and Water Elements associate with the Upper and Lower Heater respectively, while Wood and Earth Elements associate with the Middle Heater.

This chapter's title mentions the dual functions of the Extraordinary Meridians, as there are what I call secondary functions of the Extraordinary Meridians, which constitute the next topic to address. I think of the primary functions as the physiological potentials and the secondary functions as the pathological potentials of the Extraordinary Meridians. The *Nánjīng* presented the first systematic discussion of these secondary functions as reservoirs for absorbing the overflow of pathogenic factors (xié qì) in Difficulties 27 through 29. I think of absorbing and containing xié qì as one way the organism limits the damage that might occur otherwise, but limiting damage is not avoiding it entirely. The typical symptomatology of the Extraordinary Meridians, described in the *Nánjīng*, is the price paid for not being able to simply expel or transform the xié qì in the first place, which is why these descriptions are of the excess states of the Extraordinary Meridians. Opening up the Extraordinary Meridians in such cases provides an opportunity for a return of proper functioning to both the Extraordinary Meridians and the Regular Meridians, which in CCA is part of the process of transforming the Conditional state back to the Constitutional state. However, CCA is not based on treating xié qì, but rather on restoring normal functioning of the zhèng qì, so I prefer to think of abnormalities of the Extraordinary Meridians as being either hyperactivity or hypoactivity (Excess or Deficiency). That is, they may need dispersing treatment or tonifying treatment, which is a distinction I have not encountered in other Oriental medical presentations. In 2019, before the Corona virus pandemic, I was invited to teach a workshop in Shanghai covering my usage of the Eight Extraordinary Vessels (EEVs) via pulse diagnosis. The hosts were students from my prior classes in China who had seen me use EEV treatments with impressive results. These students also were followers of the lineage of Jacques Pialoux, so they asked me to study Pialoux's teachings about the EEVs so that my contribution to their understanding would lead to an enhancement, without creating a conflicting set of teachings. For this reason (even though the pandemic prevented the Shanghai class from taking place as scheduled), I studied Pialoux's textbook and found some useful material on the EEVs that I will include here. Pialoux did not use pulse diagnosis to choose EEV treatments (which is why the Chinese students wanted my guidance), but rather relied on symptomatology from classical sources, along with palpatory information along the trajectories of the EEVs. Once a particular EEV was chosen for treatment, Pialoux's protocol was to tonify the Master Points (Opening and Closing) of the EEV in question, and to disperse the Master Points (Opening and Closing) of its coupled EEV of opposite polarity, all on the same side of the body. The reader should recognize this selection of Points from my earlier protocols for treating imbalanced EEVs. Pialoux's treatment protocols were designed to address cases of excess EEVs, since the symptomatology he used in choosing an EEV reflects excess states. In his approach, as an example, Chōng Mài is coupled with Dài Mài, and Yáng Qiāo Mài is coupled with Yīn Qiāo Mài. Pialoux additionally needled one or more sensitive Points along the EEV's trajectory. Initially I was a bit skeptical of this approach, since in my experience, and that of my teachers, the Opening Point of an EEV was treated by dispersing it, while the Closing Point was treated by tonifying it. I experimented with all the possible combinations of tonifying and dispersing these Points, and concluded that my original understanding of how to treat excess EEVs was typically effective, but that adding tonification treatment of the coupled EEV was helpful, with this coupled EEV having its Opening Point tonified and its Closing Point dispersed. As an example of this confusing clinical strategy, let's assume the excessive Yáng Qiāo Mài needs treatment on the left side. The treatment involves dispersion of the Yáng Qiāo Mài and reinforcement of the Yīn Qiāo Mài. The first needle is inserted at UB 62 against the direction of flow (i.e. upwards) and the second needle is inserted at SI 3 with the direction of flow (i.e. upwards toward the head). The third needle is inserted at KI 6 with the flow (upwards) and the forth needle is inserted at LU 7 against the flow (upwards toward the head), all on the left side. Finally, a needle can be inserted along the trajectory of the Yáng Qiāo Mài (e.g. LI 15) perpendicularly. This protocol reflects my understanding that when excessive Extraordinary Vessels need treatment, it is because they need dispersion. It is easy to understand this in considering the Yáng Qiāo Mài, where a common symptom is insomnia. In such a case it is logical to disperse the Yáng Qiāo Mài and tonify the Yīn Qiāo Mài. This combination acts synergistically to draw the qì and Blood away from the head by dispersing Yáng and tonifying Yīn through use of these two EEVs. The traditional pairing of UB 62 and KI 6 to treat insomnia is easily understood as reflective of the activation of these two Extraordinary Meridians, and adding LI 15 focuses this treatment on the excessive Yáng Qiāo Mài as the source of the pathology.

When EEV presentations of pulses and symptoms appeared in my patients, I often noticed a different EEV signal in the pulse on the contralateral wrist. To use the previous example of a left sided excess Yáng Qiāo Mài pulse, I often found the right side to show a Dū Mài pulse. Dū Mài uses the same Master Points as Yáng Qiāo Mài, but with opposite needling polarity, however dispersing either one has a similar effect of dispersing yáng. I subsequently confirmed that this pattern of EEV pulses is a universal finding, and is now a crucial step in applying my criteria for diagnosing EEV imbalances. Whenever I felt this pulse configuration I used the same thinking to choose Points on the right side, based on a modification of Pialoux's teachings. Here the first needle on the right side is inserted at SI 3 against the flow (downwards, away from the head). The second needle is inserted at UB 62 with the flow (downwards). The third needle is inserted at LU 7 with the flow (downwards) and the forth needle is inserted at KI 6 against the flow (downwards). The effect of this group of needles is to disperse the Governing Vessel and tonify the Conception Vessel, again dispersing Yáng and tonifying Yīn. Treating both sides together leads to a coherent message that uses both the First and Second Ancestry EEVs to correct the organism's overall balance of yīn and yáng. This is a very powerful treatment, as it influences half of the EEVs. The reader should recognize that all the Extraordinary Meridians treated in this example are in the first group that I identified above as a useful application of duality in understanding the Extraordinary Meridians. Since the wrist pulses are typically reflective of complementary Extraordinary Meridian signals on opposite sides, it seems logical that there might be a RYCK imbalance signal on both sides, just as I have indicated as a criterion I use for diagnosing an EEV imbalance. The protocols above are easy to remember, since on one side of the body all the needles are directed upwards while on the other side, all the needles are directed downwards. Curiously, there is a strong similarity between this needling arrangement (upwards on one side and downwards on the other side, arranged on the four "limbs") and the symbol known as the swastika, which is commonly found in both

Buddhist and Hindu artwork (see Fig. 38).[222] William Ceurvels has pointed out that the traditional Chinese astronomical association related to the direction of the handle of the Big Dipper constellation can be thought of as depicting a swastika, and might be an easily observable illustration of natural law (Fig. 38 C).[223] As the previous Footnote indicates, not only is the swastika historically an auspicious symbol, but its literal meaning is "conducive to well being". This symbol long predated the classical Chinese medical texts, and it is not inconceivable that it represented an ancient recognition that this pattern of stimulation might restore the primitive balance mechanism established by the Extra Meridian system as a whole, based on resonance theory of the relationship of humans to Heaven. The model of following the Dipper handle illustrated suggests an explanation of why only four EEVs are treated in the complex protocol, rather than all eight. The most important things to bear in mind in carrying out this type of treatment are the order in which the needles are placed, their direction (polarity) and the resulting rectification of pulse signs and patient symptoms. I believe the organism reads the message from this type of needling much as we understand written words by seeing the letters. The order appears to be equally as important as the letters (Points) themselves. I have used this protocol on many patients, and it has been quite helpful. Generally when using this 4 EEV treatment, the needles are inserted quite shallowly, so patients report how calming such treatment is, and yet it produces profound beneficial effects on both pulses and symptoms. Typically, in my current practice, I will follow this EEV treatment with the indicated Regular Meridian/Organ balancing treatment described in previous chapters. The initial pulse depths are reliable indicators of which Regular Meridians need balancing treatment, however the Subdoshas must be rechecked after the EEV treatment, as they most likely will have changed once the Extraordinary Meridians are functioning physiologically.

222 Excerpted from Wikipedia: The swastika symbol, 卐 (*right-facing* or *clockwise*) or 卍 (*left-facing* or *counterclockwise*), is an ancient religious icon in the cultures of Eurasia. The left-facing version may also be referred to as sauwastika. It is used as a symbol of divinity and spirituality in Indian religions, including Hinduism, Buddhism and Jainism. In the Western world it was a symbol of auspiciousness and good luck until the 1930s when the right-facing tilted form became a feature of Nazi symbolism as an emblem of the Aryan race. As a result of World War II and the Holocaust, many people in the West still strongly associate it with Nazism and anti-Semitism. The swastika continues to be used as a symbol of good luck and prosperity in Hindu and Buddhist countries such as Nepal, India, Mongolia, China and Japan. The word *swastika* comes from Sanskrit, Romanized as: *svástika*, meaning 'conducive to well being'. In Chinese it is called 萬字 (*wànzì*) meaning 'all things symbol'.

223 Facebook post of a YouTube video lecture posted 10/21/21.

A

B

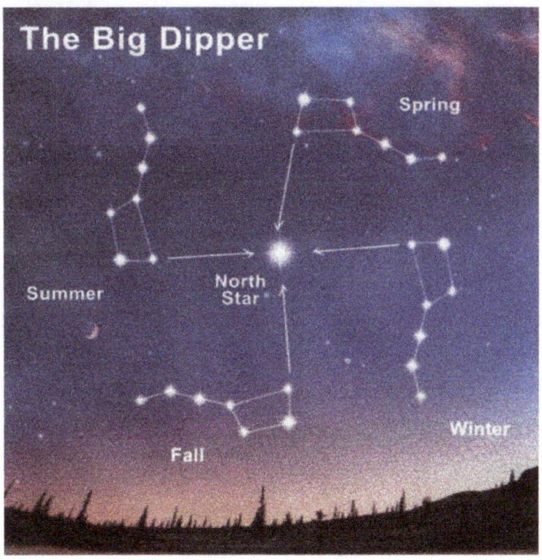

C

Fig. 38 A) Guanyin Statue, Hanshan Temple, Suzhou, China (counter-clockwise)
B) Tibetan Buddhist Sigil, (clockwise)
C) The Big Dipper in the Four Seasons, (clockwise) [224]

224 Fig. 38 A) Jonas Ericsson/Wikimedia Commons. B) Michel Royou/Wikimedia Commons. C) Facebook post by William Ceurvels. The handle of the Big Dipper points East in Spring, South in Summer, West in Autumn and North in Winter.

In summary, the group of EEVs appears to act as an integrated system to keep the organism oriented properly. Each Extraordinary Vessel is supported by three others: its yīn yáng opposite (e.g. Chōng and Dài), both of which could be treated on the same side of the body, plus its inverse of the same polarity with that EEV's yīn yáng opposite (for the same example, Yīn Wéi and Yáng Wéi), both of which could be treated on the opposite side from the target EEV[225]. These protocols always involve two First Ancestry Meridians and two Second Ancestry Meridians. The result of such treatment is theoretically a very well supported framework for orienting the individual in three dimensions, however, in some situations, I might simply treat the target EEV by using the ipsilateral Opening and Closing Points of the Vessel affected. I do not always treat the Extraordinary Meridians on the other side of the body, but I do use their pulse patterns to help confirm the diagnosis.

Finally I should remind the reader that the *Nánjīng* ostensibly discusses the EEVs as a mechanism for containing excessive pathogenic factors (xié qì). In that text however, no explanation is given for why one particular EEV becomes involved, nor for what pathogenic factor triggers its activation. Once again, I think that the conventional interpretation of *Nánjīng* chapter 27 misses its main point. Chapter 27 starts by stating that the circulation of qì in the EEVs is not part of the sequential circulation of qì in the Regular Meridians (jīngmài). As such, this statement fully agrees with my presentation so far. The EEVs and the Regular Meridians are concerned with separate issues of functional coordination. The EEVs deal with the primary energetic architecture, while the Regular Meridians deal with the functions of the zàng fǔ, including their 12 pathways of qì circulation. It is only at the end of Chapter 27 that the *Nánjīng* text mentions the metaphor of ditches and reservoirs to contain overflow. Nowhere is the term pathogenic factor (xié qì) mentioned. Thus the only significant term here is "overflow" which must

225 Readers familiar with Japanese acupuncture might recognize a similarity between these EEV protocols and Kawai's "infinity treatment" protocols. Kiiko Matsumoto has taught Kawai's methods in a number of seminars, and the details of the infinity protocols can be found online at acupuncturemedical.org>uploads>2015/07. There is a big difference however between our protocols. Kawai always uses Opening and Closing Points for any EEV on opposite sides of the body, whereas in CCA the Opening and Closing Points are always on the same side of the body. Furthermore, the Japanese styles of treatment of the EEVs typically are based on symptoms and tender areas, while CCA treatments prioritize pulse diagnosis. Matsumoto and Birch, in their book on the Extraordinary Vessels, describe several Japanese methods of diagnosis and treatment, but none of them refer to EEV pulse diagnosis.

refer to an excess. Furthermore, this excess cannot be handled (corrected) by the system of Regular Meridians, so the EEVs are called into play, and are themselves subject to this excess. The ditches and reservoirs mentioned in the *Nánjīng* are thus metaphors for how to deal with excess qì that is beyond the capacity of the Regular Meridians to handle via their normal (not extraordinary) faculties. I believe that it is for this reason that the EEVs have been primarily treated by dispersing techniques, such as sedating their Opening Points. Logically, this guideline makes sense in the context of the *Nánjīng* presentation. So where does this excess qì go when an EEV is treated? As the *Nánjīng* hints, it does not go to the Regular Meridians, which have an entirely separate circulation. The only other possibility that makes sense is that the system of EEVs as a whole is stimulated to rebalance, so that the excess qì in one Vessel is distributed to other EEVs. To continue the analogy used by the *Nánjīng*, these ditches and reservoirs are not filled during normal circumstances (good health), and thus do not show up in pulse or symptom findings. It is only following unusual circumstances that the EEVs reveal themselves in the pulse. I think this is the reason that acupuncture is ordinarily started by addressing the Regular Meridians (as has been the approach in CCA so far), but should an Extraordinary Meridian pulse already be present, it appears to be important to address it first. In such a case, the EEVs rebalance by shifting the excess qì in one EEV to the deficient EEVs. I think such an explanation is actually very much in tune with the *Nánjīng*'s goal of leveling the qì.

The following is a hypothetical example in order to illustrate the approach I have described. Other examples may be found in the case histories. Suppose that someone is diagnosed as having a RYCK pattern where RY is found to be wider than CK on both sides. In addition the pulse quality on the right is zào and the right side Subdoshas reflect the Fire Element, while the pulse quality on the left is jìng and the left side Subdoshas reflect the Water Element. Thus the patient is displaying a state of yáng excess with yīn deficiency. When we check the pulses to decide what kind of EEV problem exists, we must focus on the yáng EEVs (because RY>CK): Dū Mài, Yáng Wéi Mài, Dài Mài and Yáng Qiāo Mài. The jìng quality on the left suggests that if there is an EEV imbalance on the right, it will be an EEV with its Opening Point on a foot Meridian, leaving only Yáng Qiāo Mài or Dài Mài as possibilities. These EEVs are easily distinguished by their pulse signatures. Suppose we feel a Yáng Qiāo Mài pulse on the right wrist. Then we can check the left side pulses. If we are correct, and my understanding is also correct, then there should be a Dū Mài pulse on the left side. These two EEVs have opposite polarities of Opening and Closing Points (which can balance each other), but they both indicate a state of Yáng excess. Having confirmed the diagnosis we can disperse Yáng Qiāo Mài on the right with UB 62- and SI 3+. Next we must ask ourselves what Yīn EEV should be tonified in order to balance Yáng Qiāo Mài? Clearly it would be Yīn Qiāo Mài, which we can tonify with KI 6+ and LU 7- on the right. Depending on the individual's Constitution, it might be preferable to disperse the Yáng EEV first or to tonify the Yīn EEV first. What about the left side? Dū Mài can balance Yáng Qiāo Mài with SI 3- and UB 62+, while Rèn Mài can balance the excess Yáng tendency in Dū Mài, so we could tonify Rèn Mài on the left side with LU 7+ and KI 6-. By attending to the structure of the posited complete four EEV treatment, we can now understand why the common practice of treating the Opening and Closing Points of EEVs on opposite sides has historically been a standard recommendation. In the illustrative example above, UB 62 was dispersed on the right and SI 3 was dispersed on the left, which can be thought of as either dispersing Yáng Qiāo Mài on the right side or dispersing Dū Mài on the left side, depending on whichever choice better matched the symptoms. This combination of Points (UB 62 and SI 3) on opposite sides in dispersion) is effective for dispersing excess yáng in the Extraordinary Meridians of this patient, but without the rationale I've given, it becomes a recipe (i.e. treat the Opening and Closing Points of an Extraordinary Meridian on opposite sides) rather than a rational approach to treatment. One of my goals in formulating CCA is to provide such a rationale for the various ways acupuncture can be practiced in an Eastern scientific manner. It may be possible in the future to describe typical clinical symptoms of deficiency in each of the EEVs, in which case this last example could be possibly classified instead as either deficiency of Yīn Qiāo Mài or deficiency of Rèn Mài. So far I have only tentatively identified a way of using the pulses to determine if one particular Extraordinary Meridian caused the others to lose their normal state of balance. If confirmed, such an approach will strengthen the practice of CCA, an ongoing field of scientific inquiry into the nature of life's processes.

CHAPTER 13

Updating the Treatment Protocols in CCA

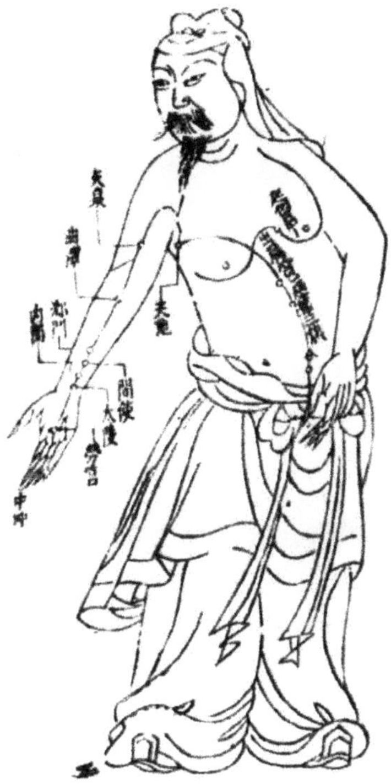

Figure 39: Ming dynasty chart of the Hand Jueyin Meridian [226]

All treatments in CCA depend on first identifying the Constitution, which I will call the Root Organ (or Official or Meridian). This Root Organ/ Meridian must belong to one of the "formed" Organs, not to one of the

226 Source: Wikimedia commons. Imagery From the History of Medicine (original in Bibliothèque nationale de France, Paris).

Ministerial Fire functions (depicted in Fig. 39). In most patients, the Root Organ of their Constitution will be found to be out of coordination with one or more of the other Organs. This Root is responsible for supporting all the other functions in the organism, so it is most often directly treated by a Root Formula, with the addition of treatment directed at the Conditionally imbalanced function(s) by Auxiliary Formula(s). The goal of treatment is to return all the pulse findings from their current Conditional state to their original Constitutional state. The treatment protocols for individuals with Zàng or Fŭ Constitutions are similar, but not identical. In many cases they are mirror images of each other. Fig. 40 shows the family of functional relationships in someone with a Zàng Constitution.

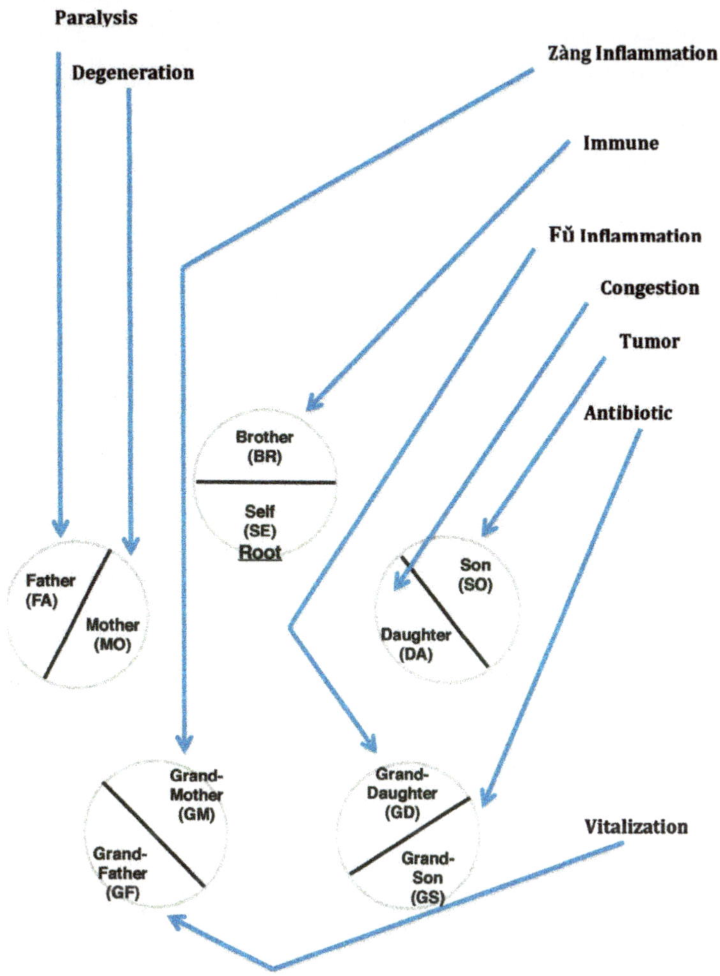

Figure 40: Functional Relationships in Zàng Constitutions [227]

As indicated in Fig. 40, the nomenclature used has been updated from that in my previous publications. The Constitutional Formula (SE or Self) is also called the Root Formula, and is designated by either R or K in this and prior publications. Originally Kuon Dowon described Auxiliary formulas according

227 Source: Peter Eckman. Several new Auxiliary Formulas are introduced here. The Immune Formula is really only applicable to individuals with Zàng Constitutions. It is the yáng Meridian combination for treating the Constitutional Element. In individuals with a Fǔ Constitution, the Root Formula already treats these yáng Meridians. This terminology was used by one of my Korean teachers, Puramo Chong. The other new Formulas, not discussed in my prior publications, are the Tumor Auxiliary Formula, the Degeneration Auxiliary and the Congestion Auxiliary Formula.

to the typical Conditions they were able to treat, but he also used Auxiliary formulas without indicating their functions or giving them names. In Fig. 40, the Grandson (GS) is the Antibiotic Formula, the Granddaughter (GD) is the Fǔ Inflammation Formula, the Grandfather (GF) is the Vitalization Formula, the Grandmother (GM) is the Zàng Inflammation formula, the Brother (BR) is the Immune Formula, the Son (SO) is the Tumor Formula, the Daughter (DA) is the Congestion Formula, the Father (FA) is the Paralysis Formula and the Mother (MO) is the Degeneration Formula (some of the last 5 were used, but not named by Kuon. The Constitutional Root Formula is always the Self (SE) Formula. The functional relationships for people of Fǔ Constitutions are related in a mirror image way as follows (Fig. 41):

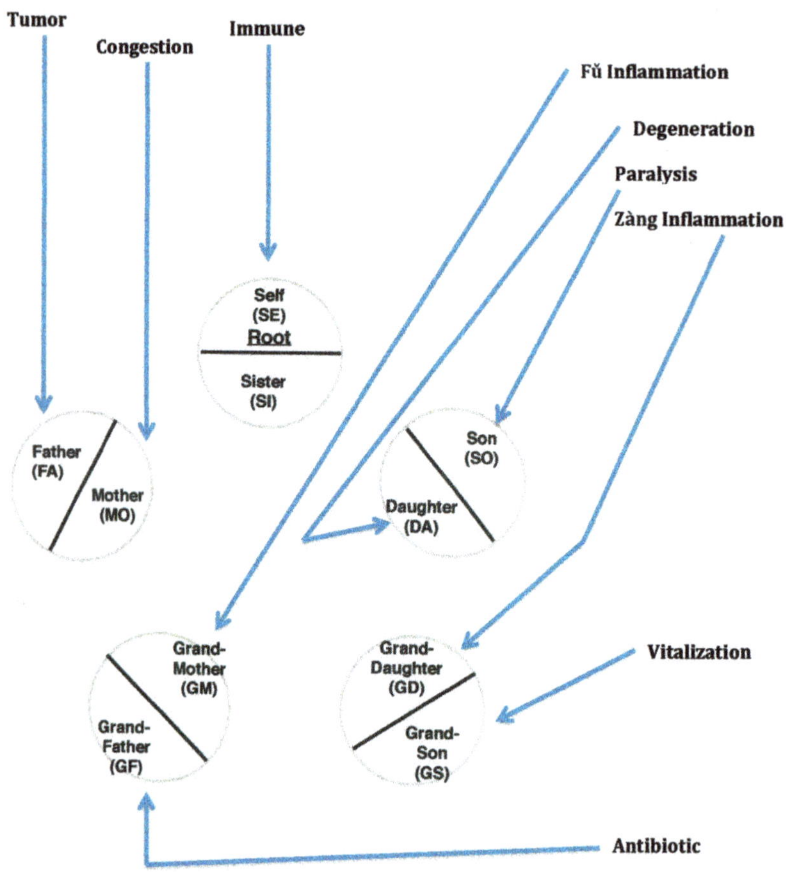

Figure 41: Functional Relationships in Fǔ Constitutions [228]

228 Source: Peter Eckman

The Grandson (GS) is the Vitalization Formula, the Granddaughter (GD) is the Zàng Inflammation Formula, the Grandfather (GF) is the Antibiotic Formula, the Grandmother (GM) is the Fŭ Inflammation formula, the Father (FA) is the Tumor Formula, the Mother is the Congestion Formula, the Son (SO) is the Paralysis Formula, and the Daughter (DA) is the Degeneration Formula (some of the last 4 were used, but not named by Kuon). Kuon did often substitute the Sister (SI) for the Root Formula, but I have not found it indicated by the pulses. If we think of the sequence of Auxiliary Formulas in Zàng Constitutions as following a clockwise order[229], then the sequence of Auxiliary Formulas in Fŭ Constitutions follows a counter-clockwise order.[230] This is an example of both nature's symmetry and also of the opposite natures of yīn and yáng.

Patients typically present with multiple complaints, so how do we know what to treat first? I believe that in formulating KCA, Kuon chose to treat the most serious problem as his default mode, but there are drawbacks to such a method. Often it is difficult to decide which problem is most serious. In CCA however, there is a more objective way to determine the applicable treatment, and as usual, the determinant is the pulse pattern. My belief is that the human organism always tries to marshal its forces to deal with the most pressing impediment to its functioning, and in so doing it creates a pulse pattern revealing this focus of struggle. Therefore, while it can be suggestive to let the complaints or even the pathology guide one's thinking about treatment focus, ultimately the choice in using CCA must be confirmed by what the body is saying through the pulse. In my experience, the first priority in treatment should be to remove any "blocks (EEV pulses, stagnation at any depth of the yáng position pulses as in "possession" or cancer, Akabane imbalances, Entry/Exit Blocks, or possibly retained xié qì). More commonly treatment begins with addressing imbalances involving the Five Elements, which manifest as Organs that are not properly coordinated with each other. When major Organ systems are in conflict, the organism as a whole cannot function normally. Both the pulse depths and the Subdosha patterns will

229 Tumor, Antibiotic, Vitalization, Paralysis for the Fŭ Meridians and Congestion, Fŭ inflammation, Zàng inflammation, Degeneration for the Zàng Meridians.

230 Paralysis, Vitalization, Antibiotic, Tumor for the Fŭ Meridians and Degeneration, Zàng inflammation, Fŭ inflammation, Congestion, for the Zàng Meridians.

indicate that this dynamic is where the body's attention is being directed, and a RYCK ratio imbalance will be found to be present, as described in the *Nèijīng*. The pulse quality (solid vs. hollow) at any Organ's correct pulse depth can be helpful in making an accurate Conditional assessment regarding excess vs. deficiency.

In the following paragraphs it is assumed that RYCK imbalances are found to be unilateral presentations, unless stated otherwise. Kuon originally discovered that individuals with Zàng Constitutions tended to develop more symptoms on their left side, but were most effectively treated on their right (Constitutional) side. Contrariwise, individuals with Fŭ Constitutions tended to develop more symptoms on their right side, but were most effectively treated on their left (Constitutional) side. This approach, of contralateral (to the symptom side) needling, is an important, but not universal, aspect of CCA treatment formulation. Pre-testing with press pellets is my recommendation for confirming the accuracy of any questionable Conditional treatment choices. In any case, it appears that the *Nèijīng*'s implied dictum, to treat the side of the body that displays a RYCK imbalance, is an accurate guide to the choice of treatment laterality.[231]

The simplest type of pathology commonly treated with acupuncture is exemplified by a case of minor acute trauma, such as a sprained ankle. Applying the Root formula on the opposite side will usually be effective in speeding the healing process. However, in my experience most patients with acute injuries have underlying chronic problems, so it is always important to do a thorough history and examination before reaching a diagnosis or deciding on a treatment.

The next most complex type of pathology is one where the two Organs of the Constitutional Element are not in harmony with each other. If one is excessive while the other is deficient, then simply tonifying the Luò Point of the deficient Meridian should restore balance, but in my experience this is not a very common situation. There is a very close connection between the two Meridians of the same Element (via the Luò Mài), so nature is pretty

231 The *Nèijīng* did not explicitly refer to laterality with regard to the RYCK pulses, but I have never encountered a source that did not assume that RY should be compared to CK on the same side of the body.

effective at keeping them from polarizing in opposite directions. That situation (opposite polarities) should always result in a RYCK imbalance, and if the Luò Point treatment isn't sufficiently effective, it could be treated more strongly by the RYCK protocols described in the *Nèijīng*. It is not always the case however that both Meridians of the same Element experience opposite polarities when out of balance. They both may be either deficient or excessive, but to different degrees. Take the example of a person with a Kidney deficiency Constitution who feels unwell, perhaps with urinary symptoms among others, but who has no RYCK imbalance, and all of whose Organs are at their proper depths. They also have the same Subdoshas bilaterally, i.e. Pitta/Kapha of the Water type. On careful examination of the 12 pulses, it will often be found that the Bladder pulse has the hollow shape of deficiency. The Kidney pulse will be even more deficient, reflecting its status as the Constitutional Organ. Such a situation might be treated by using the Root formula followed by the brother formula that tonifies the Bladder, but such cases have not been encountered enough to establish the proper treatment protocol. Possibly just tonifying the Source Point of both Meridians will turn out to be ideal. This is a better approach than either treating a Luò Point or the RYCK treatment, both of which will make one of these Organs even more deficient, while helping the other one. The brother/sister combination of 4 needle formulas was not described by Kuon, but it does offer a speculative answer as to why he changed the Root formula for individuals with a Fu Constitution from the self (Fǔ Meridian) to the sister (Zàng Meridian). After all other imbalances are corrected, it is possible for this type of situation to remain. All of the treatments earlier will have been based on the Root Meridian, but its sibling Meridian has been untouched and might also need adjustment. I believe this is one reason that Worsley's Five Element approach has been so successful. It almost always treats Brother and Sister Meridians together, using the same polarity of treatment, in the same session. On the other hand, the reason it is not usually necessary to treat the Brother or Sister Meridian in CCA and other styles of acupuncture, is that the coupled Meridians of the same Element are naturally strongly interactive, so that treating one of them usually affects both in a beneficial manner.

Many common pathologies can be effectively treated by the Root Formula plus the Fǔ Inflammation formula, with both formulas applied on the

Constitutional side. The applicable problems include most digestive complaints, urinary disorders, sinus inflammation, gynecological dysfunctions and skin diseases, but other issues are also treated with this protocol if the pulses show that these are the Meridians that are out of coordination. One standard treatment is to employ the Four Needle technique to tonify or disperse the two target Meridians (Root and Fǔ inflammation). In KCA, Kuon used repeated shallow punctures of the four Acupoints in each Formula (called touch treatment in CCA), and I have found this to be a very effective and rapidly completed treatment approach. One advantage of this method is that by varying the number of repetitions of either the tonified or dispersed points in the Auxiliary Formula, the treatment effect can be directed to either the upper or lower body. If there is no intention to direct the treatment up or down, the Root Formula is punctured four times in a row, followed by puncturing the Fǔ Inflammation formula two times in a row. Such treatments are described as 4/2 formulas. Repeating the dispersed Points in the Auxiliary Formula sends the effect to the upper body (Fǔ-a), while repeating the tonified Points sends the effect to the lower body (Fǔ-c)[232]. If the practitioner has difficulty identifying a RYCK imbalance, another way to determine which side to treat is by identifying the Subdoshas on both sides. Treatment for Fǔ Inflammation Conditions is best given on the opposite side of the body from the one showing the Subdosha pattern of the Constitution. In a similar fashion, the pulse quality (zào or jìng) corresponding to the Constitution will also be found on the side opposite to the treatment side. Both the Subdoshas and pulse quality of the Fǔ Inflammation Meridian will be present on the side with the RYCK imbalance. This rule appears to pertain to any Condition reflecting only one other Organ being out of balance with the Constitutional Organ. An easy guideline to remember in such cases is that treatment opposite to the symptomatic side usually shows pulses of the Conditional Subdoshas and the Conditional pulse quality (zào or jìng), accompanying a RYCK imbalance, all on the treatment side.

There is a certain logic behind treating the aforementioned conditions with the Fǔ inflammation Auxiliary Formula. The skin, sinuses and even the Fǔ Organs themselves are all comparatively superficial structures that have

232 This protocol and nomenclature were originated by Kuon Dowon.

contacts with the outside world. What we eat and excrete follow this relationship, and this is also true for the uterus and other gynecological and urogenital problems. The Zàng inflammation formula, on the other hand, targets the deeper structures such as the Zàng Organs themselves, but also inflammation of the bones (arthritis). These structures have no direct contact with the outside, except for the Lung. While the Lung challenges this simple classification scheme, I think it is possible to account for this discrepancy. The Lung functions have to do with the internal behavior of qì and Blood, including the gaseous exchange of oxygen and carbon dioxide.[233] The parts of the respiratory system that actually connect to the outside (bronchioles, bronchi, trachea and sinuses) all belong to the Fǔ category in CCA theory. These respiratory structures act as associated Fǔ for the Lung in an analogous way to what the Urinary Bladder does for the Kidney. I find that asthma most often is a Fǔ inflammation condition, resulting from bronchial constriction, and is not a problem of the Lung Organ itself.

CCA treatment of Zàng inflammation conditions is done in a similar manner as treating Fǔ Inflammation problems. The Root Formula is given on the side of the body that shows a RYCK imbalance, usually the dominant side for that Constitution, opposite the symptomatic side. The Zàng Inflammation Auxiliary Formula is also needled on the same side as the Root Formula. Again, the Subdoshas and pulse qualities (zào or jìng) usually follow the rule just given for Fǔ Inflammation problems. First the Points of the individual's Root Formula are needled, then the Points of the Zàng inflammation Formula are needled. If repeated shallow punctures are used, the Root points are punctured 5 times and then the Zàng inflammation Points are needled either 1 or 3 times. For cases presenting with Organ inflammation or fatigue, needling the Zàng inflammation Formula 1 time has been effective, while in cases of arthritis, needling the Zàng inflammation Formula 3 times is found to be more effective. The rules for directing the effect to the upper or lower body are the same: In the Auxiliary Formula, repeating the Points that are tonified directs the effect to the lower body, while repeating the Points to be dispersed directs the effect to the upper body. Most commonly, individuals

233 LU 9 is the Meeting Point of the Vessels (mài), indicating this important interface between qì and Blood. In TCM theory LU 9 "stimulates circulation, tonifies LU Qi, harmonizes the Qi and Blood in the chest, and stops bleeding," according to https://theory.yinyanghouse.com/acupuncturepoints/theory_huimeeting, accessed 4/19/19.

with Zàng Constitutions will need both the Root Formula and the Zàng inflammation Formula on the right side, while those with Fǔ Constitutions will need both treated on the left side.

Regardless of the Constitution, inflammatory conditions of all types are treated with Points on the Zàng Meridians for the Auxiliary Formula. I interpret this situation as indicating that inflammation is a reaction from the deeper Organ systems that control the behavior of qì and Blood in their attempt to regulate the internal milieu. By contrast, the antibiotic and vitalization Auxiliary formulas (to be described) employ Meridians of the Fǔ Organs. How might we understand this clinical finding? Infections are certainly a result of invasion of the body by organisms that come from outside it. Likewise, the conditions treated successfully with the Vitalization Formula appear to me to be those where the body cannot properly adapt to outside forces, especially gravity. The most typical problems treated by the Vitalization Formula include prolapses and ptosis of all kinds, or numbness and paresthesias. These problems typically become more frequent with aging, as the body's yáng qì naturally declines. Wèi 衛 qì is a form of yáng qì, so it is understandable that both infection and vitalization would resonate with the yáng Organs and their Meridians (i.e. the Fǔ), while yíng 營 qì, representing the yīn components of qì and Blood, resonates with the yīn Organs and Meridians (i.e. the Zàng). The Antibiotic Auxiliary Formula is typically used following the Root Formula in a ratio of 5/1 or 5/3, while the Vitalization Auxiliary Formula is typically used following the Root Formula in a ratio of 4/2. The two components of these treatments are given on the same side of the body and in the same manner as the inflammation treatments just described.

A dramatic example of the practical application of the forgoing remarks about these Auxiliary Formulas is found in the treatment of Chronic Fatigue Syndrome (CFS), mentioned earlier. It is also known as Myalgic Encephalomyelitis (ME). This illness is quite prevalent in the United States, but as yet is very poorly understood or effectively treated, even by other styles of acupuncture (as I have been informed by my patients, many of whom have previously been treated with acupuncture by other practitioners). Commonly, I find that the most effective treatment for these individuals is Root formula 5 times and Zàng inflammation formula 1 time, both on the dominant side[234]. However, I have recently observed even better results in some patients from using the Root formula 4 times and the Fŭ inflammation formula 2 times, also on the dominant side. Although there is evidence that many of these CFS patients have experienced viral infections, most do not experience significant lasting improvement from pharmaceutical anti-viral treatment or from application of the Antibiotic Auxiliary Formula. Researchers currently are studying an apparent abnormality in mitochondrial function in these patients. The mitochondria are the "energy" generators of the body's cells, and thus must have some relationship to the yíng qì, and the metabolism of qì and Blood. The Zàng Organs are responsible for the transformation of nutrients, supplied by the Fŭ Organs, into new qì and Blood, so either of these systems (Zàng or Fŭ) might be affected in a given patient. Naturally, disruption of these functions can lead to profound fatigue. While both the Zàng inflammation and Fŭ inflammation treatments have improved the health of many of my CFS patients, I do not believe they have provided a cure. It appears that there are additional abnormalities present in the pulses of these patients, and further study is underway to see if there are other typical imbalances common to this cohort. I currently suspect that a block at the level of the Extraordinary Meridian system may be a necessary target of treatment before a cure can be achieved, and this level of treatment has been addressed in the previous chapter. Also, the Neuro Auxiliary Formula has occasionally been helpful when the autonomic nervous system is malfunctioning, and autonomic imbalances are commonly present in severe cases of CFS, including Postural Orthostatic Tachycardia Syndrome (POTS) in severe cases. Repeating what has been said before, the decision of which Meridians to treat in any given clinical situation might be suggested by the theoretical analyses in this chapter, or by the symptomatic presentation, but ultimately must be confirmed by specific abnormalities detected in the Constitutional and Conditional pulses.

Probably the most common reason people seek acupuncture treatment is for the relief of pain. I did not start this chapter with this topic however, because pain treatment represents a more complex pathological and treatment issue

234 Repeating for clarity, in Zàng Constitutions, the right side is the dominant one, while for Fŭ Constitutions the left side is dominant. These designations follow Kuon's observation that acupuncture treatment is usually needed on these dominant sides, which are typically opposite to the symptomatic side, but not always.

than the topics I have just covered. In cases of inflammation, infection and devitalization, although there may be some degree of discomfort present, the Constitutional Organ is still able to play its leading role in supporting the other Organs, and acupuncture treatment as outlined above for these Conditions will usually be effective in treating the accompanying discomfort as well. We can see the results in both the pulse depth exam and the Subdosha exam returning to normal. In all of these situations, the depth of the pulse at the Constitutional Organ's location before treatment will have changed to the depth of the Element associated with the Conditional Organ, and in addition the Subdosha pulses on one side will be found to match those of the Constitution, while on the opposite side the Subdosha pulses will be found to match those of the Conditional Organ. This simple relationship breaks down in many cases where pain presents as the predominant problem. Kuon Dowon used a number of different protocols in the KCA treatment of pain, but always chose the left side for treatment. I have confirmed the accuracy of this approach, but will discuss its implications further along. It is my experience that by using the Conditional pulse exams, it is possible to determine the appropriate treatment protocol in each clinical case. What seems to be common when certain types of pain such as all radicular syndromes including sciatica become the focus of the body's imbalance, is that the Constitutional Organ is no longer able to fully support the other Organs, and turns to either its Mother or Daughter Organ for help. This pathological change is reflected in the pulse depth of the Constitutional Organ moving to the proper depth of the Element of its Mother or Daughter (the Degeneration Auxiliary Formula), and vice-versa for the depth of the Mother or Daughter Organ, which moves to the Constitutional Organ's proper depth. This Degeneration Organ now assumes responsibility for supporting all the other Organs. The pulse will show that either the Zàng inflammation Organ or the Vitalization Organ's pulse has also moved to the proper depth of the Degeneration Organ, confirming this physiological response. In addition, the Ministerial Fire pulse, of the same polarity as the Constitutional Organ, will be found to have moved from the Water depth to the Fire depth. As usual, the Subdoshas and RYCK pulse qualities will confirm this pathology of the Zàng inflammation Organ or the Vitalization Organ on one wrist, and the Degeneration Organ on the opposite wrist. Which wrist shows which Auxiliary Formula Subdosha, and also which order to use for treating the Auxiliary formulas depends on whether the individual has a Zàng or a

Fŭ Constitution. For people with Zàng Constitutions the order of treatment components is D/Z/N or D/V/N, and the left Subdosha and RYCK quality will correspond to the Degeneration Organ. The right Subdosha and RYCK quality will correspond to either the Zàng inflammation or the Vitalization Organ. This assignment is opposite for people with Fŭ Constitutions. The left side for these individuals will show the Subdosha and RYCK quality of either the Zàng inflammation or the Vitalization Organ, while the right side will show the Subdosha and RYCK quality of the Degeneration Organ. The treatment order for these individuals is Z/D/N or V/D/N. Regardless of the Constitution, this pain protocol is always administered on the left side, and should result in normalization of the pulse depths, Subdoshas, and RYCK size and quality abnormalities, in addition to at least some immediate pain relief. Treatment of pain thus involves three Formulas and their Meridians: Degeneration, Zàng inflammation or Vitalization, and Neuro. The Constitutional Meridian is not treated directly in this case. The types of pain I am referring to here seem to involve neural components coming from the spinal cord, suggesting some malfunction of the nervous system as well. There is always an interactive relationship between the central and autonomic nervous systems, so it would be fair to assume that both are involved in the genesis of pain. Two Auxiliary Formulas thus become involved in treating pain: the Degeneration Auxiliary Formula and the Neuro Auxiliary Formula.[235] Looking at Figs. 40 and 41, one can see that in Zàng Constitutions, the Degeneration Auxiliary Formula treats

235 These Auxiliary Formulas are given new names here. Some Korean practitioners who studied with Kuon coined the term Degeneration. Many cases of osteoarthritis are painful, and in those cases the three-part pain prescription may show itself as the present Condition needing treatment. Kuon also has referred to the Ministerial Fire Formulas as the "Psycho" Auxiliary Formulas. I prefer to call them Neuro Formulas because the implication of "Psycho" is somewhat pejorative. Other authors have commented that there appears to be a correlation between the Ministerial Fire functions and those of the autonomic nervous system. See: Woojin Choi, Seunggi Lee, Seunghun Cho, and Kyungmo Park.The Journal of Alternative and Complementary Medicine, Oct 2012, pp. 959-964. http://doi.org/10.1089/acm.2010.0751 The authors' conclusions were that the wood points PE 9 and TH 3 increase sympathetic activity; and the metal points TE 1 and PE 5 increase the parasympathetic activity. The problem with this study, as I see it, is that it didn't differentiate tonifying or sedating techniqies, so it was not really possible to tell which branch of the Autonomic Nervous System was being targeted at each Point that was treated. One finding of this physiological study was that skin conductance and peripheral skin temperature, which are indicative of sympathetic activity and blood flow respectively, were significantly augmented at PE 9, during acupuncture compared to pre-acupuncture, in the test subjects. Many types of pain do appear to reflect abnormalities in the nervous system, so "Neuro" is a preferable term for these Auxiliary Formulas. It is also commonly believed that pain inherently reflects an abnormality of the Fire Element (see *Sù Wèn* Chapter 74, which discusses disease triggers). CCA would support this idea, as it treats Ministerial Fire in most cases of pain syndromes. No other Elements in my experience have quite such a universal relationship to pathology, again demonstrating the atypical nature of Ministerial Fire compared to the other Elements. The Neuro Formula selected to treat pain is of the same yīnyáng polarity as the Constitutional Organ.

the Mother Meridian, while in Fǔ Constitutions the Degeneration Auxiliary Formula treats the Daughter Meridian. The Neuro Formulas are not shown in those figures at all, because they treat Meridians of Ministerial Fire, which are outside the Five Element cycles depicted in those figures. The original rules that I have reported for the Neuro Auxiliary Formulas are applicable as follows: In Zàng Constitutions the treatment Points are PE 7 and PE 3. For Taeumin and Soeumin types, first tonify PE 7 then disperse PE 3. For Taeyangin and Soyangin types first disperse PE 7 then tonify PE 3. In Fǔ Constitutions the treatment Points are TH 1 and TH 3. For Taeumin and Soeumin types first tonify TH 1 then disperse TH 3. For Taeyangin and Soyangin types first disperse TH 1 then tonify TH 3. Fig. 42 shows which Constitutions fit these four Sasang types.

- Taeyangin-1 (Fǔ): LI+, SI-, UB+ (All lack Vata)
- Taeyangin-2 (Zàng): LV-, SP+ (All lack Vata)
- Soyangin-1 (Fǔ): ST+, GB- (All have Pitta)
- Soyangin-2 (Zàng): KI-, HE+, LU- (All have Pitta)
- Taeumin-1 (Fǔ): LI-, SI+, UB- (All have Vata)
- Taeumin-2 (Zàng): LV+, SP- (All have Vata)
- Soumin-1 (Fǔ): ST-, GB+ (All lack Pitta)
- Soumin-2 (Zàng): KI+, HE-, LU+ (All lack Pitta)

Figure 42: Correlation of SCM, Organs and Ayurvedic Doshas

In summary, several different protocols may be indicated in treating pain, depending on the pulses. The actual treatment protocol for Zàng Constitutions is either Degeneration/Zàng inflammation/Neuro or Degeneration/Vitalization/Neuro, and for Fǔ Constitutions it is either Zàng inflammation/Degeneration/Neuro or Vitalization/Degeneration/Neuro. One difference between treating pain in individuals of Zàng and Fǔ Constitutions is thus the order in which the different components of the whole protocol are optimally needled. I use these ordering rules because they were developed by Kuon, and I keep his methods as part of CCA unless I find that they are not supported by my own clinical results. I suspect that further study of the Subdoshas and RYCK pulse qualities on each side will lead to a rational guide for ordering the Auxiliary components in these pain treatments, but for now I am relying on Kuon's experience as my guide. This approach (always treating the left side for pain) was observed at Kuon's clinic in the 1980s, and has shown itself to be important for getting optimal results.

One can only speculate as to why this tripartite treatment is effective for painful Conditions. I see pain syndromes as a pathological progression from syndromes due to Zàng inflammation or Devitalization alone. When the Constitutional Organ is unable to control the metabolism of qì and Blood within the limits of physical comfort, it calls on either its Mother or Daughter Organ for support, but once the Constitutional Organ has lost this degree of control over the other Organs, the autonomic and central nervous systems are also less properly controlled, and Ministerial Fire cannot function properly as a result. Such a description represents the mechanism at play, for example, in the progression from proliferative osteoarthritis to significantly painful osteoarthritis.[236] Similarly, in cases where Vitalization has been compromised, when the degree of discord between the body and its environment passes a certain threshold, both the Degeneration Organ and the Ministerial Fire Function are recruited to support the Constitutional Organ.

The type of pain can be a helpful guide for which treatment protocol is likely to be indicated by the pulses. Pain that is characterized by tingling, paresthesia or numbness is often an indication that the Vitalization Formula is needed as part of the protocol. Pain that is sharp and clearly defined, especially if it is a result of structural abnormalities such as intervertebral disc hernia, is often an indication that the Zàng Inflammation Formula is needed as part of the protocol. However, the Conditional pulse findings are the most reliable indicator of which protocol to choose. Following Kuon, I usually treat these painful Conditions with the 5/5/5 ratio of repetitions for individuals with Zàng Constitutions and 4/4/4 repetitions for those with Fŭ Constitutions, when using touch treatment technique. Kuon's approach of treating all three parts on the same side has been found to be the most effective approach, and has been confirmed by pre-testing. The treatment side, as usual, is the side with the RYCK imbalance, which as I just mentioned is always the left side.

236 For example, many of my patients have displayed relatively non-painful bone enlargement such as bunions and Heberden's nodes on their fingers. These are cases of simple osteoarthritis, treated by the K/Z protocol.

Why is the Degeneration Formula used, and the Root Formula omitted in the treatment of pain? Kuon discovered this approach, but he never explained his reasoning, leaving me to assume that his treatment protocols were based on trial and error. Because Kuon didn't use classical radial pulse diagnosis methods, he was not able to formulate the following hypothesis, which is my own attempt to provide a rational basis for the protocols useful in treating pain. In such cases, I find that, as stated, the pulse depth of the Constitutional Organ shifts to that characteristic of the Degeneration Organ. Additionally, the pulse depth of the Degeneration Organ shifts to that characteristic of the Constitutional Organ. Finally, the pulse depth of either the Vitalization Organ or the Zàng Inflammation Organ shifts to that characteristic of the Degeneration Organ. My interpretation of these findings is that in these cases, the Constitutional Organ is no longer able to maintain the proper relationships between the other Organs/Officials, and is now calling on either its Mother or Daughter Organ/Official for support, or rather to temporarily assume overall responsibility for inter-Organ coordination. Both Mother and Daughter Organs always have the same polarity (strong vs. weak) as the Constitutional Organ, so they are best suited to carry out this homeostatic role. Another question that occurs to me is why the Zàng Inflammation Formula is indicated in these cases of severe pain. Before offering an explanation, I need to say a few words about the Neuro Auxiliary Formula, as it is used in the composite approach to treating pain.

I have repeatedly mentioned that Ministerial Fire behaves in a manner that is outside the general rules applicable to the 10 formed Organs. It is my belief that the Neuro Auxiliary Formulas always treat Ministerial Fire, contrary to Kuon's original theory. As mentioned, the Neuro Formulas only contain 2 Points, rather than the 4 Points used in both the Root and the other Auxiliary Formulas. Although Kuon reported that the Neuro formula was effective for a number of Conditions when used with only the Root formula, I have rarely encountered such cases. When I have suspected possible Neuro only Conditions, it has been my experience that the RYCK pulses were equal, and the *Nánjīng* depth exam showed that the Constitutional Organ was correctly widest at its "home" depth. The only consistent pulse abnormality was that one of the Ministerial Fire pulses was not at its proper depth, but rather changed from the Water depth to the Fire depth. Because this is the only pulse abnormality I have occasionally observed when the Neuro Formula was apparently needed without other Auxiliary formulas, it could easily be missed when checking the pulse. However, the Neuro Formula is quite frequently needed in combination with other Auxiliary Formulas, as in the treatment of pain, cancer and occasionally chronic fatigue syndrome. When used without any other Auxiliary Formulas, it was Kuon's protocol to simply combine it with the Root Formula in a ratio of 4/2 (for touch treatment) as Root 4x/ Neuro 2x, all on the same Constitutional side. I am reserving judgment on whether there is a real need for this protocol. Kuon's original indications, including insomnia and neuro-psychiatric abnormalities, have not consistently shown the above pulse patterns, nor the expected response to such treatment, in my patients. It is possible that I might have missed the need for other Auxiliary formulas in such cases. To summarize my thoughts about the Neuro Formulas and the Ministerial Fire Meridians in general, I would propose that embryologically, Ministerial Fire is involved with the organism's primitive system for internal communication. Thus in some patients, I have found that after all the pulses of the formed Organs have returned to their normal depths, the Subdoshas reflect only the Constitution, and the RYCK pulses are all equal, yet there still might be an abnormal pulse depth on Ministerial Fire, albeit this is a rare finding. I believe this situation would indicate that the imbalances which were responsible for the presenting complaints have been corrected, but that underneath them there still remains a need to correct this part of the primitive signaling system emanating from Ministerial Fire. These signals could relate to purely neural dysfunctions, but also psycho-emotional issues that the patient still needs to address. Although following Kuon's lead, I have incorporated ideas from Western medicine here, I should point out that the passages cited earlier from the *Nánjīng*, referring to the role of the Ministerial Fire functions in the process of the unfolding of the Constitution, provide an Asian medicine basis for my hypotheses.

The question of why pain treatments should always be done on the left side has been a nagging concern of mine for a long time. Nature generally does not exhibit such unbalanced symmetry. I believe that I have rather recently discovered at least a partial answer to this mystery. I have noticed that some patients exhibit a pulse pattern that appears to be symmetrically opposite to the pain pattern of pulses. What I mean is that instead of the Constitutional

Organ's pulse shifting to the Mother Organ's natural depth, as in Zàng Constitution pain Conditions, it shifts to the normal depth of the Daughter Organ (Congestion Organ/Meridian), and the Daughter Organ's depth shifts to that normally found for the Constitutional Organ. Similarly, in individuals with Fŭ Constitutions, the Constitutional Organ and the Mother Organ (Congestion) sometimes exchange their pulse depths. This situation is a mirror image of what happens in painful Conditions. Once again, the Mother or Daughter Organ/Official is being recruited to temporarily take over responsibility for maintaining the proper relationship between the 12 Officials. What confirms this interpretation for me is the additional finding that in such cases, the Fŭ Inflammation Organ's pulse or the Antibiotic Organ's pulse also shifts to the depth that is natural for the affected Congestion Organ/Meridian. I have named this latter Organ the Congestion Organ, as Kuon never gave it a name, nor used it when I observed his practice. My reasons for choosing this terminology are as follows: From symmetry considerations, I think the designation should be opposite to the term Degeneration, but still imply a type of pathology. It is not easy to find a proper antonym, since degeneration of joints (as in degenerative joint disease) usually leads to abnormal deposition of boney tissue, so it may involve a quantitative abnormality in the body's structure. Consideration should also be given to the opposite natures of the Zàng and Fŭ Organs' functions with regard to these two clinical situations, implying that yīn yáng theory should be considered. The Fŭ Organs, being yáng, should be more involved with the functional communications in the body, while the Zàng Organs, being yīn, might be expected to be involved with more physical, structural abnormalities, as already indicated in degenerative joint disease. Obstruction of communication suggests the term congestion to me, and fits the often mentioned claim that the Zàng Organs tend towards Deficiency, while the Fŭ Organs tend towards excess (congestion or blockage). While I do not believe such a rule is universally accurate, it does seem to fit in the present case. Such "congested" patients have typically shared one thing in common: they have had chronic difficulties with the functioning of their Fŭ Organs, including digestive, excretory, respiratory, reproductive or other systems, with complaints including bloating, sinus congestion, dermatitis, infertility, etc. These are the reasons I have chosen the term Congestive Organ or Formula, but I am open to a better choice of terminology should anyone come up with a good suggestion. The final aspect of the diagnosis and treatment of these chronic congestive Conditions is the finding that that one of the Ministerial Fire pulses is also displaced, analogous to what happens in pain Conditions. In Congestion the Ministerial Fire Official involved is of the same yīn yáng polarity as the one involved in treating pain. That is, the yīn yáng polarity of the Ministerial Fire Official involved in these cases is the same as the polarity of the Constitutional Organ. The depth of this Ministerial Fire Organ's pulse shifts from the normal Water Element depth to the Fire Element depth. There has long been an argument in Chinese Medicine about whether Ministerial Fire functions relate more to the Fire or the Water Element. This pulse finding supports the contention that both Elements are mutually involved. Also, pain resonates with the Fire Element, while congestion resonates with the Water Element, so the pathological presentations also support this conclusion. Treatment of these congestive disorders is given on the patient's right side, opposite to the side used for treating pain, thus restoring natural law's expression of symmetry. The complete treatment protocol for these Congestion Conditions in patients with a Zàng Constitution is as follows: the Congestion Auxiliary Formula, then the Fǔ Inflammation or Antibiotic Auxiliary Formula, then the Neuro Auxiliary Formula (C/F/N or C/B/N). The Subdoshas and RYCK quality on the right side reflect the Congestion Organ, while on the left side they reflect either the Fǔ Inflammation Organ or the Antibiotic Organ. For patients with a Fǔ Constitution, the order is (F/C/N or B/C/N). These components are all treated on the right side, which will show a RYCK imbalance. The Subdoshas and RYCK quality on the right side will reflect either the Fǔ Inflammation Organ or the Antibiotic Organ. The touch treatment number of repetitions is the same as in the pain protocols. This congestion pattern appears to be present in many chronic problems of the Fǔ Organs. My awareness of the importance of symmetry in natural law was a crucial factor in discovering this method of treating congestive disorders. Once again, the Subdoshas and RYCK pulse qualities will reflect whether a hand or a foot branch of a Great Meridian is the place to start treatment regardless of the Constitution. I had previously wondered why the order of components in pain treatments (and now in Congestion treatments) followed different orders in those individuals with different Constitutions. The explanation appears to be that the organism tells us through the pulses which Organ system or its Meridian needs to be addressed first by showing the Subdosha and RYCK quality on the left side for pain syndromes and on the right side for Congestion syndromes. This discovery of Congestion syndromes fills a rather large gap in Kuon's attempt to create a comprehensive system of acupuncture (KCA), and I hope it will bring CCA much closer to success in that regard.

All the recommendations regarding laterality of treatment discussed in this chapter reflect both my own and Kuon's experiences, however they should not be considered invariable. I have found rare occasions where the treatment side needs to be opposite to the standardly chosen side. The only guide that seems to always be accurate regarding laterality is to treat the side with a RYCK imbalance. All the Conditional pulse findings should be found to conform to the unusual choice of laterality indicated by the RYCK imbalance.

After any of the treatments described in this chapter there is often at least some immediate improvement in the presenting symptoms and well-being of the patient, if the diagnosis was correct. More importantly there should be observable signs of better neuro-muscular coordination (O-Ring test and range of motion) and return of the pulse depths and Subdoshas to their Constitutional norms. In my experience, the ideal treatment ends with the RYCK balance showing a ratio of 1/1, at least temporarily. The *Nèijīng* states that treatment should continue until RY equals CK, and I agree with that goal based on experience, but I think there is an additional criterion that will help avoid future relapses. I will explain it, in the chapter on cancer, as something I call "prodromal" signals in the pulse, even when RY equals CK. Continuing treatment until even these prodromal signals disappear seems to bring more lasting results in all types of pathology.

The preceding protocols are ones that are mostly similar to those used in KCA by Kuon. I have developed protocols for other Conditions such as cancer, inspired by Kuon's treatments that I observed in his clinic. I would also reiterate that many of the Conditions just described can be equally well treated either by Worsley style qì transfers, or by using the Four Needle techniques developed by Kuon. In fact, sometimes the qì transfer treatments are preferable, as they may involve significantly fewer needles. For some patients this latter consideration is a major issue.

All the foregoing Conditions mentioned in this chapter result from Organ malfunctions due to the inability of the Constitutional Organ to maintain the proper harmonious relationship of the Zàng Fǔ constellation. This statement expresses the essence of a Constitutional approach to acupuncture. There are many other styles of acupuncture that are tremendously effective in treating all kinds of symptomatology without adopting such a Constitutional approach. I have long wondered how to reconcile this seemingly contradictory situation. While I cannot present a comprehensive answer, I have most recently found that the pulses point to a promising avenue of inquiry. After there is no longer a RYCK ratio imbalance and no longer any depth abnormality in the pulses, there can still be important symptoms remaining. In most such cases I have found that the RYCK pulse quality and the Subdoshas on left and right sides reflect opposite branches of the Six Great Meridians (Liù Jīng). Earlier I discussed the Liù Jīng in relationship to pathogenic factors, but the Meridian system has inherent physiological functions that are simply expressions of biological natural law, unrelated to pathogens. Each Great Meridian has two branches: a hand branch associated with the zào pulse quality and either the Fire or Metal Element, coupled with a foot branch associated with the jìng pulse quality and either the Water, Wood or Earth Element. These two branches of the same Great Meridian must also function harmoniously to maintain good health, and when the Organs are not showing signs of imbalance on the pulses, these two parts of the Great Meridians should be evaluated. Here is an example of what I have observed: Let us suppose that someone has a Small Intestine Excess Constitution, but no longer has any RYCK ratio imbalance or abnormal pulse depth; however they still get pain from longstanding arthritic joints. This is actually what I found on patient number 7 in the case history chapter to follow. On the right side her pulse quality was jìng with a Kapha (Earth type) Subdosha. Since the Earth Element tends towards excess in her Constitution (child of Fire), these pulses correspond to Spleen excess (not Stomach deficiency). On the left side her pulse quality was zào and the Subdosha was Pitta. Since the Metal Element (grandchild of Fire) tends towards deficiency in her Constitution, these pulses correspond to Lung Deficiency (not Large Intestine excess). Spleen and Lung are connected to the foot and hand branches of Taiyin. On that visit she had mild discomfort in the left wrist and both knees. Pressing along these Meridians showed tenderness at LU 9 on the left and SP 9 on the right sides. I inserted needles in these two points and left them for 10 minutes. After withdrawal, all the pain was gone and now the pulse quality was zào bilaterally with Fire Subdoshas on both sides, reflecting only her Constitution. Clearly this was a case of Taiyin Meridian imbalance, even though her Fire Element's Small Intestine Constitution was functioning normally. I have had a number of similar cases of imbalanced Great Meridians, and this 2 needle treatment has produced dramatic symptom improvement in each case. However, it is important to note that such treatment was given after a period of CCA treatment that resulted in returning functional control to her Constitutional Organ. The general rule seems to be that the two branches of an imbalanced Great Meridian should be treated on opposite sides, each one corresponding to the pulse quality and Subdoshas on its own side. The specific Points to needle are best chosen by sensitivity to pressure, and may or may not be near areas of physical pathology. Distal Points seem to be typically reactive in this manner. Earlier I asked if there was any way of distinguishing problems of the Organs from problems of the Meridians. It is possible that this latter type of pulse pattern is specific to the Meridians, while the previous pulse patterns described in this chapter are specific to the Organs. This hypothesis also addresses the question of how to reconcile the lens of 5 (Elements and Organs) with the lens of 6 (Meridians). I should add that this pattern of treatment was a major component of Richard Tan's balance methodology, which in turn was derived from the Tung Family style of acupuncture. What I have added, as a component of CCA, is the specific pulse findings that inform the practitioner about which Great Meridian to treat, and which side to needle on either its arm or leg branch. As usual, this diagnostic assessment starts with identifying the individual's Constitution.

CHAPTER 14

Advanced Constitutional Conditional Acupuncture: Cancer Diagnosis and Treatment

Fig. 43: Cancer Research Apparatus [237]

As Fig. 43 nicely illustrates, on every level cancer has been viewed as a complex disease requiring the most advanced experimental treatments, but still carries a high degree of mortality and morbidity despite the best of conventional

237 Source: commons.wikimedia.org. From the Burrows Wellcome Exhibition in Turin (1911). An early experiment in synthesizing a chemotherapeutic drug from the fungus Ergot. Donated by the Wellcome Trust. Image L0012221.

medical treatment. There have been many reports on the use of acupuncture in the treatment of cancer, but for the most part these have been focused on alleviating the side effects of Western medical treatment, or else on alleviating the primary symptoms cancer patients report, such as pain, nausea and fatigue. Almost no research has focused on using acupuncture to address the pathomechanism of the neoplastic process itself. A recent issue of the North American Journal of Oriental Medicine was devoted to the topic of treating cancer, and included 11 articles on this subject.[238] Sadly, none of the articles indicated that acupuncture was effective in treating the mechanism of cancer; rather the focus was exclusively on supportive treatment while Western medical treatments were being used. Indeed, the very first article stated, "I have never administered treatment with the goal of eliminating or reducing cancer." The next article added that, "studying in an acupuncture college in Japan, we were taught that cancer patients were contraindicated for acupuncture and moxibustion." One more quote, "I've learned that the best approach is to be clear with the patient that I cannot treat their cancer. What I can treat are some of the symptoms associated with the disease, or symptoms that arise from conventional cancer treatments." There are at least four more statements in the other articles reinforcing this kind of thinking.

Is this attitude towards cancer treatment in Oriental medicine particularly a Japanese belief? Actually the answer is no. In a 600 plus page book on the TCM approach to treating cancer, translated and published in English in 2003, very similar statements can be found in the Foreword, Preface and text itself.[239] From Giovanni Maciocia's Foreword, "It is generally recognized in modern China that the techniques of Western biomedicine (essentially surgery, radiotherapy or chemotherapy) are more effective than Chinese medicine in eliminating cancer and restricting the growth and metastasis of tumors. However, Western treatments of cancer are frequently toxic and have adverse side effects (particularly when chemotherapy and radiotherapy are used)...This is where Chinese medicine comes into its own: when used in conjunction with Western medicine, it has an essential role to play in restoring the body's balance, strengthening the immune system, reducing the side effects of these more drastic treatment methods...." The author's Preface

238 North American Journal of Oriental Medicine, Vol.24, No. 73 2018.

239 Li Peiwen, *Management of Cancer with Chinese Medicine*, Donica Publishing, St. Albans, U.K. 2003

sums up his point of view thusly, "Traditional Chinese medicine and Chinese materia medica have produced extremely promising results in in dealing with the adverse side effects and complications occurring during radiotherapy and chemotherapy. The rich treasury of knowledge and experience bequeathed to us has spurred me on to compile this book to present current approaches involving the use of Chinese medicine in the treatment of cancer complications and commonly seen side-effects occurring as a result of radiotherapy and chemotherapy." Finally, the text itself includes the following statement: "Supporting Vital Qi (Zheng Qi) and cultivating the Root is the basic principle for treating tumors in TCM and has proven to be very effective. However, practitioners should not work under the illusion that Chinese materia medica are likely to be as effective as surgery, radiotherapy or chemotherapy in curing cancer."[240] Although the text covers herbal medicine and acupuncture, nowhere is there a consistent mechanism presented that explains why the standard etiological factors in Chinese medical theory (exogenous, endogenous and miscellaneous) lead to cancer, as opposed to other illnesses produced by these same pathogenic factors. I might add that there is also no discussion of pulse findings that are unique to cancer patients. Whatever it is that is uniquely common to all forms of cancer is left as an unaddressed mystery. What I found most interesting in this book, on the other hand, was its historical review of writings about tumors in the classical Chinese literature.

The author Li, reports that the term liú 瘤 (in modern Chinese meaning tumor) can be found on Shang dynasty (16th-11th century BCE) oracle bones.[241] Ancient dictionaries including the *Shuo Wen Jie Zi* explained this character as a synonym for its homophone liú 流, meaning "flow". Why these terms might have been considered as synonyms is not readily apparent, but several possibilities come to mind. If blood or other substances overflow into any place in the body, a mass will begin to develop. Similarly, if there is an obstruction to flow, commonly described as stagnation or stasis, then once again this blockage can lead to a mass or accumulation. Hua Tuo (d. 208)

240 ibid. p. 47.

241 The modern Chinese term for cancer, 癌 ái, first appeared in 1171 in *Wei Ji Bao Shu (A Treasury of Relief and Treatment)* by Dongxuan Jushi. The character signifies the unevenness of the surface and consistency of cancers, which are hard as a rock. Another term 岩 yán, meaning "rock" was also used for cancers during the Song and Yuan dynasties according to Li, ibid. page 7.

commented that dysfunction of the zàng fǔ plays a role in the formation of tumors, leading to an accumulation of toxins, which inhibits the movement of Qi and Blood. He added that "when accumulations in the Interior lead to illness and cannot be reached by needles or herbs, they must be cut out,"[242] a surprisingly modern statement, recognizing that Hua Tuo was reputedly an adept surgeon. It appears that Hua Tuo understood cancers as systemic illnesses (dysfunctions of the zàng fǔ) in addition to having a localized manifestation. My own experience based on pulse diagnosis corroborates that line of thought. Further on I will describe what I feel in the pulses of all of the cancer patients I have examined, and the pattern I find is also one that is characteristic of a state of stagnation. Li cites a Song dynasty reference that supports the stagnation hypothesis: "the character liú 瘤 signifies stagnation and lack of movement. The normal flow of Qi and Blood is impaired and the body loses its harmonious equilibrium. When there is stagnation, binding, congestion and obstruction, pathogenic factors take advantage of the Deficient condition and illness occurs."[243] Analysis of the character liú 瘤 shows the radical meaning illness on the left and the phonetic liú 留, meaning kept in place, to the right and underneath, providing insight into the choice of this character for such tumors.

My Korean mentor, Kuon Dowon, had a diametrically opposite approach to cancer treatment than the Japanese and Chinese references cited above. In fact, he refused to treat patients who opted for any kind of Western medical treatment, and yet I saw clear evidence of improvement in several cancer patients during the short time I was able to observe his clinical practice in the 1980s. Kuon eventually established the Dawnting Cancer Research Institute in Seoul, and collaborated with biomedical scientists in research that showed his treatments led to normalization of biomarkers present in various types of cancer that previously had not normalized even when Western medical treatment had successfully induced remissions in these cancer patients.[244] Unfortunately, Kuon never published his method

---

242 *Zhong Zang Jing* (*Central Viscera Classic*) quotation of Hua Tuo cited by Li, page 5.

243 *Sheng Ji Zong Lu (General Collection for Holy Relief)*, cited in Li, page 7.

244 Paik, M., La S., Lee Y., Kim J., Choi, Y., Kuon D., Kim I., Kim K. "Therapeutic monitoring on urinary nucleosides and polyamine levels of cancer patients by CE and GC under acupuncture treatment." Proceedings of the Convention of the Pharmaceutical Society of Korea, Vol. 1 (2003).

for treating cancer, but my colleague Stuart Kutchins and I were able to develop a tentative understanding of Kuon's approach from our observations at his clinic. Over the years I had only a few cancer patients with whom to try out my interpretation of Kuon's treatment protocols, with encouraging, but inconclusive results. Then, in 2017 I began to attract a number of clients with cancer and noted a curious finding in their pulses: In every case, at least 4 out of the 6 Fŭ Organ pulses were widest at the Fire Element depth, instead of at their healthy depth, and frequently 5 or even all 6 Fŭ Organ pulses were widest at the Fire Element depth. Based on this discovery, I was able to adapt Kuon's treatment protocols into a standardized approach, similar to those for pain, inflammation, infection and devitalization, but one that is considerably more complex, as befits such a serious illness[245]. One of the first patients I tried this new protocol on was a 76 year old man whom I had treated eight years previously for sciatica and two years previously for capsulitis of the shoulder, with complete resolution in both cases. During the past six months he had been feeling unwell, with chest congestion, loss of appetite, weight loss of 10-15 pounds, fatigue and a feeling that he wasn't his normal self anymore. Noting the abnormal pattern of Fire level pulses, together with his history, I shared with him my suspicion that he had some kind of cancer, and that he needed to get it evaluated by his regular physician. He related that he had already come to the same conclusion, but that he hoped I could treat him with acupuncture until a definitive diagnosis was confirmed. His case history (#25) will be found in chapter 16, but what is important to relate here is that after only two treatments he reported that he was "96 percent" back to normal in all ways except his weight and the persistent chest congestion. Most importantly, he felt like himself again. A subsequent bone marrow biopsy confirmed that he actually had a large B cell lymphoma as his primary diagnosis, but as the case history shows, he has continued to feel well after the initial series of weekly acupuncture treatments followed by about 2 years of monthly maintenance acupuncture, which was interrupted by the

245 One additional finding in all the cancer cases I've encountered, is that the RYCK pulse quality corresponding to the Constitution is on the same side as the RYCK imbalanced width ratio, while the RYCK quality on the rpposite side is always zào. I believe this zào quality represents the Fire Element, which plays such a crucial role in cancer diagnosis. The Fire Meridians are all Hand branches of the Six Great Meridians, and naturally produce the zào quality in the RYCK pulses, but this manifests contralaterally to the needed treatment side. The treatment side itself appears to reflect the expected RYCK quality associated with the individual's Constitutional Meridian.

Covid-19 pandemic. It is now more than 3.5 years after his first treatment, and he continues to be cancer free without further treatment. I reported my methods of using the pulse for diagnosis and acupuncture for treating cancer, including this case, in the Journal of Chinese Medicine, so this chapter will cover much of the same material, but expand upon it, while there will be additional case histories in chapter 16.[246] So far the results of this approach have been astonishingly successful.

The starting point for treating cancer is in making a correct Constitutional diagnosis, as usual. Two patients with identical types of cancer will need different treatments if they have different Constitutions. Next, the presence of 4 to 6 Fǔ Organ pulses at the Fire depth must be confirmed.[247] Even with this minimal amount of information, I have found that cancers of many different kinds (breast, colon, prostate, melanoma, lung, tongue, neuro-endocrine, bladder, meningioma and lymphoma so far) can be treated with fairly rapid clinical improvement. I should point out that the abnormal Fire level pulses are not found in stage 0-1 tumors (carcinoma in situ), so I interpret this pulse pattern as a marker of metastatic progression, as it has appeared in all my cancer patients with stages two through four, but not in those with stage zero to one (2 cases). With consistent treatment however, the Fire level pulse pattern will disappear, as long as there are no aggravating factors. Unfortunately, most chemotherapy, radiation therapy, and possibly surgery, do at times seem to act as aggravating factors, so in treating patients who are simultaneously getting conventional cancer treatment there is usually a back and forth pulse picture. The metastatic signal I have described usually goes away right after the acupuncture treatment, and

246 Eckman, P. "The Diagnosis and Treatment of Cancer with Constitutional Conditional Acupuncture" J. Chin. Med. (U.K.) Issue 125, 2021, pp. 3-16.

247 I mentioned earlier that cancer pulses implied stagnation as a mechanism. I am indebted to Igor Siminov for suggesting in a personal communication that when several pulses under different examining fingers are felt at the same depth or pressure level, stagnation is indicated. I am also indebted to Lu Yubin who posted the following information on Facebook 5/26/20: "Xia Ku Cao literally means the grass that will become withered when summer arrives. In Chinese medicine, it functions to purge fire (liver fire) to improve vision, detoxicate and dissipate nodules… the spike will dry up and become shrinking quickly when summer arrives. This suggests that the fire stagnates in a local part. This phenomena includes three aspects, the fire, the yin deficiency and the nodule-like results. This is why it can detoxicate and dissipate nodules in the human body. Although this herb can be used to treat carbuncles, etc., it is particularly applied to treatment of … cancer." Although I don't posit a total commonality of theoretical bases for herbology and acupuncture, Lu's comments do bring together my ideas about the importance of stagnation and Fire in the mechanism of cancer.

will usually stay away if the patient follows an appropriate Constitutional diet, but at some time interval after Western medical treatment the metastatic pulse might reappear and must be reversed by repetition of the same acupuncture protocol. I never advise my patients to refuse or terminate Western medical treatment, but simply share with them my experience in treating other cancer patients. I always encourage patients to make their own decisions about the best course of therapy, and so far virtually all of my patients have opted to combine both Western medical treatment with acupuncture treatment, although they often choose the shortest duration of Western therapy agreeable to their oncologists. Since my impression is that the pulse marker I've discovered correlates with metastatic activity, I don't know if existing tumors would always shrink from the acupuncture protocol I've developed, but tumor shrinkage or disappearance has occurred in a number of patients. With that in mind, it may be a good idea to use Western medical treatments to destroy those existing tumors, while adding acupuncture to inhibit their growth or the development of new lesions, however I feel it is important to recognize the impressive results Kuon published regarding the results of treating cancer with acupuncture and diet alone. Kuon only treated patients who did not receive Western medical therapy, and his patients did show shrinkage of pre-existing tumors, plus general health improvement. The optimal approach is still a matter for further study.

I would also like to add that following any apparent "remission" or even "cure" of a malignancy while undergoing acupuncture, it seems to be important to give a booster preventative treatment approximately once a month as follow-up. The required duration of these booster treatments is unknown at this point. I make this suggestion based on my experiences with cancer patients who abstained from acupuncture during the Covid-19 pandemic. All of my cancer patients did well for the first 2 months of isolation when my clinic was closed, but then there were a few relapses. One patient resumed acupuncture upon relapse and is now apparently cancer free. Two other patients had relapses 4 or more months into isolation having initially completed less than a year of acupuncture treatment, and they have chosen to avoid possible Covid exposure from resuming acupuncture. One of these patients has not yet been able to clear their malignancy, even with resuming allopathic treatment, while the other achieved remission at least. I'm hopeful that with the Covid vaccine becoming available, they might return for acupuncture treatment. All the rest of my cancer patients have stayed cancer free once the initial series of treatments abolished the "metastatic signal" in their pulses, and they have elected to receive booster treatments at least once a month.

Before presenting the treatment protocols, I would like to address the question of how one might evaluate progress in the course of the acupuncture treatment of cancer patients. The most important finding is the disappearance of the "metastatic signal" previously described. However, this is not an "all or nothing" change. It is not unusual after successful treatment, for the abnormal Fǔ Organ pulses to return to their normal Elemental depths, but still there might remain a slight widening of these pulses at the Fire Element depth in addition. By this I mean that these pulses are slightly wider at the Fire Element depth than they are at the Metal and Earth Element depths. I call this slight widening at the Fire Element depth a "prodromal signal". My suspicion is that in such cases, the cancer has changed from a state in which it is growing and spreading (metastasizing) to one in which it has been inhibited from doing so. I consider this new state to essentially be one of remission, even if there is still cancerous tissue present in the body. If it is not growing and spreading, at least for the time being it is not doing further damage. During this "remission" state, it is important to continue acupuncture treatment on a regular basis (I suggest once a week treatment). Another parameter to follow is the RYCK imbalance. If this imbalance stays on the side that has been treated, then the illness is definitely not yet cured. With ongoing treatment I have found that at some point the RYCK imbalance will either switch back to the normal side for that person's Constitution, or else become equal, an even better finding. Finally, there is the matter of the Subdosha pulses and RYCK pulse qualities in cancer patients. Typically the Subdoshas and pulse quality on the RYCK imbalanced (treatment) side will reflect the individual's Constitutional Organ, while the Subdoshas and pulse quality on the opposite side will reflect the Fire Element, i.e. a zào quality. At some point, the Subdoshas and pulse quality will change, and reflect those of the Constitution on both sides. When all of these indicators have returned to normal (no metastatic signal, no prodromal signal, RYCK not imbalanced on the treated side, and the Constitutional Subdosha appearing bilaterally), then I think there is a strong possibility of cure having been achieved. However, I definitely want to add that it is possible that even with all these positive findings, it is possible that there could be some residual cancer cells remaining. In a few patients I have found that although these criteria have been met, there have been lumps persisting at the initial site of the cancer. These were then removed surgically and showed a small cluster of tumor cells surrounded by a much larger area of dead tissue and scar tissue. I think that this situation reflects the poor circulation created in the area of the dying cancer tissue, and it is thus hard for the body to get its immune mechanisms to reach this "war zone". I don't know if these residual cells are still capable of initiating a recurrence of cancer or not, so it seems appropriate to surgically remove any remaining lesions after all the pulse markers have returned to normal. What I can say is that in several of my patients treated in this way, their oncologists have declared that they are cancer-free or cured. I do still recommend that they receive a booster treatment once a month, as cancer is a mortal affliction, although I have no data to suggest how long to continue this preventive approach. Before finishing this paragraph, I would like to suggest that the "prodromal" pulse findings may not be unique to the recovery process in cancer patients, but might also be a way the body informs us that any kind of pathology has been suppressed, but not yet cured. If so, it could offer a valuable guide for determining how long to continue any form of treatment even after the presenting symptoms are gone.

The following protocols are the ones I have developed for treating all kinds of cancer. As stated, they were inspired by what was observed in Kuon's clinic, but they're not identical to his protocols. They target, in fact, the Meridians/Organs whose pulses have shifted pathologically to the Fire Element depth. Kuon used the repeated shallow "touch treatment" needling on all of his patients, whereas I have experimented with both touch treatment and deeper stimulation of the same points without needle retention, and without the repetitions characteristic of touch treatment. I have found both techniques to be equally effective. Most of the components of treatment have already been presented in discussing other types of Conditions, but cancer additionally needs treatment of what Stuart

Kutchins and I named the Tumor Auxiliary Formula as shown in Figs. 40 and 41. The protocol I use for patients with a Fŭ Organ Constitution is (Antibiotic) (Tumor) (Neuro) (Antibiotic) (Tumor) (Immune/Root) (Vitalization). For patients with a Zàng Organ Constitution the protocol I currently use is (Vitalization) (Immune) (Tumor) (Antibiotic) (Neuro) (Tumor) (Antibiotic).[248] I previously used a slightly different protocol in patients with Zàng Organ Constitutions, which was helpful, but not as dramatically as the one treating people with Fŭ Organ Constitutions, so I researched a more effective protocol. It turns out that the best Zàng Organ Constitution protocol uses exactly the same Auxiliary Formulas as the Fŭ Organ Constitution protocol, but the order of components follows an opposite sequence. These treatments can be performed with either shallow touch treatment delivered 4 times for each component or with single penetration at each Point to get dé qì and then removed without needle retention. Also bear in mind that a Five Element diagram such as is illustrated in Fig. 15 must be worked out for each patient's Constitution in order to properly select the treatment Points of the Four Point Auxiliary Formulas. In most cases these Points differ somewhat from the classical Four Needle technique Points, which only apply strictly to the Root Formula.

Each of these components has a rationale for why they might be useful in treating cancer. The Tumor Auxiliary Formula is obvious. While no clearly established pathomechanism in cancer has been proven, suggestions have been put forward that problems exist with immune surveillance, and as a result some viral infections are associated with certain cancers. Thus the Immune and Antibiotic Formulas are also understandable. Immunity is a responsibility of the yáng qì, so the Vitalization Formula, which targets this as previously discussed, also makes sense. Finally we have the Neuro Formula, which I believe is closely connected with adjusting the functioning of the Autonomic Nervous System (ANS). This primitive integrating mechanism controls all the involuntary functions of the organism, and needs to be working properly to combat the development of cancer. The only yáng Auxiliary Formula that is not part of the cancer protocol is the Paralysis Formula. As far as I know there is no particular tendency for cancer to be associated with

248 The immune Auxiliary Formula targets the Meridian of the Fŭ Organ of the Constitutional Element in people with a Zàng Constitution.

paralytic symptoms. For that reason I believe the Fǔ Organ of the Paralysis Formula does not shift its depth to the Fire Element level, as is verified by pulse diagnosis.

Individuals with a Zàng Constitution are treated on the left side with the following sequence of Auxiliary Formulas (Fig. 40): GF/BR/SO/GS/Neuro/ SO/GS. Individuals with a Fǔ Constitution are treated on the right side with the following sequence of Auxiliary Formulas (Fig. 41): GF/FA/Neuro/GF/FA/SE/GS. These laterality choices are consistent with the side showing a RYCK imbalance. As one can see, these protocols are similar, but they rotate the sequence of Auxiliary Formulas in mirror image directions from each other due to their yīn and yáng natures and the directions of the Zàng and Fǔ Organs' constellation of relationships. Interestingly, this sequence of Auxiliary Formulas, when translated into Five Element corrolates, always follows a Creative Cycle order. The Creative Cycle has often been viewed as physiological and life enhancing (Natural Law), while the opposite sequence can be thought of as pathological and life opposing (Anti-Natural Law). All of the target Meridians of these Auxiliary Formulas are of the Fǔ Organs, which perfectly matches the locations of the pulses with abnormal Fire level depths[249], but opens a question about the pathophysiology involved in cancerous metastatic behavior. I do not have any firm conclusions about this issue, but I suspect that the trigger for cancer progression might very well be something, or rather somethings, in the environment to which the organism cannot adapt (a yáng responsibility). To reiterate, these could include viruses (Antibiotic Formula and Immune Formula), failure of the yáng qì to carry out its role (Vitalization Formula) and deregulation of the Autonomic Nervous System (Neuro Formula). Another finding to keep in mind is that cancers frequently run in family lineages, and thus appear to reflect some genetic component. My best guess is that these genetic predispositions are normally not expressed without there also being a trigger that switches on these pathological genes. This is a topic that is being studied as part of the theory of epigenetics, for as we know, an individual's health is not purely a question of their genetic

249 There is one exception: As usual, Ministerial Fire does not follow this rule, but simply uses the Neuro Formula for the Constitution, even if that involves PE instead of TH. This is just one more example of how Ministerial Fire seems to function as part of an independent mechanism from the other Elements.

inheritance. I believe that epidemiological studies have suggested that cancers have become more prevalent in modern society than they were in the past, and there is no doubt that modern society has experienced a much greater environmental change than was true in earlier times. I might even say that we have created a crisis with our impact on the environment, and it is more urgent than ever that we study ways to mitigate this change.

CHAPTER 15

Stages in the Process of Acupuncture Treatment in CCA

Figure 44 [250] *"Is all of that in my pulse?"*

Although I have never seen it stated as an axiom of acupuncture theory, it is my impression that all living organisms strive to maintain their health in order to be able to accomplish whatever their destiny may be. This is not merely an abstract philosophical principle, but has very practical implications for thinking about treatment strategies in acupuncture practice. The human organism, whether consciously or not, must be able to recognize whatever poses the most serious threat to its ability to stay on its destined path (inscribed at conception in its Constitution, and stored in Mìng Mén

250 Source: wikimedia commons. Image after Moliere's play, Le Malade Imaginaire, This 17th century work is in the public domain. The original is in the Bibliotheque nationale de France.

as its original conglomerate of primordial jīngshén). Furthermore, if there is more than one threat to the organism, and if the most serious one is not dealt with first, other treatment targets may not benefit the goal of achieving one's destiny. Thus, I think that at any moment, each person experiences the biggest threat as the target of its innate homeostatic response mechanisms. And, it is precisely this response mechanism that we are able to perceive in the pulse patterns that reflect the struggle going on, whether between organism and environment, organism and pathogen, or simply malfunctions of the organism within itself. If the pulse patterns are truly a reflection of this homeostatic activity, then they should be a very useful guide to choosing the acupuncture treatment at each visit, as that should be whatever will strengthen the organism's momentary focus of its resources (Fig. 44).

When we check the pulses in CCA, there are many possible findings. I believe a correct Constitutional diagnosis should be the first goal, if simply because it clarifies the meaning of almost all the other pulse findings. However, the Constitutional Organ/Meridian itself may not be involved in the initial treatment selection! Why do I say that? Because there may be "blocks" to the patient's homeostatic functions which simply stop them from working properly. One whole chapter of this treatise was devoted to imbalances of the Extraordinay Meridians, which, if present, may limit the degree to which other treatments might be fully effective. For that reason, I think it is a good idea to re-establish the "primitive Constitutional grid" by treating the Extraordinary Meridian system first if it clearly shows imbalanced pulse signals. Such a situation should be suspected whenever the pulses at Cùnkǒu show a different deviation pattern than those further proximally at positions 4, 5 and 6. If Extraordinary Meridians are malfunctioning, then there will also be RYCK width imbalances bilaterally, with opposite pulse qualities and Subdoshas associated with the Fire and Water Elements on the two sides. An example of other types of blocks to consider is the state of "possession," that Worsley taught should be diagnosed and treated before Five Element approaches could be effective. Worsley was certainly not creating a new category of pathology by his inclusion of possession. It was a recognized form of pathology throughout the history of Chinese medicine, from its inception until modern times. I have published (in *The Compleat Acupuncturist*) the methods I have discovered for diagnosing possession via the pulses[251], my interpretation of what possession is really about in Chinese pathophysiological terms, and the treatments Worsley transmitted from his Taiwanese teacher Hsiu Yang-chai, for eliminating this type of block. Specifically, if all the yáng Meridian pulses are at either the Earth Element depth or the Water Element depth, then "possession" is preventing the zàng fǔ from self-regulating homeostatically. Although these "possession" blocks are not very frequent in my experience, they are not rare either. On a single day in April 2021 I found one case of internal and one case of external "devils" in my clinical practice. Both presented with persistent anxiety, one severe and incapacitating, the other merely preventing a normal quality of life. Both responded well to the treatments Worsley transmitted for such conditions. The pulses tell us we need to treat this situation first (which is not to say that there are no other ways of diagnosing possession). The Extraordinary Meridian imbalances and both types of possession implicate problems in the functioning of an individual's yuán qì and jīngshén, so that this person cannot fully recognize their Original Nature, and thus these blocks must be removed in order for treatments based on their Constitution to be fully effective.

The prior chapter on cancer reported my finding that malignant diseases present with Fǔ Organ (yáng Meridian) pulses displaced to the Fire Element depth. The similarity of this finding with that of the Fǔ Organ pulse displacement to the Water Element depth in "External Devil Possession" and of the Fǔ Organ pulse displacement to the Earth Element depth in "Internal Devil Possession" leads to a consideration of whether or not cancer is a type of "block" that needs to be treated as an initial step in acupuncture when the pulses indicate its presence, regardless of whether or not there are clinical signs and symptoms of malignancy. One aspect of the pulses in cancer that I did not emphasize is that Ministerial Fire is the only Element that is imbalanced in <u>all</u> cases of cancer. I have repeatedly pointed out that Ministerial Fire behaves differently than the Elements of the "formed Organs", and plays a vital role in connecting the "Primitive Constitution" of the Extraordinary Meridians to the Organ-based Constitution emphasized in CCA. The pulse

251 Following the recognition of similarities with the pulses of cancer patients, I have refined the pulse diagnosis of "possession". Instead of needing all 6 Fǔ Organ pulses to be at the same abnormal depth, I now believe that only 5 of these pulses at such a depth is typical in diagnosing the presence of these blocks.

patterns in cancer are thus similar to those found in the blocks due to possession, all manifesting on the Fǔ Organ pulses. Treating these findings seems to result in opening up these blocks, suggesting that blocks are a form of stagnation. Stagnation prevents a normal response to balancing treatments, as these latter responses depend on the ability to move and change (the opposite of stagnation). One might ask: Is there any generalization that can help us to understand why these blocks occur at the levels I've described? Cancer is a condition in which the organism no longer recognizes abnormal cells as "other" than one's self. The Heart (Fire Element) is the Organ that houses our Spirit (shén 神), which is reflective of our unique identity. If we cannot access the information carried by our shén, then we will not be able to recognize malignant cells as abnormal, and marshal our immune systems to eliminate them. Fixed Fǔ Organ pulses at the Earth Element depth have been explained as interfering with the production of post-natal essence, leading to Internal Devil Possession. The essence is the most refined and subtle part of the food consumed, whose somewhat grosser parts are transformed into qì, but this subtle essence joins with its Spirit partner, the yì 意, whose function is to keep our intention consonant with what is in our unique shén. Thus we can understand fixed pulses at the Earth depth as a form of Spiritual stagnation. We no longer devote our most precious resources to our unique destiny, but rather seem to be guided by another plan (so called "demonic possession"). The Earth Element also reflects the nutritional aspect of our being, which is more "internal" than our more trauma-related "external" experiences, whether physical or social in nature. These latter relate more to the fixed Fǔ Organ pulses at the Water Element depth that I have previously indicated as correlating with External Devil Possession. This affects our inherited essence, which like our acquired essence is of great subtlety. It is connected to our will (zhì 志) without which we cannot follow our destined path in life. Although the Water Element pulses are located at the deep level, as we might classify the DNA in the nucleus of all of our cells, it is recognized that the expression of our DNA can be turned off or on by outside influences, a field of study called epigenetics. Perhaps this is the mechanism whereby External Devil Possession occurs. Seen in this light, all these fixed pulse patterns express stagnation (partial or complete) at the Spirit level and so must be the first place to address in treatment. As the *Nèijīng* tells us, all acupuncture must be based upon the Spirit. These Spirit level blocks seem to interfere with the ability of the Organs and Meridians to properly adjust yīn yáng and wǔ xíng balance mechanisms. It is important to remember, however, that treating these blocks should be followed by the usual CCA treatments in order to help address any Constitutional Conditional imbalances. Such treatment positively reinforces an individual's ability to adhere to their unique destiny. Thus cancer also appears to exert its block at a level that is deeper than that of the CCA treatments that I have been describing in my previous publications. Cancer is indeed a very deep-seated disturbance of the organism that is both systemic and difficult to cure if there is no attempt to engage the person's inborn Constitution, a concept that is foreign to conventional medicine. Other types of blocks can result in stagnation in the flow of qì and blood, which can limit the response to CCA treatment, although they are not as serious a class of pathology. These include Akabane imbalances and Entry/Exit blocks. If found, these should also be treated before trying to balance the Organs or Elements. Quite recently, I have recognized a type of Entry/Exit block that I have not seen described before. In the usual type of Entry/Exit block the preceding Meridian is in an excess (solid) state and the following Meridian is in a deficient (hollow) state, both determined by pulse quality diagnosis. Such a Condition is treated by tonifying the Exit Point of the first Meridian and then tonifying the Entry Point of the second Meridian. This maneuver has the effect of opening the connection between these Meridians, like opening an overly constricted valve, thus allowing greater qì flow to equilibrate the excess and deficient states. However, often I find that the pulse shows that the preceding Meridian is deficient (hollow) and the following Meridian is in excess (solid). In such a case dispersing both the Exit Point of the first Meridian and the Entry Point of the second Meridian appears to equilibrate the two. I conceptualize this as analogous to partially closing a leaky valve. The leakage from the first Meridian causes it to become deficient, which in turn allows a "flooding" of the second Meridian, causing it to accumulate an excess of qì. This scenario appears to be quite common, especially if one of the two Meridians belongs to the Root Organ of the Constitution. An unusual finding in such cases is that often the Constitutional Organ's pulse is found to deviate opposite to its natural tendency. By that I mean that if the Constitutional Organ tends towards deficiency, in this situation its pulse may display an excess quality, and similarly, if the Constitutional Organ tends towards excess, its pulse may display a deficient quality. Also, the RYCK imbalance appears to occur on the non-dominant side for the person's Constitution. I have only found this unusual type of pulse change in these cases of "reverse" Entry/Exit blocks. Another common finding in any Entry/Exit blockage is that the RYCK width ratio is expressed on the opposite side as that which is typically found for individuals of that Constitution. This finding suggests that indeed there are differences between Meridian states (Entry/Exit blocks) and Organ states (Zàng Fǔ imbalances), an issue that I left unanswered earlier. Finally however, these Entry/Exit blocks are easy to treat, and often provide immediate symptomatic relief in addition to promoting a generalized healing response.

Once blocks have been excluded, there is a sequential ordering that I have found both effective and logical in selecting acupuncture treatments. In a way, the whole process of how the organism strives to restore a state of health seems to follow the rule that the most recent changes in the pulses reflect relatively more recent, but also potentially more serious pathological developments, if not resolved. The blocks mentioned above are examples of that hypothesis, and I have frequently heard patients with these blocks say, "I have never felt well since..." in conditions brought on by severe physical injury (motor vehicle accidents, for example, which often disrupt the Extraordinary Meridian grid) or psycho-emotional injury (possession following non-physical shocks, for example). When no blocks are present, the most common situation is one where the 12 Officials are not coordinated with each other. To treat this situation, it is necessary in CCA to diagnose both the Constitution and the Condition.

A common pulse finding occurs in many Conditions characterized by pain or other altered neurological symptoms such as tingling, numbness or any kind of dysesthesia. These Conditions are frequently similar in the following ways: 4 Organ systems are found to be out of coordination in the Conditional pulses (with Ministerial Fire being one of them)[252], and the Constitutional

252 *Su Wen* 74 states, "All pain, itch and sores, without exception, they are associated with the Heart." Unschuld (p. 627) cites a commentary on this chapter by Gao Shishi (1980),"Old editions have Heart instead of Fire. I have changed this (to Fire) now. All pain, itch and ulcers are associated with the Fire of the hand Shao Yang (conduits of) the Triple Burner." Essentially this Chinese scholar has come to a similar conclusion as I have in implicating Ministerial Fire in painful Conditions, where it serves as one of the 19 disease mechanisms (bìng jī 病機) discussed in *Su Wen* 74.

Organ will have enlisted its Degenerative Organ for help in controlling the Organs of the other Elements. The Meridian of either the Zàng inflammation or Vitalization Auxiliary Formula is usually needed for treatment, in addition to those of the Degeneration and Neuro Auxiliary Formulas. Treatment is given on the left side of the body, which is the side of the body that shows a RYCK imbalance. Whenever treating the Degeneration Organ is indicated by the pulses, there is no need to also treat the Constitutional Organ. Properly treating the Degeneration Organ will restore control to the Constitutional Organ, as these Organs are always in a Mother/Child mutually beneficial relationship. The Zàng inflammation or Vitalization Auxiliary Formula is applied on the same side as the Degeneration formula. The last component of this treatment is the Neuro Auxiliary formula.

Pathologies due to congestion are diagnosed and treated in a manner similar to that for pain. The Constitutional Organ calls on its family member, the Congestion Organ to assume responsibility for inter-Organ coordination. This can be observed in the pulse, as the Constitutional Organ and the Congestion Organ having exchanged their usual depths. In addition, either the Fǔ Inflammation Organ or the Antibiotic Organ will have moved to the usual depth of the Congestion Organ. These Congestion Conditions present with RYCK imbalances on the right side, and are treated on the right side. In addition to the Congestion Auxiliary Formula, the Fǔ Inflammation or Antibiotic Formula is treated, as is the Neuro Formula, all on the right side. The pulse depths, Subdoshas and RYCK qualities will confirm each of these components in the treatment of congestive disorders.

Like peeling off the layers of an onion, painful Conditions are often preceded by simpler pathologies such as arthritic changes or prolapse tendencies that can be visualized by imagery studies, but that are not severely uncomfortable. These pathologies show in the pulse as reciprocal imbalances between the Constitutional Organ and the Organ of the Zàng inflammation or Vitalization Auxiliary Formula, but there is no problem showing on the pulses of the Degeneration Organ or the Neuro Ministerial Fire function. Some patients seem prone to develop this kind of pathology in different locations in their bodies, so the same acupuncture treatment often is effective for multiple symptoms. Similarly, other patients tend to develop problems requiring treatment of the Fŭ inflammation Auxiliary Formula or else the Antibiotic Auxiliary Formula, but there is no abnormality in the pulses of the Congestion Organ or Ministerial Fire function. All the pathologies in this paragraph involve imbalances between 2 Meridians only, and can be treated by a similar rule: The Constitutional formula is treated on the side of the RYCK imbalance, which is also the side that shows the Subdoshas and RYCK pulse quality of the Auxiliary Organ. The Auxiliary formula is then treated on the same side as the Constitutional formula.

If there are no imbalances detected in the depths of all 12 Meridians, and the Subdoshas show only the pattern of the Constitution bilaterally, but there is still a RYCK imbalance, then patients can be treated on the side of that imbalance by the *Nèijīng*'s RYCK protocol. Next, if even the RYCK balance is equal, but the Subdoshas show opposite patterns of the Constitutional Element on left and right sides, that seems to indicate that there is a condition of a latent pathogen present, as hypothesized earlier, which will be illustrated in the case histories. In such cases it is necessary to find out which of the Organs has shifted its pulse depth to that of the Constitutional Organ, and then to disperse its Luò Point. This is different than the case of Akabane imbalances, which involve alterations in the Organ's medial or lateral deviation on the pulse, rather than its depth. Akabane imbalances are treated by tonifying the Meridian's Luò Point on the deficient side. Often when patients have responded well to CCA treatment, but some symptoms remain and there are neither RYCK width imbalances nor abnormal pulse depths present, the reason has been a Liù Jīng imbalance. Such Meridian abnormalities appear to be an indication that any remaining pathologies are relatively superficial and easily treated by the Liù Jīng protocol I discussed earlier. This Six Great Meridians balancing technique is implicated when the Organs show no abnormal pulse depths, but the RYCK pulse qualities and Subdoshas both reflect one of the Great Meridians on opposite sides, i.e. the arm branch on one side and the leg branch on the other side. Finally, I want to remind readers of Simonov's method of macrocosm/microcosm balancing, which can be applied even if there is a RYCK imbalance. I often find that using this method to harmonize the Constitutional Organ/Meridian with the environment can solve problems that don't seem to respond to any other strategies I've tried. There is a simple hypothesis that can explain this finding.

Many of the previous strategies are based on the assumption that in a given Element, the yīn and yáng Organs/Meridians deviate from normal in opposite directions. E.g. if KI becomes deficient, then UB will become excess. That is the basis of the *Nèijīng*'s RYCK diagnostic and treatment strategy. What would that accomplish if both yīn and yáng Organs/Meridians (e.g. KI and UB) are in a deficient state? The RYCK treatment for tonifying the KI would further weaken the UB, which is inadvisable. Worsley's Five Element style is based on the assumption that in most cases, the coupled yīn and yáng Organs deviate in the same direction, so that in this case both KI and UB would be tonified in the same treatment. Simonov's method comes to the same conclusion, but with a rational explanation provided. Keeping the same example, he would say that the KI has become yīn deficient, while the UB has become yáng deficient. This explanation allows for the two coupled Organ systems to keep their opposite polarities, but to be treated with the same intention of either tonifying or dispersing them both. Obviously it is not reasonable to say the *Nèijīng* is correct and Simonov is wrong, or vice-versa. Nature offers us many complex situations to understand, and no single rule is applicable to all possible circumstances. The mark of a good clinician is the flexibility to individualize the treatment in every case, based on all the information available. In CCA that is principally via the pulse, and both the RYCK and Simonov's approach are pulse based methods that can be verified in advance, as to applicability, by the careful practitioner.

The expectation in all of the foregoing treatments is that all of the abnormalities in the Conditional pulses will disappear (at least temporarily), and be replaced by signs consistent with the Constitution. In addition, the RYCK pulses should become equal (at least temporarily) and the O-Ring kinesiology test should show improved neuromuscular coordination. Most often there is also a symptomatic improvement right after treatment if it was chosen correctly, especially in the general sense of well-being of the patient. The case histories following will give examples of these different kinds of treatment and the outcomes I have observed. The optimal response I've found in my patients has been when they spontaneously report, "I feel like myself again for the first time in ages." Treatment can be continued until no pulse abnormalities are present, including what I have termed prodromal pulse signals. These indicate that the individual is in the recovery phase of whatever problem they had, but is still capable of experiencing a relapse if treatment is stopped prematurely. On the other hand, treatment when all the pulses indicate normal functions is not generally helpful, and could even provoke new imbalances. Because cancer is such a devastating illness, and prodromal signals are difficult to entirely rule out, at this time I recommend treating all cancer patients once a month after achieving remission. I have not found any adverse effects of such a protocol, and in fact have found that patients report benefits in their well-being. Whether ongoing monthly treatments are beneficial for for cancer patients who are cured, rather than simply in remission, remains a question to be studied over a much longer time frame.

I wish to emphasize that acupuncture cannot eliminate illnesses if the precipitating causes are still present in a patient's life. Whether it is a question of improper diet, uncontrolled lifestyle or recurrent emotional trauma, these circumstances must be addressed and corrected. The practitioner can help patients discover where these pathological situations exist in each case, and discuss strategies to eliminate them, but in the end each person must take on the responsibility for the choices they make in this regard.

CHAPTER 16

Case Histories from San Francisco to Beijing

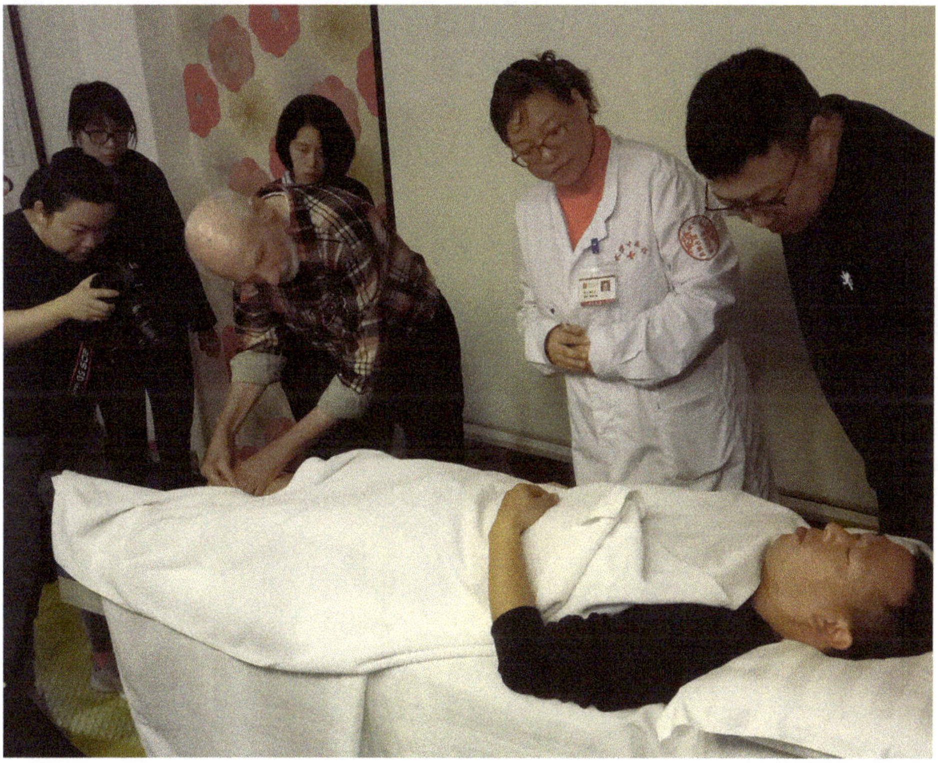

Figure 45: CCA Treatment in Kunming [253]

The following case histories have been selected to illustrate the various types of diagnoses and treatments used in CCA, together with their clinical

253 The author's Kunming clinical workshop in 2017. Source: Olga Fedina, permission granted by all participants for public viewing.

outcomes. Since *The Compleat Acupuncturist* (with a selection of case histories) was published in 2014, I was invited to conduct workshops in China where the Constitutional Conditional methodology was put to the test, as each day I was the attending physician for all the morning patients, followed by theoretical and practical training each afternoon (Fig. 45). These workshops were held in Kunming in 2017 (3 days) and Beijing in 2018 and 2019 (5 days each). Case histories from these workshops are included in this chapter.[254]

The diagnosis and treatment of cancer has been emphasized by its much greater number of cases described. This choice does not reflect the frequency with which this pathology has been seen in my practice in the past, but the dramatic improvements I've introduced with CCA cancer treatment has markedly increased word of mouth referrals from other patients. These cases have been chosen because this mortal illness typically has not been successfully treated by other approaches, and I have included the cases I previously reported in my JCM article to provide the latest data on the longer-term effects of the CCA treatment of cancer. I have also included a disproportionate number of cases of chronic fatigue syndrome because, although not fatal, this condition can be a severe impediment to those suffering from it, and my patients have told me that previous trials of acupuncture were not as helpful. CCA has provided a unique understanding of the diagnosis and treatment of this illness, and my patients keep referring others from their support networks.

It would be misleading to claim that all my patients respond as well as the ones illustrated here. It can take several treatments for me to ascertain that I am correct about the Constitution, but I have found that it is rare that CCA does not provide some significant degree of benefit to my patients, regardless of their Constitution or Condition. In addition to pulse findings, I presently use a kinesiology test to identify the Sasang classification of the

254 A full presentation of all the patients treated in the Kunming workshop can be found in "Constitutional/Conditional Acupuncture: Basic Theory and Clinical Practice", Peter Eckman, The Journal of Chinese Medicine (England), Issue 117, pp. 45-53, 2018. A shorter summary was also published as "Constitutional/Conditional acupuncture." Peter Eckman, Chinese Medicine and Culture, Volume 1 Issue 1 (Chin Med Cult) 2018; pp. 46-48. Shanghai, China.

patient's Constitution.[255]

Banana (Napa cabbage) strong = Taeyangin

Cucumber (Melon) strong = Soyangin

Carrot (Radish) strong = Taeumin

Potato (Orange) strong = Soeumin

Fig. 46: Four Foods Sasang Correlates[256]

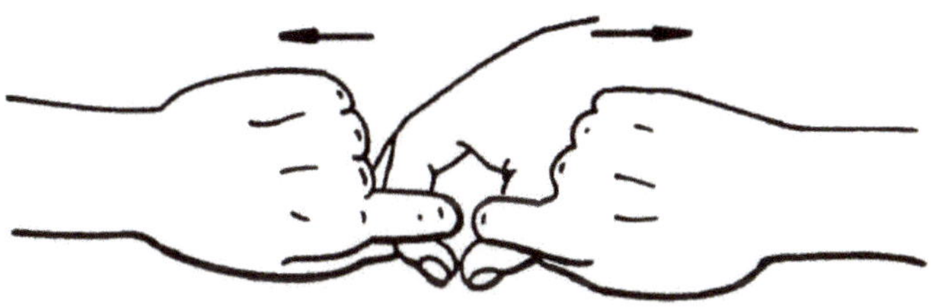

Figure 47: Omura's Bi-Digital O-Ring Test[257]

---

255 The 4 food tropisms as Constitutional indicators (Fig. 46) show up in some of these cases, but I have not previously published this method. Omura's Bi-Digital O-Ring kinesiology test (Fig. 47) was described in *The Compleat Acupuncturist* (pages 193-194). It was only after the publication of that book that I discovered how to use the Four Foods kinesiology test to identify the Four Sasang Constitutional types. I based my method on the work of Jihae Ko described in his graduation thesis from Shenzhou Open University, Amsterdam, 2014. (http://www.iwooridul.com/english/sasang-constitution-type). I substituted carrot for pumpkin in diagnosing Taeumin because pumpkins are not so readily available for testing. When the patient holds these foods one at a time, only one of them will result in increased muscle strength. I use the following foods, but others (tentatively those in parentheses) may prove equally accurate: Banana indicates Taeyangin, cucumber indicates Soyangin, carrot indicates Taeumin and potato indicates Soeumin. All the patients I treated in China over the course of 3 years were diagnosed using this method, and the uniformly positive treatment outcomes convinced me of the accuracy of this technique, however it takes practice to develop the required skill in performing this type of examination. Using organic produce is recommended to minimize errors.

256 Source: Peter Eckman

257 Source: en.wikipedia.org

The cases are generally arranged from the less complex (pathologically) to those more complex, with an occasional bit of randomness thrown in. In relating these cases, my intention is to show how diagnoses and treatment formulations are established, but the interpersonal part of treatment, i.e. the practitioner/patient relationship, is definitely a crucial part of every therapeutic encounter, although that aspect of treatment does not lend itself as easily to this type of case presentation. And so, the reader should keep in mind that CCA is not merely a mechanical process, but involves helping the patient explore the possible reasons they have developed the problems they experience, and guide them in their search for ways to resolve the underlying roots of their difficulties in life. I have organized the cases by the most significant treatment formula (with their abbreviations listed below) that was used in each patient, in order to show as clearly as possible how these formulas are applied, however each case is different and might have included different treatments on successive sessions. Since I could not include every visit's treatment, only the most significant ones have been selected for presentation. The following abbreviations are used to indicate Formula components:

K= Constitution
F= Fǔ inflammation
Z= Zàng inflammation
V= Vitalization
B= Antibiotic
D= Degeneration
N= Neuro
I= Immune
P= Paralysis
T= Tumor
C= Congestion
CA is used as an abbreviation for cancer, so the reader can identify cases concerning this diagnosis for easier comparison.
AKA indicates an Akabane imbalance.
EPF indicates an Exogenous Pathogenic Factor.
IPF indicates an Endogenous Pathogenic Factor.
ID indicates "Internal Devils," a form of "Possession."
8E= Extraordinary Meridian Treatment

50 Cycles= Contact needling to balance Róng (Yíng) and wèi qì.
6L= Great Meridian imbalances.
E/E= Entry/Exit

1) K/F, 8E, EPF, 50 Cycles: A 63 year old male building contractor was referred to me by another acupuncturist in January of 2017 for treatment of a long history of skin problems. His main complaint was a frequent feeling of heat in his skin all over, with red bumps coming and going. It felt to him like he was attacked by a porcupine. He had suffered from asthma since childhood and at 18 had a combination of mononucleosis, strep throat and pneumonia. Dermatologists did not agree on what kinds of skin disorders he had, but had treated him with steroids, antihistamines and antibiotics for possible folliculitis, and possible eczema. Nothing gave him more than temporary symptomatic relief, and he was currently taking antihistamines 3 times a day and an albuterol inhaler 3 times a day for asthma. There seemed to be a pattern of aggravation of his skin issues after any respiratory infection, which he was prone to develop. His pulse diagnosis frequently showed the following findings: Left side CK>RY with a 2/1 ratio and the zào quality present on the right side, but not on the left side. The *Màijīng* deviation was radial at cùn, indicating a Constitution of SI Deficiency (Fire Taeyangin). The Conditional pulses showed the SI at the Water depth and the KI at the Fire depth. Left Subdosha was Vata (Water type) and right Subdosha was Pitta/Kapha (Fire type). The treatment was Root/ Fǔ inflammation (K/F), all on the left side. He was mostly treated with regular needling to obtain dé qì, and then the needles were removed at each point as follows: (UB 66, SI 2- and GB 41, SI 3+) plus (LU 8, KI 7- and HE 8, KI 2+). Following this treatment approach he has not experienced any significant skin symptoms or asthma attacks for the last 5 years, and he stopped all medications for both. He has never felt this healthy before and continues to come for treatment of musculoskeletal pains brought on by heavy lifting at work and vigorous basketball games (pre-pandemic), so I have good follow-up examinations. At one visit all his pulses were normal, except the left sided CK>RY, however he was experiencing a generalized achy feeling, most noticeable in his sacral area. He was very tense in the pelvic region, so I dispersed Rèn 4, and supplemented SI 4 and HE 7 (the Fire Source Points). This immediately resolved all his symptoms. I often look to disperse a yīn Meridian (either on Rèn Mài or a Mù Point) when CK>RY (which means yīn excess and yáng deficiency) or a

yáng Meridian (either on Dū Mài or a Back Shū Point) when RY>CK (which means yáng excess and yīn deficiency). In this case the Front Mù Point of SI was the obvious choice. The logic here is to treat the front for problems of the back, but always in accord with the Constitution. I'd like to mention a few other treatment approaches used over time, which illustrate how the different components of CCA can be sequentially applied or combined. On one visit, although there were no depth abnormalities in his pulses, I did feel that there was a Yīn Qiāo Mài pulse on the left and a Rèn Mài pulse on the right. Since his Constitution is one of yáng deficiency (SI deficiency), I tonified Yáng Qiāo Mài and dispersed Yīn Qiāo Mài on the left side (his Constitutional side) as follows (UB 62+ and SI 3- plus KI 6- and LU 7+). Although he had not previously mentioned these symptoms, he reported that following this treatment his sleep was markedly improved, as was his general mood. I think it is important to check for Extrameridian pulses in anyone who is prone to repeated musculoskeletal traumas, even if they are relatively minor. More recently, I again felt the same Extraordinary Meridian pulse signals in addition to pulse depth reversals on SI and KI. I gave him the full 4 limb version of the 8E treatment: Left side UB 62+ and SI 3- with KI 6- and LU 7+. Right side SI 3+ and UB 62- with LU 7- and KI 6+. I followed the 8E treatment with a transfer component by simply tonifying SI 2 on the left side (Control Cycle transfer from KI to SI). Right after this treatment he said, "I hate putting out fires." The context was that he was taking care of crises in family members that others had left for him to deal with. Given his deficient Fire Element Constitution, I found his remark quite poignant. As usual, he felt better in himself following this treatment. On another visit, I noticed that his Subdoshas were exact opposites, both reflecting his Constitutional Fire Element. The spikes on his left side (on my index finger) were at positions 1 and 2, while on his right side they were at positions 3 and 4. In searching his pulse depths at all positions, I discovered that his UB pulse was at the Fire depth, but there were no other depth abnormalities. I believe this finding indicates a retention of xié qì associated with the Operator of Taiyang, which is UB. Curiously, his hands were always cold when I examined him, but I was never clear on how to explain that, as his Constitution was Fire. He then reminded me that at age 18, after "burning the candle at both ends" he had come down with a fever of 104 degrees that landed him in the hospital for two weeks with strep throat, pneumonia and mononucleosis. It appears he was left with a retained pathogen (EPF) that resulted from a shāng hán attack on tài yáng that was never fully resolved. Ever since then he has worn insulated underwear and extra layers of clothing, as he was always colder than the people around him. This made sense as a case of a Cold pathogen lodged in Taiyang. I tested silver press pellets on the Hand Acupuncture Points corresponding to UB 58 on both sides. Only the right side restored a normal depth to his UB pulse, so I dispersed UB 58 on the right for 10 minutes. With this one needle in place, his hands became warm, his pulses normalized and all his musculoskeletal discomforts disappeared. It is astonishing to think that an EPF (Cold in this case) could stay trapped for 45 years and still be expelled to some degree by one correctly placed needle. He no longer wears extra layers of clothing to keep warm. Finally, it is interesting to note that this case of skin disease and asthma involved someone with a Taeyangin Constitution. Taeyangin was originally described by Li Jema as typified by a strong Lung and a weak Liver. Small Intestine deficiency satisfies this general relationship.[258] Li believed that Taeyangin was a rare Constitutional type (less than one in a thousand), but my experience contradicts that, as I find many Taeyangin patients. Although the Lung is relatively strong in this Constitution, and the symptoms in this case affected the skin, bronchioles and lungs, the successful treatment approach did not routinely involve treating the Lung directly. It was more important to recognize and treat his Constitutional Organ (SI), the Fŭ inflammation pathomechanism (KI), the EEV disharmony (Deficient Yáng Qiāo Mài and Excess Yīn Qiāo Mài, and the EPF (UB 58, Luò Point of the Operator responsible for treating Cold), all situations that were reflected in his pulses. Now that there are no ongoing pulse abnormalities, he requested that I treat him to prevent the development of future problems. My most successful approach for preventive treatment is to use contact needling (no penetration) aimed at harmonizing macrocosm/microcosm, as discussed in Chapter 7. I would like to show at least one example of using this strategy. On his visit on 6/9/21, he arrived at 8:30 A.M. I set the slide rule (Fig. 24) for solar noon at 1:09 P.M. Since his visit was in the morning, that is a time when males experience "victory of yīn", meaning that solar energy was available for him to absorb as yīn qì. The slide rule showed that he was experiencing a yīn period at that time, so to treat his Constitution (Fire Element deficiency) I could direct this yīn qì to strengthen his HE Meridian (the open yīn Meridian of his Fire Element CF). At 8:39 A.M.

258 The order of Elements from weakest to strongest is Fire<Wood<Earth<Metal<Water.

shaoyin opened for about 5 minutes as part of the second cut (kè 刻) of this yīn period, during which the Meridians of the right hand are in a Róng (荣) phase. The treatment indicated was to simply touch as many Points as possible on the HE Meridian on the right side with a filiform acupuncture needle during this 5 minute time span. Each Point was contacted for about 7 seconds. The needle simply acts as an antenna to amplify the absorption of the ambient solar yīn qì, and direct it to the HE Meridian, in this case by reinforcing the Róng movement into the HE Meridian and Organ, thus strengthening them. Following this treatment the kinesiology (O-Ring test) was markedly stronger, and the minor amount of stiffness in his hands (as a manual worker) had disappeared. I do not expect the reader to be able to understand all the details of this kind of treatment, but at least this aspect of what I incorporate in CCA is now documented. Once again, I encourage interested readers to contact Dr. Igor Simonov if they wish to learn how to apply this method themselves.

2) K/F: In October of 2017 I was teaching a workshop on CCA in Kunming, China. There were a number of European and Chinese acupuncturists present, and many asked for treatment after seeing the results in class, so the evenings were spent treating those who could not be treated in class. A 49 year old Chinese woman had been diagnosed with a problem in one of the blood vessels supplying her brain. This abnormality had been treated successfully by surgery, but afterwards her tongue had become totally numb at the tip, and this condition had lasted for a whole year already, with no treatments providing any relief. On examination, she had a positive tropism for potato indicating a Soeumin Constitution. On her left side RY equaled CK, but on the right side CK>RY with a ratio of 2/1 and the pulse deviation was ulnar at the chǐ position. These findings indicate that her Constitution was KI Excess (Water Soeumin).[259] Examination at cùn, guān and chǐ showed the HE pulse was widest at the Water depth and the KI pulse was widest at the Fire depth. The Conditional diagnosis was therefore Fǔ inflammation (HE deficiency), with the Heart being unable to properly control the Fǔ Organ system. The tip of the tongue is associated with the Heart, and the numbness indicated Heart deficiency. Because we were not in clinic at the time I treated her with gold (tonifying) and siver (dispersing)

259 The pulses are consistent with either KI excess or HE excess, however the former is a Soeumin type while the latter is a Soyangin type.

press pellets at these corresponding hand acupuncture points on the right side: (SP 3, KI 3 gold and LV 1, KI 1 silver), plus (SP 3, HE 7 gold and KI 10, HE 3) silver. As soon as these pellets were in place, she reported that her tongue was back to normal with no numbness. This case illustrates that the Extraordinary Fǔ are also treated by the Fǔ Inflammation Auxiliary formula. Blood vessels (血脈) are classified as Extraordinary Fǔ Organs, and her Condition was initially clearly a blood vessel abnormality. Surgery had prevented a possible stroke, but had not addressed the disease mechanism, which CCA was able to accomplish.

3) K/F: On the same visit to Kunming, I was asked to treat one of the Spanish attendees. This 52 year old woman was severely intolerant of lactose, fructose, gluten, histamines and other food components, leading to a very restricted diet. Exposure provoked headaches and nausea, which were both present when I examined her. She had a positive tropism for cucumber, which indicated that her Constitution was Soyangin. The presence of a RY>CK pulse and a radial deviation at the guān position specified her Constitution as ST Excess (Earth Soyangin). Conditionally, her LV pulse was widest at the Earth depth and her ST pulse was widest at the Wood depth, so she had a LV deficiency Condition, an indication of Fǔ Inflammation for her Constitution. Again, I was only able to use press pellets for treatment, as in the preceding case, so the pellets were applied to the corresponding KHA Points as follows: (LI 1, ST 45 silver and GB 41, ST 43 gold), plus (SP 3, LV 3 silver and KI 10, LV 8 gold), all on the left side, and kept in place for 30 minutes. At that time she no longer had any headache or nausea. Fortunately she joined our tour through Yunnan Province the next week, so I was able to monitor her afterwards. She discovered that apparently she could eat anything without any of her previous symptoms, and she indulged herself every day. This improvement was still maintained when she returned to Spain. Clearly the hollow Organs of her digestive system were the site of her Fǔ Inflammation Condition.

4) K/F: One final example from that memorable Yunnan tour. The wife of one of the French acupuncturists was experiencing a number of health issues. This 45 year old woman was chronically constipated, only able to have a bowel movement every other day at most. This resulted in abdominal pain

if there was any pressure on the area, and she also suffered from severe low back pain every night in bed, such that she was unable to turn her body over in order to change position. These problems had been present for many years already, were worse towards the end of menses, and she had only gotten temporary relief from osteopathy and acupuncture. Medically, she was diagnosed as suffering from endometriosis. On examination she had a positive tropism for banana, indicating a Taeyangin Constitution. RY>CK and ulnar deviation at cùn confirmed her Constitution as LV Deficiency (Wood Taeyangin). Her LV pulse was widest at the Earth depth and her SP pulse was widest at the Wood depth, indicating the she had a Fǔ Inflammation Condition. Press pellets were again the available treatment, applied as follows: (LU 8, LV 4 silver and KI 10, LV 8 gold), plus (LU 8, SP 5 silver and LV 1, SP 1 gold), all on the right side. This treatment led to a stronger kinesiology test, normal pulses and the absence of any symptoms. For the first time in ages she slept through the night afterwards, without any pain. As discussed in the text, gynecological problems are usually treated as Fǔ Inflammation Conditions, which makes sense of her accompanying constipation, which was aggravated towards the end of her menstrual cycle. Again we see that this protocol can address problems of both the Regular and the Extraordinary Fǔ Organs, including the uterus.

All these K/F cases illustrate the common pathomechanism of Fǔ inflammation, which is shared in problems of the skin, bronchioles, intestines, uterus and blood vessels. The treatments depend however, on first making a Constitutional diagnosis. Although the pathomechanism was the same for all four cases, their Constitutional Elements were different (Fire, Water, Earth and Wood) as were their Conditionally affected Elements (Water, Fire, Wood and Earth).

5) K/Z: A 37 year old Chinese woman attended my workshop in Kunming. Each evening I met with the students and this woman asked for treatment of her case of rheumatoid arthritis, which had been present for over 4 years. She had only received moderate relief from a combination of Western medications, Chinese herbs and TCM acupuncture. At the time she was experiencing multiple joint pains, with shoulders, neck, wrists and knees being the worst. On exam she had a positive tropism for potato. Her pulses showed

right side CK>RY with a ratio of 3/1, and the deviation was ulnar at the guān location. These findings indicated that her Constitution was LU Excess (Metal Soeumin). Her Conditional pulses showed both Fire and Vata/Kapha (Metal) Subdoshas, with the LU and HE pulses at each other's proper depth. I treated her on the right side with silver and gold press pellets at the Korean Hand Acupuncture (KHA) Points that corresponded to the following prescription (HE 8, LU 10+ and KI 10, LU 5-) plus (LV 1, HE 9+ and LU 8, HE 4-). These were left in place for 20 minutes. At that time she said, "This is amazing. The pain is so much better." She continued to report her status in class for the next 3 days, and she was very happy that most of her pains were markedly diminished, with only her knee pain starting to return. Even though her Condition was characterized by pain, her Ministerial Fire pulses were normally placed, so this was a simple case of Zàng inflammation affecting the bones and joints. Rheumatoid arthritis is a disfiguring illness, and it is remarkable that such physical Conditions can stop being painful as soon as the pathomechanism is addressed.

6) K/Z, EPF: A 78 year old man was referred to me by his podiatrist for treatment of hip pain that had lasted for 3 years, and hadn't responded to conventional therapies. An MRI had shown spinal stenosis, with diffuse degenerative disc disease in the lumbosacral spine, however his complaints were typically described in his visits as right sided hip pain. As usual, I suggested he try 6 treatments to see if acupuncture could help. Initially I thought his Constitution was GB Excess, perhaps lead astray by the location of his pain, but after several treatments there had been no significant improvement. Whenever I encounter such a result, I assume that I must have missed the correct diagnosis. On reexamination I found the following: Left side RY>CK with a ratio of 2/1 and the presence of the zào quality, but on the right side RY equaled CK without the zào quality. The deviation was radial at the cùn position. So, instead of GB Excess, the correct Constitution seemed to be UB Excess. Conditionally his pulses showed UB at the Fire Element depth and HE at the Water Element depth. This pattern reflects Zàng inflammation. Acupuncture was then given as touch treatment at the following Points: (GB 41, UB 65- and ST 36, UB 40+) 5 times plus (KI 10, HE 3- and SP 3, HE 7, SP 3, HE 7+) 3 times. This is the K/Z-c formula that directs the effect of treatment to the lower body (hips). He returned 2 weeks later to report that

his pain had completely disappeared. His pulses at that time were all at the normal depth and showed no other abnormality, so I did not do any further treatment. I told him to call me if the pain recurred, but he replied, "when, not if." He subsequently had surgery for benign prostatic hypertrophy, and the hip pain recurred soon afterwards. He decided that regular acupuncture was preferable to waiting for something to go wrong, and has scheduled treatments for twice a month. Occasionally I have changed his treatment, depending on the pulse readings on a given day. These have included both D/Z/N touch treatment on the left side (KI 10, LV 8- and SP 3, LV 3+) plus (KI 10, HE 3- and SP 3, HE 7+) plus (TH 1- and TH 3+) and D/V/N touch treatment on the left side (KI 10, LV 8- and SP 3, LV 3+) plus (UB 66, SI 2- and ST 36, SI 8+) plus (TH 1- and TH 3+). The D/Z/N treatment was given when, as is often found in pain cases, 4 pulse locations were not at the expected depths as follows: UB was at the Wood depth, HE was at the Wood depth, LV was at the Water depth and TH was at the Fire depth. The left Subdosha was Pitta/Vata (Wood type) and the right Subdosha was Fire. The only difference when D/V/N was indicated was when the SI was at the Wood depth instead of the HE pulse having shifted there. When I observed at Kuon Dowon's clinic in Seoul, I tried (along with my colleague Stuart Kutchins) to determine when Kuon chose to use D/V/N treatment and when to use D/Z/N treatment in patients with pain. My own method is to let the pulses determine the best treatment, but our feeling was that Kuon used D/Z/N when there was structural damage present and D/V/N when structural damage was absent, however the quality of pain was also a factor in Kuon's approach. Sharp pain was treated with D/Z/N, while dull pain, tingling or numbness was treated with D/V/N. This case showed me that such a "recipe" approach is not always accurate. It is also often said that acupuncture is useful in treating functional problems, but not once they become structural. I think this case refutes that idea as well. It appears that painful arthritis can be reversed in terms of symptoms, even if the boney abnormalities persist. Most often, when this patient comes for treatment now, his pulses show the K/Z pattern as initially described. I interpret this as reflecting the underlying arthritic abnormalities, which are almost devoid of painful symptoms now, but which are still present on a structural level. K/Z-C treatment in a ratio of 5/3 has kept him virtually pain free. In September of 2021 he developed an egg sized swelling at his left elbow, which his regular doctor diagnosed as bursitis. The fluid was drained, but quickly returned again. This condition remained unchanged for the next 2 months. On his next visit I noticed that his pulses were close to normal, but that the Subdoshas were Pitta Kapha (Water type) on the left and Vata (Water type) on the right. In addition, his SP pulse (Operator of Taiyin/Dampness) had changed to the Water depth. I interpreted this as a case of retained damp pathogen, and treated him by dispersing SP 4 on the left side (his Constitutional side). When seen 2 weeks later the swelling was about 90 percent diminished.

7) K/Z, ID: I have treated this 79 year old female chiropractor for the past 10 years, starting in 2011. She is still working full time, but it has become somewhat more difficult over time, due to painful joints. X-rays show typical findings of osteoarthritis in several places, the worst being in her thumbs bilaterally, which are visibly deformed. In spite of this long-term systemic problem, she has had a relatively minor degree of discomfort, and is more commonly bothered by fatigue. These are both typical symptoms of Zàng inflammation. She happens to be needle-shy, so I try to use the simplest and gentlest treatments. Typically a transfer treatment fits this requirement. Her Constitution is SI excess (Fire Taeumin), with left side showing RY>CK and both sides typically having the zào quality. Often the SI pulse is at the Metal depth and the LU pulse is at the Fire depth. The following Points typically relieve all of her symptoms by transferring qì from SI to LU on the left side: LU 9, SP 2 and HE 5, with LU 9 tonified and the other Points with neutral insertion. The shallow use of only 3 needles is greatly appreciated by her. During the pandemic she became depressed, a reminder of how she felt around age 11 when her family moved, and she began to cope by overeating, with resultant weight gain. After vaccination, she resumed acupuncture treatment. I noted that her Fǔ Meridian pulses at that time were unusually all displaced to the Earth Element depth. Such a finding is diagnostic of "Possession" by "Internal Devils", whose pathophysiology I hypothesized in my prior publications. The internal devil could be thought of as either endogenous worry over loss of her center (her work), or exogenously as improper eating behavior. The Subdosha pulses on that occasion were Pitta/Vata of the Earth type bilaterally, so the treatment was to gently disperse the Master Point below Ren 15, plus ST 32, the Master Point below ST 36, and ST 41. This treatment restored normal pulse depths and eliminated her depression,

while keeping her arthritic complains at bay. She remains in good health, only coming in for maintenance treatments periodically.

These 3 patients all had demonstrable arthritic pathology. Arthritis is obviously an inflammatory condition ("itis"), and is often benefitted by the K/Z treatment formula. Compared to the patients with Fǔ inflammation Conditions (involving superficial tissues near the interface between the body and the environment), these patients with arthritis were experiencing inflammation of deep bone tissue. My impression is that the Zàng inflammation formula targets the deeper tissues, as opposed to the Fǔ inflammation formula that targets the more superficial tissues. I might take this analysis one step further, and propose that even on a cellular level the Zàng inflammation formula acts deeply, as these findings are simply reflections of the law of resonance (gǎnyìng 感應). The mitochondria are intracellular (deep), but not located in the cell's nucleus (superficial part of the intracellular space), so either the K/Z or K/F treatments might be indicated in a given case of mitochondrial malfunctioning, as in chronic fatigue syndrome. I should point out that these 3 arthritis patients had different Elemental Constitutions: Metal, Water and Fire. While Conditional states do appear to reflect similar pathomechanisms, more study is needed to examine the role Constitution plays in susceptibility to various kinds of pathology. Kuon was convinced that there was a strong correlation of specific illnesses with certain Constitutions in many cases, but this is clearly not true for arthritis in general.

8) K/Z: A 34 year old woman had suffered from chronic fatigue syndrome (CFS or ME) for more than 10 years. I began treating her on 10/15/15. Her symptoms included exhaustion, brain fog, left sided chest pain, joint pains and insomnia. She used a wheel chair and a scooter to get around at home because she could not climb stairs without aggravating her symptoms. Her illness had started insidiously, with no prodromal infections. The only significant finding in her history was that she had previously been an avid dancer and volleyball player, and had pushed herself to the point of fatigue repeatedly. She had been treated with moxibustion by a Japanese practitioner, which helped some, but not as much as she expected. Her experience with acupuncture led her to feel that she could not tolerate needles being retained for any length of time. On exam she almost always has the same pulse

pattern: right side RY>CK in a ratio of 3/1 without the zào quality on either side. The deviation is ulnar at the guān position, so her Constitution is SP Deficiency (Earth Taeumin). Her SP pulse is usually at the Wood depth and her LV pulse is usually at the Earth depth. On the right side her Subdoshas are usually Pitta/Vata (Wood type), while on the left side they are Pitta/Vata (Earth type). Various treatment strategies to address this situation have been successful at improving her symptoms and correcting her pulse signs. These include the transfer protocol from LV to SP (SP 2, HE 9+), and the 4 needle protocol (HE 8, SP 2+ and LV 1, SP 1-) plus (SP 3, LV 3+ and KI 10, LV 8-), all on the right side with simple shallow insertion, but no retention. She usually feels better for several weeks after treatment, but the pulses and symptoms revert after a while. Her life is very stressful, as her husband also has the same illness, and is in worse condition than she is, although he has never been my patient. As a result she has to be the primary caregiver, which can be a particular challenge for someone with a deficient Earth Constitution. As long as she receives acupuncture, she feels like she can cope with her illness, and was able to enjoy a 2 week vacation in Paris, during which she felt reasonably well. As yet, after more than 5 years of treatment, I have not figured out any way to cure her, especially given the recurrent stressful life she leads, but treatment has been successful in helping her maintain functionality. She has now moved to Los Angeles, so I will not be able to continue searching for an acupunctural cure of her illness, but I hope future clinical research will provide a more lasting solution for her.

9) K/Z: A 62 year old woman had a case of chronic fatigue syndrome that began 3 years previously after a series of respiratory infections. Her symptoms included exhaustion, insomnia, headaches, joint pains and swollen lymph nodes. At first she was not even able to leave her bed, and she ended up getting depressed as well. At the time of her first visit with me she was on multiple medications including bupropion, lyrica, Ritalin and trazidone, all of which have strong neuropsychiatric effects in addition to a broad range of possible side effects. I mention this because she experienced "brain fog," grogginess and jitteriness, but wasn't clear if these resulted from her illness or her medications. Her pulses showed left side RY>CK with a ratio of 3/1 and the presence of the zào quality on the right side. The deviation was radial at the guān position, so her Constitution is LI Excess (Metal Taeyangin). Her

Conditional pulses showed LI at the Wood depth and LV at the Metal depth. The left Subdosha was Kapha (Wood type) and the right Subdosha was Pitta. This finding is typical of a Root/ Zàng inflammation Condition and was treated with 20 minute retained needles at the following Points on the left side: (UB 66, LI 2- and SI 5, LI 5+) plus (LU 8, LV 4 - and HE 8, LV 2+). After this treatment she felt well (good energy, no brain fog or pain), but her doctor found a questionable mass on a mammogram and recommended an excisional biopsy. I did not feel any indication of what I call the "metastatic signal" on her pulses, and reassured her that the likelihood of a tumor greater than stage 1 (treatable by excisional biopsy) was remote, in my opinion. The biopsy confirmed a benign growth. Kuon's research indicated that Taeyangin individuals are the most likely types to have negative reactions to pharmaceuticals in general, so it is possible that her medications were exacerbating her underlying illness. I have not seen her since the Covid-19 pandemic.

10) K/Z, EPF: I began treating this 66 year old man over 20 years ago for chronic fatigue syndrome. It apparently started after a viral upper respiratory illness. He is very motivated to use all types of treatment to maximize his health, and to look at him one would not suspect he was suffering from a debilitating illness. I have tried many different approaches over the years, but he responds best to the CCA Root/ Zàng inflammation protocol. He had previously been treated by both Five Element practitioners and TCM practitioners with good results, but has chosen to continue with CCA as his only acupuncture therapy. His pulses generally show left side CK>RY with a ratio of 3/1 and the zào quality is present on the right side. The deviation is radial at the guān position, so his Constitution is LI Deficiency (Metal Taeumin). A typical treatment includes the following Points: (ST 36, LI 11+ and SI 5, LI 5-) plus (LU 8, LV 4+ and HE 8, LV 2-) all on the left side. In the early days of his treatment I thought his Constitution was LV Excess, so I dispersed the LV Meridian many times with good results, but since formulating CCA and reclassifying his Constitution, he has done even better. What I was treating before was his Conditional Organ and Element, which is not a bad approach in any case. Recently I noticed opposite Metal Element Subdoshas on one of his exams. All his pulse depths were normal except for the LV, which was at the Metal depth. LV is the Operator Meridian of Jueyin-Wind. On questioning, he mentioned that the most severe symptom when he originally got

sick was intense vertigo, which is a classic Wind pathogen indication. He also admitted that he had been a frequent user of recreational drugs before he became ill. This leads me to think that his LV was affected by the drugs, perhaps acting as a miscellaneous pathogenic factor, further weakening his resistance to viral attacks. It remains to be seen if dispersing the Luò Point, LV 5, will result in a greater degree of healing.

11) K/Z: A 28 year old woman was referred by her mother, who was also a patient of mine. This young woman had been a vigorous athlete in school, participating in 4 different sports, doing double practice workouts and running several miles every day. After one such run she became totally exhausted and couldn't get out of bed the next day. Subsequently she began to get recurrent colds, sinus pain, with sore throat and swollen lymph nodes. This went on for 2 years. She was then diagnosed with chronic fatigue syndrome. I have been treating her now for almost 20 years, and she is able to lead a very demanding life, having gotten married and raised 2 children. She credits acupuncture with keeping her functional. Her pulse diagnosis and treatment is essentially identical to the previous case. The only difference seems to be in the precipitating cause of illness. In her case it appears to have been extreme exercise.

12) K/Z, AKA, CFN: This, at that time, 50 year old doctor was first referred to me by her previous acupuncturist 30 years ago, upon his retirement. He had been treating her for CFS with good results, based on his diagnosis of KI deficiency as her major problem. I concurred with his diagnosis, as her pulses showed right side RY>CK and ulnar deviation at chǐ. Most commonly her KI pulse was at the Earth depth and her SP pulse was at the Water depth. The Subdoshas at those times were Pitta/Kapha (Water type) on the left and Kapha (Earth type) on the right. These findings indicate that her Constitution is KI deficiency (Water Soyangin) and her Condition is SP excess, representing Zàng inflammation. Although acupuncture has not cured this illness, she has functioned at a normal level, still seeing patients, now at 80 years of age, and highly respected by her colleagues. Neither of these groups are even aware of her diagnosis, but she compensates by sleeping more than 12 hours a night whenever possible. The treatment for the K/Z pattern in her case is (SP 3, KI 3- plus LU 8, KI 7+) and (HE 8, SP 2- plus KI 10, SP 9+) on the right side. I decided to include this long-term patient

to support the contention that CFS is a K/Z type of pathology, but also to show how treatments can be fine-tuned beyond this introductory level. On a recent visit she was both exhausted and suffering from right elbow pain of unknown etiology. Her pulses were as described above, but in addition, I noticed that her HE pulse was deviated to the left. I made 2 adjustments to her usual treatment: First I tonified HE 5 on the left to correct what I felt to be an Akabane imbalance, then I modified the above K/Z protocol by using touch treatment, in a 5/1 ratio. The SP dispersion was directed to the upper body by repeating the HE 8 and SP 2 sedation. In getting off the table she remarked that not only was she much less tired, but that her elbow pain was gone. The reason I directed the treatment to the upper body is that I believe that CFS is an illness of the brain's functioning, often labeled as myalgic encephalomyelitis, or ME. This case report supports such a conclusion. On a recent visit she informed me that her balance had gotten worse. Her pulses did in fact reflect a different pattern than usual. The RYCK exam showed right side RY>CK, but her KI pulse was at the Wood depth rather than its typical Earth depth for her. The Liver pulse was at the Water depth, the Heart pulse was at the Wood depth and the PE pulse was at the Fire depth. This presentation reflects the Congestion pathomechanism, and I noted that she had a long history of alleries and asthma. I treated her with the CFN protocol on the right side using simple stimulation without needle retention at the following Points: (SP 3, LV 3- and KI 10, LV 8+) plus (SP 3, HE 7- and KI 10, HE 3+) plus (PE 7- and PE 3+). Right away she reported that her balance was back to normal and that her energy level was much better. These findings lead me to consider the possibility that CFS may involve a combination of several pathomechanisms, with Congestion being one possibility.

13) K/Z: Over 30 years ago, a 39 year old woman was referred to me by her physician for "chronic fatigue syndrome." At that time hardly any physicians recognized CFS as a real illness. Perhaps this is because all laboratory tests, as with this woman, are typically within the normal range. Her main symptoms were extreme fatigue, insomnia, "brain fog," and low grade fevers. She had been diagnosed with early stage breast cancer 3 years before her current illness, and had bilateral mastectomies as a preventative measure. She then had reconstructive surgery with silicone implants within the year before she got sick, and subsequently did some research showing me that others

had reported chronic fatigue after similar silicone exposure. There were no other precipitating factors that I suspected as being associated with CFS, and she decided to have the implants removed. At the time of her initial visit, I was measuring the Ryodoraku electrical signals on all my new patients. Her readings showed Deficiency of SI and TH, plus Excess of UB. Her initial treatments were Five Element style, aimed at tonifying the Fire Element. She responded well to these treatments, but needed follow-up sessions about once every two weeks to maintain the improvements. I treated her in this way for many years. Since developing CCA, I have diagnosed her Constitution as SI Deficiency (Fire Taeyangin). Her pulses showed left side CK>RY with the zào quality present bilaterally, and radial deviation at the cùn position, which supported that diagnosis. Her last treatment was almost a year ago, and she has not needed to come in for treatment since then. That treatment used the Root/Zàng inflammation protocol as follows: Left side (UB 66, SI 2- and GB 41, SI 3+) plus (HE 8, LU 10- and KI 10, LU 5+). Because she was "needle shy" all her treatments were done using the Korean Hand Acupuncture equivalent Points, by the application of gold and silver press pellets for tonification and dispersion respectively.

In my experience, there are 3 different clinical scenarios in people who develop chronic fatigue syndrome. The first is with the onset of fatigue following an infection, often viral in nature. The second is with the onset of fatigue following physical exhaustion in people who over-exercise, as in marathon runners, triathlon participants, or even avid joggers. The third group experiences the onset of fatigue following exposure to a toxic environment, such as silicone implants, heavy metal exposure, overuse of psychedelics, or possibly too much electromagnetic radiation exposure. What I believe to be common to all three variations is that the intracellular mechanism for energy production has been damaged. Viruses are known to hijack the genetic code in order to switch cells from performing normal tasks, to making more viruses instead. Toxic exposure is also a possible cause for altered genetic expression of a person's DNA, as a result of induced mutation. It is not as clear how overwork would fit this pattern other than the possibility that the individual's jīng (also related to their genetic endowment) has been damaged by serious depleting behaviors over time. The level of one's DNA corresponds to (resonates with) one's jīng, and this analysis lends support to my hypothesis that ultimately it may be necessary to treat the Extraordinary Meridian system in order to finally be able to cure this group of illnesses. The presence of both yīn and yáng Constitutional types in these cases of CFS would support the conclusion that any particular Constitution is not a necessary aspect of susceptibility to this illness.

14) K/V: A 42 year old woman had an 18 year history of migraine headaches, starting after an injury to her left side. They occurred about 4 times each month, mostly in the premenstrual phase of her cycle. Her symptoms at that time included severe anxiety, irritability, depression and paranoid thoughts, in addition to the headache which radiated from the front of her neck to her jaws and head bilaterally. She was happily married with a 4 year old child, and owned a bakery/restaurant with her husband. Her pulses showed CK>RY with a 2/1 ratio on the right side. The *Màijīng* deviation was ulnar at the chǐ position. These findings indicate a HE Excess Constitution (Fire Soyangin). Her Conditional pulses showed the HE at the Water depth and the UB at the Fire depth. The left Subdoshas showed Pitta/Kapha (Fire type) and the right Subdoshas showed Vata (Water type) indicating UB Deficiency. This pulse relationship is typical of Root/Vitalization imbalances, which happen to be quite common with migraine headaches in my experience. I asked her if she ever felt like things were dropping down in her body when she got the headaches. Her answer was, "How did you know that? It's like I'm in an elevator and suddenly someone pushes the stop button, and all my insides drop down." I treated her with touch treatment on the right side as follows, doubling the tonified Points to direct the effect to the ovaries and uterus, which I believed to be the source of the headaches: (SP 3, HE 7- and KI 10, HE 3+) 4 times plus (SI 5, UB 60- and LI 1, UB 67, LI 1, UB 67+) 2 times. She has since experienced almost no premenstrual symptoms and no migraines. The Vitalization Auxiliary Formula for all Constitutions are on the yáng Meridians, and I believe that an abnormality of the yáng's function of lifting the qì is the pathological mechanism in all the different manifestations of devitalization Conditions.

15) K/V: A 35 year old Chinese woman in Kunming had a 2 year long illness that left her completely unable to stand or walk without holding on to somebody or something. The symptoms started after she felt shocked by a sudden

increase in work stress at her job as a hairdresser. Originally presenting as a headache, her symptoms quickly changed to a constant feeling of dizziness, like she was drunk all the time. She felt worse in rainy or damp weather. For 2 years she had been treated with Chinese herbal medicines and acupuncture (twice a week) for a diagnosis of damp invasion and yáng deficiency, but that had not been effective in restoring her ability to stand or walk unaided. On examination she had a positive tropism for carrot. Her pulses showed RY > CK, but this was before I developed a way to tell if the pulse was zào quality or not. However, the deviation was clearly ulnar at the guān position, so her Constitution had to be SP Deficiency (Earth Taeumin). I was able to treat her twice, and on the second visit her Conditional pulses showed the Spleen pulse to be at the Wood depth and the Gall Bladder pulse to be at the Earth depth. The left Subdosha was Pitta/Vata (Earth type) and the right Subdosha was Kapha (Wood type). This indicated a Condition of Devitalization, so I treated her on the right side with simple insertion of needles (getting dé qì, but without needle retention) at the following Points: (HE 8, SP 2+ and LV 1, SP 1-) plus (ST 36, GB 34+ and UB 66, GB 43-). Right after this treatment I asked her to try getting up and walking. To the astonishment of both the class participants and myself, she practically bounced up and strolled back and forth in the room and down the hallway without any help! It is interesting that her TCM diagnosis was much the same as my CCA diagnosis. As I explained, devitalization is often a manifestation of yáng deficiency (she would fall down without support), and dampness is understandable as a result of SP Deficiency. The involvement of the GB suggests that her dizziness was also expressing some wind pathology. The CCA treatment was not aimed at expelling pathogenic factors, however, but only aimed at returning her Conditional Organ functions to their Constitutional state.

16) K/B: A 54 year old man was seen for a chronic illness that had lasted for more than 15 years, and was both variable in presentation and difficult to diagnose. Symptoms included tinnitus, dizziness, palpitations, headache and such severe "brain fog" that he could not function. After extensive testing, it was found that he had signs of an infection with Ehrlichia, a germ that is often detected in patients with Lyme disease. He had been treated with multiple antibiotics without any sign of improvement, however. On examination he showed a positive tropism for banana. His pulses showed RY>CK

on the left side in a 3/1 ratio, with the zào quality present. The deviation was radially at the guān position, thus his Constitution was LI Excess (Metal Taeyangin). The Conditional pulse exam showed LI at the Fire depth and SI at the Metal depth. His right Subdosha was Pitta and left Subdosha was Fire. This combination indicates the need for the antibiotic Auxiliary Formula. The following Points were treated on the left side with simple stimulation to achieve dé qì, but without needle retention: (UB 66, LI 2- and SI 5, LI 5+) plus (LI 1, SI 1- and GB 41, SI 3+). Since using this type of treatment he has felt essentially back to normal, with only very mild headaches occasionally. Of course it was impressed on him that he should be consuming a "pescatarian" diet, quite different than his prior omnivorous approach. He was also encouraged to avoid all alcohol and other intoxicants. He has been very happy to make these changes, seeing the dramatic improvement in his health over the 4 months since starting acupuncture.

17) D/V/N: A previously healthy 61 year old woman was hit by a car as she was crossing the street 6 months prior to consulting me. Since the injury she experienced constant right-sided neck and shoulder pain with some restriction of movement. An MRI was negative. She had received a lot of acupuncture for prior problems, which had always helped, but this current problem was not helped by her previous acupuncturist, physical therapist or chiropractor. As a result, she felt afraid when crossing the street, and generally was in poor spirits. Her podiatrist insisted she try CCA treatment. On examination she had a positive tropism for carrot, her left side CK was wider than RY with a ratio of 2/1, and there was radial deviation at the cùn position. These findings define her Constitution as UB Deficiency (Water Taeumin). The Conditional exam showed the left Subdosha as Kapha (Wood type) and the right Subdosha as Fire. Her UB pulse was at the Wood depth, and LV pulse was at the Water depth. In addition her SI pulse was at the Wood depth and her TH pulse was at the Fire depth. These findings indicated that her pain syndrome was reflecting a need to treat the Degeneration, Vitalization and Neuro Auxiliary Formulas, all on the left side. This was done using touch treatment with a ratio of 4/4/4 for each part, at the following Points: (KI 10, LV8+ and SP 3, LV 3-) plus (UB 66, SI 2+ and ST 36, SI 8-) plus (TH 1+ and TH 3-). She felt some relief right away and was next seen 6 days later. At that time she reported that she had slept 11 hours after

the initial treatment, and when she woke up almost all the pain was gone. Her spirits were markedly improved. Several months later she was still free of pain from that single initial treatment, and she has stayed pain free during the ensuing years. With such sustained dramatic improvements, I usually refrain from further acupuncture, allowing nature to continue the healing process. In retrospect, I would change the order of treatment components to V/D/N as her Constitution pertains to a Fǔ Organ, but as the results indicate, this treatment order is obviously less important than choosing the correct Constitution, Auxiliary formulas and laterality of treatment.

18) Z/D/N, IPF: A 67 year old woman was referred by her masseuse for treatment of back pain. She described herself as in great health except for "mechanical" problems. For a long time she suffered from low back pain and 9 years previously had a lumbar discectomy after an MRI showed a number of lumbar vertebral abnormalities. After surgery she was able to manage the pain with physical therapy and massage. However, in the last 2 months she had developed similar types of pain in her shoulders and arms, which she described as "nerve-like". These pains were intermittent, but seemed to be aggravated by physical therapy. Her past medical history was presented as completely unremarkable, but on questioning she mentioned that she had fallen down some stairs, not long before the upper body pains began. On exam, her left side showed CK>RY with a ratio of 2/1. The deviation was radial at cùn, so her Constitution was either UB Deficiency (Water Taeumin) or SI Deficiency (Fire Taeyangin), pending confirmation by her Conditional pulses. The Conditional pulses showed 4 locations with abnormal depths: UB was at the Wood depth, LV was at the Water depth, HE was at the Wood depth and TH was at the Fire depth. This is reflective of the pain pattern in people with UB Deficiency Constitutions. Her Constitutional Organ, Urinary Bladder, was calling on its daughter, Liver, for support. The Zàng inflammation Organ, Heart, was therefore reacting to the Liver rather than the Bladder. This is typical of the Z/D/N pattern, so touch treatment was given on the left side as follows, with 4 repetitions of each part: (KI 10, HE 3+ and SP 3, HE 7-) plus (KI 10, LV 8+ and SP 3, LV 3-) plus (TH 1+ and TH 3-). Prior to treatment she could not lie flat on her back without bending one of her knees up, but as soon as treatment was finished she could do so comfortably. When she got off the table she was astonished that all

the pain in both her back and arms had disappeared. I explained to her my impression that she probably had spinal abnormalities in the cervical region, similar to those in the lumbar region. The fall she sustained probably led to further compromise of these vertebrae, which resulted in her recent pains. Of note, these vertebral problems are bone level issues that suggest a need for the Zàng inflammation Formula rather than the Vitalization formula. Just one treatment resolved her several months of pain. I saw her from time to time for other problems (she fractured a bone in her foot after a fall), but she responded well to treatment. After being away for over a year due to the pandemic, she returned for treatment of a recurrence of neck pain. Her pulses were all normal as far as I could tell, except that her TH pulse was at the Fire depth and her Subdoshas were opposite types of the Water Element, which is her Constitution. Testing with silver press pellets indicated that the left side TH 5 needed dispersion. This treatment resulted in an immediate return of her pulses to normal, a marked improvement in kinesiology testing, and relief from any pain. I found this a curious situation because the TH is the Operator of Shaoyang and is associated with the EPF of Fire. She denied ever being burned, not even a sunburn, and had never experienced an electrical shock. I therefore questioned her about her emotional history. She related that even as a child she felt inadequate because she was overweight and unpopular with other children. Nobody else in her family had weight issues. My supposition was that the role of the TH in facilitating the transition from the primitive EEV Constitution to the post-natal Constitution had been adversely affected, probably by something in her mother's life while she was in utero. This would be a kind of pre-natal IPF that because it affected her yuán qì, disrupted her Ministerial Fire. There is also a teaching that the TH function is responsible for developing smooth social interaction skills, and perhaps this began to malfunction in her early childhood. It will be interesting to see if her weight changes following this treatment. Her normal demeanor displays excessive laughter (inappropriate), which would lead Five Element practitioners to diagnose Fire as her Constitution, but she actually has a Water Constitution. I find that this appearance of inappropriate emotions is quite commonly correlated with retained pathogens, rather than as a sign of one's Constitution. When the TH is holding a retained pathogen, it should reflect the level of the Constitution on the pulse. However, since her Constitution is Water, the shift in the TH pulse is to the other attribute of

Ministerial Fire, which is a shift from Water to Fire Element depth. She has been feeling well for the past 3 months and has not scheduled a visit, so I have no recent information on her current weight.

19) D/Z/N: A 39 year old man came for treatment of low back pain and sciatica on the right side, which he had suffered from for 30 years (i.e. practically his whole life). After an MRI showed a ruptured disc at L5/S1, he had discectomy surgery, which failed to resolve the problem, and a subsequent MRI showed a second ruptured disc at L4/5. He has been a patient of mine now for 18 years, believing that periodic acupuncture has kept his pain level manageable. His Constitution is KI deficiency (Left side: RY>CK and ulnar deviation at the chǐ position). Commonly he presents with a need for left side D/Z/N treatment as follows: (He 8, LU 10- and KI 10, LU 5+) plus (HE 8, SP 2- and KI 10, SP 9+) plus (PE 7- and PE 3+) using touch treatment. This has been effective at keeping his discomfort at a minimal level. He often told me that his goal was to be able to resume his favorite sport, golfing, at which he used to excel. Eventually he was able to start playing golf again without aggravating his back, and even got his first hole in one, ever. Apparently the acupuncture not only diminished his pain, but also improved his coordination.

20) Z/D/N: This 53 year old sculptress was referred to me by one of the senior members of the "Scholars of Chinese Medicine" Facebook group. Originally trained as a ballet dancer, she later developed severe back pain radiating from the lumbar area down her right leg. Pilates classes had helped control the pain, but with the Covid pandemic she was unable to continue that coping behavior. An unusual remark she shared was that in her whole life, she had never felt as if she had an internal source of energy, and that she never felt like she could lead a normal life as a result. On examination, her pulses showed RY>CK on the left with a ratio of 3/1. Her LI and LV pulses were both at the Water Element level, KI was at the Metal level and TH was at the Fire level. This pattern reflects the Z/D/N presentation in someone with a LI excess Constitution. I treated her with retained needles on the left side as follows: (LU 8, LV 4- and HE 8, LV 2+) plus (LU 8, KI 7- and HE 8, KI 2+) plus (TH 1- and TH 3+). The pain diminished by more than 70 percent while the needles were in place and stayed better until her next treatment a week later,

which was a repeat of the same points. She called to report how much better she felt, and also said that "for the first time in my life I have enough energy to lead a normal life."

21) C/F/N: A 44 year old woman started seeing me about 5 years ago for multiple medical problems that were both difficult to diagnose and only partially helped by previous acupuncture treatment. Her doctors concluded that she has Sjogren's syndrome, which causes dryness of the eyes, and mucous membranes in general. It is considered an autoimmune disorder. Her most persistent and bothersome symptom has been a feeling of fullness in her stomach that makes eating anything difficult. Her pulses showed right side RY>CK with the zào quality present, left side RY equal to CK without the zào quality, ulnar deviation at the guān position, SP and KI at the Metal depth, LU at the Earth depth and PE at the Fire depth. Her left Subdosha was Vata (Water type) and her right Subdosha was Pitta. This pulse pattern fits only one diagnosis: SP Deficiency Constitution with Congestion of the Fŭ Organ system. The treatment was (SP 3, LU 9+ and LV 1, LU 11-) plus (SP 3, KI 3+ and LV 1, KI 1-) plus (PE 7+ and PE 3-) using touch treatment repeated 5 times. Even though there are a number of Jing Well Points being used, the depth of needling was less than 1 mm, so it was not particularly painful. Following this approach, all her symptoms resolved and her pulses normalized. I am hoping that the discovery of these Congestion patterns will be the answer to keeping her well, but that seems to have been the best approach to treating her so far.

22) C/F/N: A 25 year old female was seen in 2019 for a severely itchy erythematous rash around her left axilla that had been present for 4 months, and showed no trend towards healing. It could be ameliorated with steroid ointments, but always recurred as soon as they were stopped. I had treated her off and on since she was 17, as she was the daughter of another patient. Her original problems had been asthma and obsessive/compulsive thoughts, but these had been successfully managed by a combination of acupuncture and antidepressant medication. However, when the rash appeared, she also found it more difficult to express her emotions, although she knew they were not normal for her. At the same time she became more constipated than usual. She had no idea what had occasioned this change in her state of health, but it appeared

not to be improving with time. On examination, her pulses were not entirely clear to me, other than having RY>CK on the right side, but I diagnosed her as having a Taeyangin Constitution. Her skin and bowel symptoms suggested a Fŭ inflammation Condition, and I decided to treat her as having a LI Excess Constitution (Metal Taeyangin). Thus her Condition would be HE Deficiency. Touch treatment was given on the right side as follows: (UB 66, LI2- and SI 5, LI 5+) repeated 4 times plus (LU 8, HE 4, Lu 8, HE 4- and LV 1, HE 9+) repeated 2 times, (LU 8 and HE 4 were punctured twice each time to direct the effect to her upper body). Right away all her pulses felt normal and her O-Ring test became much stronger. When seen the following week, the rash was more than 95% gone and had stopped itching (Fig. 48). She also reported that she was once again able to express her feelings normally, and that her bowel function had corrected itself. As her pulses seemed normal, no further treatment was given, and she continued to feel well over the next few years.

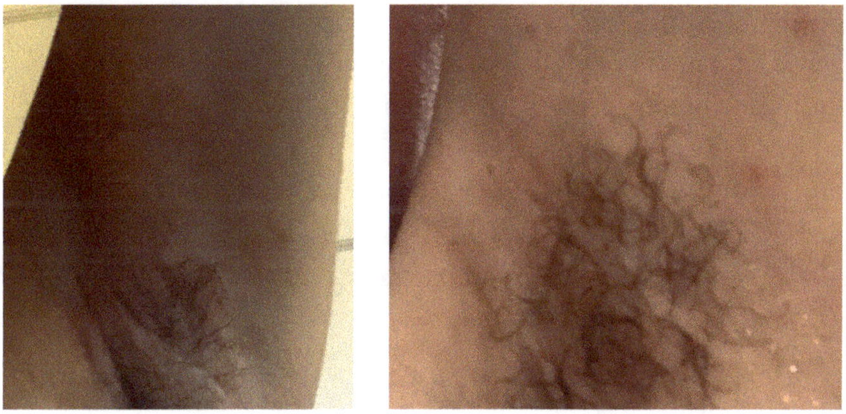

BEFORE **AFTER**

Figure 48: Axilla Rash [260]

However, in hindsight I realized that as she was still on antidepressant medication, she had not completely returned to her original nature. Patients with Fŭ Organ Constitutions usually have RYCK imbalances on the left side, but hers was on the right side, so I was open to considering if she really had a Zàng Organ Constitution. The next time I saw her was two and a half years later, just after she weaned herself off the antidepressant. At that

260 Source: Peter Eckman

time she was again experiencing irregular bowel movements and the onset of a strong case of body odor coming from her axillae. I was better able to evaluate her pulses at that time and they indicated that her Constitution was LV Deficiency (Wood Taeyangin). LV, SP and PE were all at the Fire depth, while HE was at the Wood depth. The left side RYCK quality was not zào and the Subdosha was Kapha (Earth type), while the right side had the zào quality and a Fire Subdosha. Putting all the findings together indicated a need to treat her for Congestion on the right side with simple stimulation as follows: (LU 8, HE 4- and LV 1, HE 9+) plus (LU 8, SP 5- and LV 1, SP 1+) plus (PE 7- and PE 3+). After this the body odor completely disappeared and her bowel function markedly improved. The only remaining pulse abnormality was a very small but solid LI pulse. I discussed with her the Metal Element's need to feel respected, and she immediately stated that she was struggling with that issue in her relationship. Tonifying LV 3 and GB 40 (the Wood Constitution's Source Points) helped soften her LI pulse. I explained that Wood needs to take action, and that her problems would improve faster if she brought up the respect issue in her relationship. She was very grateful for this advice, which made perfect sense to her. I find that often, when CCA treatments have eliminated any RYCK imbalance, it is often helpful to treat the Source Points of the Constitutional Element. This approach illustrates the closeness of CCA to Worsley's Five Element style of diagnosis and treatment. Both are Constitutional systems whose findings should be congruent with each other.

It is interesting to note that both case number 1 and this case of skin disease and history of asthma had Taeyangin Constitutions, but they had different Elements and Organs at the Root. Taeyangin was originally described by Li Jema as typified by a strong Lung and a weak Liver. Both of these individuals satisfied this relationship. Li believed that Taeyangin was a rare Constitutional type (less than one in a thousand), but my experience contradicts that. Using a different lens (pathological presentation) both individuals also responded to treatment aimed at Fǔ inflammation. Finally, we can see that the common association of the Metal Element (LI and LU) with the skin in Chinese medicine is corroborated in both cases, but in quite different ways. Thus three lenses, Chinese medicine, Sasang Constitutional medicine and Korean Constitutional Acupuncture were all accurate in reflecting facets of these individuals, and by combining all three viewpoints, CCA was able to establish a comprehensive diagnosis and an effective treatment for both.

23) B/C/N: I began treating a 79 year old woman in August of 2017 for a medical condition called interstitial pneumonitis. This illness has no known cause in Western medicine, but it often occurs in people who have had chronic bronchitis in the past, and this woman had bronchitis for more than 20 years. Her symptoms were mostly just a persistent cough, but listening to her lungs revealed very pronounced wet rales (a crackling/bubbling sound) in the lower half of her lung fields bilaterally. She also suffered from gastro-esophageal reflux. Her pulses showed RY>CK, radial deviation at the guān position, and the KCA pattern that Kuon called Gastrotonia. These findings indicated that her Constitution was ST Excess (Earth Soyangin). There was an unusual finding however: her RYCK imbalance usually showed up on the right side. I used various treatments based on dispersing the ST, and although her symptoms didn't improve much, they did stop getting worse. When the Covid-19 pandemic started she stopped getting acupuncture treatment, and she began to cough more as time went by. In 2021 she decided to resume acupuncture treatment, and the lung sounds on both sides were more pronounced and had the wet or bubbly sound quality. It was at this time that I began to formulate my understanding of Congestion Conditions. Her right sided RYCK presentation made sense from that perspective. Here is the complete pulse picture: right side RY>CK with no zào quality, left side RY equal to CK with the zào quality present, ST, GB and TH at the Fire depth and HE at the Earth depth. The left Subdosha was Fire and the right Subdosha was Pitta/Vata (Wood type). This pattern indicates Congestion due to infection, and it was treated with the B/C/N formula on the right side as follows, leaving needles in place for 20 minutes: (ST 36, GB 34- and UB 66, GB 43+) plus (SP 3, HE 7- and KI 10, HE 3+) plus (TH 1- and TH 3+). For the first time since starting acupuncture, the lung sounds were at least 50 percent diminished and sounded drier when seen a week later. As of her most recent visit on 10/18/21, she is continuing to improve with this approach, which targets both congestion and infection. Pulmonary function tests PFTs) now show that she has not lost any respiratory function since before the pandemic, an unexpected finding in

those with this diagnosis.

24) CA/8E: At my first workshop in Beijing, in May of 2018, I was asked to treat a 60 year old woman who had a chronic cough, and chest pain radiating to her left arm. 8 months previously she had had surgery for a tumor in her lung. She was told it was benign, but her daughter later informed me that it was actually a cancer. This was shortly before I developed my own understanding of how to diagnose and treat cancer with CCA methods. The patient had begun TCM herbal treatment, which helped for about 6 months, but in the last 2 months the pain and cough had gotten worse. A recent CAT scan showed growth of 2 nodules in her left lung. She woke every night from 1 to 3 AM with pain and shortness of breath. On exam she had a positive tropism for carrot. The deviation was not obvious because she had a Yáng Wéi Mài pattern on the left side, but otherwise I thought the deviation was ulnar at the guān position. She had RY>CK with a ratio of 3/1. I concluded that her Constitution was SP Deficiency (Earth Taeumin), but that because of the Extraordinary Meridian signal, it should be treated before interpreting the Conditional pulses. Thus the first part of treatment was to disperse TH 5 and tonify GB 41 on the left side. I added ion-pumping cords to strengthen the effect. After 10 minutes her O-Ring test was much stronger, so these needles were removed. The Extraordinary Meridian signal was no longer present. Her Conditional pulses now showed Spleen at the Fire depth, Heart at the Earth depth, Gall Bladder at the Fire depth and Pericardium at the Fire depth. The left Subdosha was Fire and the right Subdosha was Kapha (Wood type). These findings indicated pain from devitalization requiring the D/V/N treatment on the left side, which was given as follows: (SP 3, HT 7+ and KI 10, HT 3-) plus (ST 36, GB 34+ and UB 66, GB 43-) plus (PE 7+ and PE 3-). All needles were simply inserted to get dé qì and then removed without retention. Her pulses felt close to normal afterwards, and she was asked to return the following day. She did so and reported that she felt stronger with less pain, and that her breathing was 50% better. On exam her pulses had returned to the same pattern as the previous day (including the Yáng Wéi Mài pattern), so the same treatment was given again. She immediately said that there was an even greater relief of pain and shortness of breath. The improvement was reported as continuing 3 days later, when her daughter called to inform me that the diagnosis was actually lung cancer, but her mother was not told that. I've found that in Chinese culture such avoidance by doctors and family

members of telling patients about cancer is quite common. I did not have a chance to treat her again, but her daughter reported that she was maintaining the improvement 3 days later. It was only months afterwards that I initially discovered what I have been calling the metastatic pulse signal, but in hindsight I wondered why I hadn't felt it on this woman. Perhaps the Chinese herbal treatment she was receiving was effective at inhibiting progression of her tumor. If so, why was her pain and cough getting worse? It is interesting to speculate on the significance of the Yáng Wéi Mài activation. Her abnormalities were clearly one sided, all on the left, including chest and arm pain, so the trajectory involved the upper limb. Additionally, her Condition was a serious one involving multiple cancer lesions in the left lung and surgical intervention there as well. That combination of left/right asymmetry, surgical trauma, upper limb involvement, plus Organ dysfunction suggests that one of the Wéi Mài had been activated. Her presentation of RY>CK indicated to me that the Yáng Wéi Mài was in excess. Both her pulse findings and response to treatment confirmed this diagnosis.

25) CA: This is the gentleman with lymphoma whom I introduced in chapter 14. He was one of the first patients I treated with the anti-cancer protocol described in that chapter. He was a 76 year old man whom I had treated eight years previously for sciatica and two years previously for capsulitis of the shoulder, with complete resolution in both cases. He consulted me again June 7, 2018 for a different problem. During the past six months he had been feeling unwell, with chest congestion, loss of appetite, weight loss of 10-15 pounds, fatigue and a feeling that he wasn't his normal self anymore. His pulses initially were as follows: Conditionally, all the Fǔ organ pulses were widest at the fire element (blood vessel) depth except for that of the LI. His Constitutional pulses showed right side CK>RY with a ratio of 3/1 but no zào quality. The *Màijīng* deviation was radial at the guan position. The ratio indicates either taiyin meridian excess or yangming meridian deficiency. The deviation specifies that it is a yangming Constitution. Absence of the zào quality on the right side means one of the foot meridians is involved. This set of pulse findings therefore defines his Constitution as ST Deficiency (Earth Soeumin). Noting the abnormal pattern of fire element level pulses, together with his clinical history, I shared with him my suspicion that he had some kind of cancer, and that he needed to get it evaluated by his regular physician. He related that he

had already come to the same conclusion, but that he hoped I could treat him with acupuncture until a definitive diagnosis was confirmed. I treated him on the right side with simple stimulation to get "needle sensation" (de qi 得氣) at the following Points: (ST 36, GB 34+ and UB 66, GB 43-) plus (ST 36, SI 8+ and UB 66, SI 2-) plus (TH 1+ and TH 3-) plus (ST 36, GB 34+ and UB 66, GB 43-) plus (ST 36, SI 8+ and UB 66, SI 2-) plus (SI 5, ST 41+ and GB 41, ST 43-) plus (ST 36, UB 40+ and GB 41, UB 65-). This is the Fǔ organ anti-cancer protocol given in Chapter 14 for individuals with ST deficiency Constitutions. This combination targets all the yang meridians except the LI, reflecting the fact that all his fu organ pulses were at the fire element depth except that of the LI. Detecting this difference at the LI position is an extremely subtle finding, so I imagine it would only be noticed if one was looking to verify the applicability of this anti-cancer protocol, but its presence helped in formulating my approach to treating cancer. Right after the first treatment his Conditional pulses all returned to their normal depths. After only two treatments he reported that he was "96 percent" back to normal in all ways except his weight and the persistent chest congestion, but his pulse again showed a recurrence of the metastatic signal. Most importantly however, he felt like himself again. At that time a bone marrow biopsy finally confirmed that he actually had a large-B-cell lymphoma as his primary diagnosis. In response to his pulse findings I gave him the same treatment for 5 additional sessions. Although his only remaining complaints were congestion (probably unrelated to the lymphoma) and slight residual weight loss, he decided at that time to start chemotherapy, as recommended by his oncologist, in conjunction with acupuncture. By 2 months after his initial visit his pulses had consistently stopped reverting to the metastatic pattern. He completed 6 sessions of chemotherapy with only minor side effects and then terminated it. In addition he elected to receive the anti-cancer acupuncture protocol once a month, which he did until March of 2020. Due to the pandemic, he then decided to discontinue regular acupuncture treatment for the moment, as it was almost 2 years after his first treatment, and all his medical tests were normal, plus he felt well. I think a judicious combination of acupuncture and chemotherapy was the right choice for him, however I believe that he is not just in remission, but actually cured. The reason for this conclusion is that his RYCK imbalance changed from the right side back to the left side, which latter situation is the normal finding in people having a ST Deficiency Constitution and some other health issue (probably whatever was responsible for the lingering chest congestion). In my experience, the switch away from normal laterality is one key finding in patients with cancer. I spoke with him by phone in December 2020, 2.5 years after his initial cancer treatment, and he described himself as continuing in excellent health. A brief conversation early in 2021 when he brought his wife in for an acupuncture treatment confirmed his good health report.

26) CA: In October of 2019 I conducted my second clinical workshop in Beijing. One of the patients had a case of blood platelet deficiency with spontaneous bruising (idiopathic thrombocytopenic purpura or ITP) that had lasted 3 years already. His condition was refractory to both Western medicine and TCM. I was able to give him CCA treatment resulting in at least a modest increase in his platelet count, and seeing this, his 56 year old wife asked if I could treat her for pain in her legs that limited any activities involving walking or climbing stairs, which had been a problem for at least 10 years. I agreed, but when I examined her pulses I was surprised to find that all of her Fǔ Organ pulses except ST were at the Fire depth. I asked about her family medical history and was told that her father had died at 65 from lung cancer and her paternal grandfather had died from stomach cancer. Without overthinking the situation, I informed her that I suspected that her most serious condition was some kind of cancer, and that if I were to treat her, that would need to be my target. Everyone in the room, both students and the host Professor, was shocked that I would say that to a patient, as it was taboo in Chinese culture, as I mentioned. However, the patient was not upset, and informed me that her gynecologist had previously speculated about a malignancy based on some abnormal laboratory findings, so she consented to the treatment I proposed. On the left side her RYCK pulses showed CK>RY with a ratio of 3/1 and the quality was zào. The deviation was ulnar at the guān position and she had a positive tropism for potato. Thus her Constitution was LU Excess. I gave her the following treatment on the left side, using simple insertion to get dé qì. (GB 41, SI 3+ and LI 1, SI 1-) plus (SI 5, LI 5+ and UB 66, LI 2-) plus (SI 5, UB 60+ and LI 1, UB 67-) plus (SI 5, GB 38+ and LI 1, GB 44-) plus (PE 7+ and PE 3-) plus (SI 5, UB 60+ and LI 1, UB 67-) plus (SI 5, GB 38+ and LI 1, GB 44-). Right after treatment her pulses were at the normal depths, and her O-Ring test was much stronger. She returned for a second treatment the next day and reported that she coughed up a large

amount of white phlegm after treatment, but felt energized and walked for 2 kilometers without any pain. Her pulses had stayed at the normal depth, but she still showed a RYCK imbalance on the left side, so I repeated the same treatment that day and the following day. Her final report was that the phlegm production was greatly diminished, her energy was much better and she was able to fulfill her dream of walking all around the Forbidden City without any pain. Her case was instructive in showing that the pulses are an amazing guide to acupuncture treatment, and that symptoms improve from CCA treatment, even in cancer where the symptoms might seem unrelated to the underlying diagnosis. Additionally, this type of treatment seems to have resulted in the expulsion of pathogenic factors via phlegm discharge, without any treatment specifically aimed at that result.

27) CA: A 56 year old woman had been diagnosed with breast cancer in May of 2018, 5 months before coming to see me. It was probably early stage 2 as the skin was involved, but a bone scan was negative for other metastases. The primary lesion had been in the right breast, and after surgical removal she had an implant for reconstructive cosmetic reasons. Unfortunately the implant got infected and needed removal plus multiple courses of antibiotics. There was an open wound remaining at the time of her first visit. In addition to the surgery she had been given 25 radiation treatments and was on hormonal chemotherapy (Anastrozole). Her history revealed that all of her female relatives on her mother's side had also had breast cancer, and that she herself had also had uterine cancer 29 years previously, which was treated by hysterectomy. Her biggest complaint was the post-surgical pain. On examination her pulses showed RY>CK on the right side with a ratio of 3/1, and the zào quality was present. The O-Ring test showed a positive tropism for banana. The deviation on her pulse was radial at the guān position. These findings indicate that her Constitution is LI Excess (Metal Taeyangin). With this history, my first question concerning her Conditional pulses was whether or not there was the "metastatic signal" that I have found in other cancer patients. She did indeed have this signal, with all her Fǔ Organ pulses at the Fire depth except UB. I elected to treat her with simple needle insertion (getting dé qì, but without needle retention) on the right side, using the protocol described in chapter 14. The Points were as follows: (LI 1, SI 1- and GB 41, SI 3+) plus (LI 1, ST45-, and GB 41, ST 43+) plus (TH 1- and TH

3+) plus (LI 1, SI 1- and GB 41, SI 3+) plus (LI 1, ST 45- and GB 41, ST 43+) plus (UB 66, LI 2- and SI 5, LI 5+) plus (LI 1, GB 44- and SI 5, GB 38+). This treatment markedly diminished the breast pain and restored her pulse depths to normal. On all her subsequent visits the pulse depths have remained normal, but following the experience of Kuon Dowon, I gave her the same treatment 7 more times over the next 6 months. At that time, more than a year after her initial diagnosis, a follow-up mammogram was negative, suggesting that she is cancer free. It is now more than 3 years since her original cancer diagnosis, and she is probably the most energetic patient in my practice, literally running into the treatment room and jumping up on the table. She continues to feel well, working double shifts as a nurse, and has had no recurrences, but comes in for booster treatments once a month. Although her Fǔ Organ pulse depths are normal, there is occasionally a slight widening of these pulses at the Fire depth in addition. I interpret this as an indication that her body is fighting to prevent a recurrence of neoplasia, and I have referred to this pulse signal as being "prodromal." Each time the anti-cancer acupuncture treatment is administered, this prodromal signal disappears and her O-Ring kinesiology test shows markedly greater strength. Given her past medical history and her family medical history, I believe this once a month booster treatment to be appropriate indefinitely, and she agrees. I hope that others will learn enough of CCA treatment to provide this service when I can no longer practice.

28) CA: A 67 year old woman came for treatment of the side effects of chemotherapy, which she was receiving for a diagnosis of neuroendocrine cancer (carcinoid) confirmed by biopsy, CT scan, PET scan and MRI as stage 4, with metastases in the liver, lung and bone. The cancer was diagnosed two and a half years previously, but did not remit after the initial round of surgery and chemotherapy, so she was on a second, more complex combination of chemotherapy drugs. She was experiencing nausea, insomnia, fatigue and right sided hip pain. On exam she had a positive tropism for potato. On the right side RY>CK with a ratio of 1.5/1, without the zào quality. The deviation was radial in the chǐ position. Thus her Constitution is GB Excess (Wood Soeumin). All her yáng Meridian pulses were at the Fire depth, confirming for me that she was not yet reversing the metastatic process. She was treated with the standard protocol using touch treatment on the right side as

follows: (ST 36, LI 11+) and (GB 41, LI 3-) plus (ST 36, UB 40+ and GB 41, UB 65-) plus (TH 1+ and TH 3-) plus: (ST 36, LI 11+) and (GB 41, LI 3-) plus (ST 36, UB 40+ and GB 41, UB 65-) plus (LI 1, GB 44+ and SI 5, GB 38-) plus (LI 1, ST 45+ and GB 41, ST 43-). This treatment returned her pulses to their normal depths. Her hip felt some relief right away, and 4 days later she had a CT scan that showed some shrinkage of the tumors in her liver and lung. Significantly, her pulses had stayed at the normal depth. Her energy level was improved and her nausea had decreased from just that initial treatment. She subsequently had another round of chemotherapy, and when I saw her 10 days after the 2nd visit, the pulses had reverted to all at the Fire depth again. Repeating the protocol acupuncture treatment restored normal pulse depths once more. Although her hip pain had stayed gone, she was not able to maintain these normal pulse depths in the face of the new chemotherapy regimen. Unfortunately, she was unable to commit to regular acupuncture treatments. When contacted more than a year later in 2021, after the first wave of the Covid-19 pandemic, her husband related that she was doing well and in a stable state of health, reaffirmed 6 months later by her daughter.

29) CA: A 57 year old woman was referred by her doctor to see if acupuncture could help with peripheral neuropathy following cancer chemotherapy treatment. She had a family history of cancer from her father, who had skin cancers, and she had previously had 3 basal cell cancers removed surgically. In 2013 (4 years before her first appointment with me on 7/20/17) she was found to have precancerous cells in her uterus, which were removed surgically, but 4 months after that she was diagnosed with ovarian cancer. Initially she was treated with surgery and taxol, but the cancer recurred and her chemotherapy was changed to a more aggressive version, which caused neuropathy in both arms and legs, with symptoms of weakness and tingling. She was still receiving chemotherapy when I first saw her. Prior to the cancer she had developed osteoarthritis in her hands, and knees, but was otherwise healthy. I was not at all sure about her Constitution initially, as her carotid pulses were difficult to feel, and I was only beginning to investigate the pulse diagnosis and treatment protocols for cancer. Mostly I treated the yáng Meridians, as that had been Kuon's approach with cancer, and her symptoms of neuropathy resolved fairly quickly. By 2 months after her initial visit she said she was back to normal for her, the way she had

felt before developing cancer. I saw her sporadically after that, and her only real complaint was painful arthritic knees. Then 9 months after her initial visit (April 2018) she scheduled an appointment because there had been a recurrence of cancer, which presented with pelvic pain. The CA 125 cancer marker in her blood tests had increased from 4 (normal) to 13.9 (elevated), and she again had surgery plus both chemotherapy and radiation therapy. By this time I was able to make a Constitutional diagnosis of SI Excess (Right side RY>CK with a ratio of 2/1 and presence of the zào quality bilaterally. She has radial deviation at cùn, therefore SI excess, or Fire Taeumin), but her Conditional pulses did not show the metastatic pattern I had now identified in other cancer patients. Apparently the Western medical treatment had put her into remission. At present, more than 2 years after the last cancer episode, she has not shown any metastatic pulse signs, only the pattern expected for arthritis. The CA 125 marker is currently at 4.3, which is in the normal range. A recent treatment was for the arthritic knees, as her pulses showed the following pattern: SI was at the Metal level and LU was at the Fire level. The right Subdosha was Fire and the left Subdosha was Pitta. I treated her with touch treatment: Right side (UB 66, SI 2+ and ST 36, SI 8-) 5 times plus (KI 10, LU 5, KI 10, LU 5+ and HE 8, LU 10-) 3 times. This directs the Zàng inflammation effect to the lower body to treat the knees. Her knee pain was much reduced immediately after this treatment. I have since monitored her pulses every 3 to 4 weeks to see if it is possible to catch and treat any metastatic change before it develops into a physical tumor, which it has not done in over 3 years. Meanwhile I am giving her the cancer protocol acupuncture treatment as a preventive measure at these intermittent visits. At her visit on 5/20/21, I gave her the following treatment on the right side: (LI 1, UB 67+ and SI 5, UB 60-) plus (LI 1, GB 44+ and SI 5, GB 38-) plus (TH 1+ and TH 3-) plus (LI 1, UB 67+ and SI 5, UB 60-) plus (LI 1, GB 44+ and SI 5, GB 38-) plus (UB 66, SI 2+ and ST 36, SI 8-) plus (UB 66, LI 2+ and SI 5, LI 5-). One key finding is the presence of a RYCK imbalance on the right side. Most non-malignant illnesses of patients with Fu Constitutions have RYCK imbalances on the left side, so it is very important to look for metastatic signals when this imbalance is on the right side, even if it's only in the prodromal stage, which I believe to have been the case with her on most visits. Her lab tests in November 2020 showed that her CA125 cancer marker was staying in the normal range, as were all of her routine blood tests. At her visit in December 2020 her RYCK imbalance finally appeared on the left side, and her CA125 level was less than 5 (normal), so I believe that she may finally be cured from cancer. As of June 2021 there had been no increase in the CA125 test. A PET scan in February 2022 showed that there was no evidence of cancer, now more than 4 years after starting acupuncture, and currently relying only on monthly anti-cancer acupuncture treatment to prevent a recurrence, following stage 4 cancer.

30) CA: A 68 year old man sought treatment for prostate cancer. He had lived with benign prostatic hypertrophy for over 25 years, which had not been responsive to Flomax medication. Over the years he had experienced a number of bladder and prostate infections, but controlled them with herbal treatment. His PSA levels had increased over the years. In 2014 they varied between 11 and 14, but by 2017 they increased to 22, and most recently the reading was 30.2. MRI and biopsy confirmed the cancer diagnosis, but a PET scan showed no metastases to his bones. The history revealed that both his father and paternal grandfather had died from prostate cancer, while both his mother and sister had developed thyroid cancer. On exam he had a positive tropism for banana. His pulses showed right side RY>CK with the zào quality present and the deviation was radial at the guān position, thus his constitution was LI Excess (Metal Taeyangin). Conditionally, all his yáng Meridian pulses except UB were at the Fire depth, so I treated him with the cancer protocol on the right side using touch treatment as follows: (LI 1, SI 1- and GB 41, SI 3+) plus (LI 1, ST 45- and GB 41, ST 43+) plus (TH I – and TH 3+) plus (LI 1, SI 1- and GB 41, SI 3+) plus (LI 1, ST 45- and GB 41, ST 43+) plus (UB 66, LI 2- and SI 5, LI 5+) plus (LI 1, GB 44- and SI 5, GB 38+). After this treatment there were no abnormal depths on any of his pulses and the RYCK imbalance had disappeared. He reported that on the second night after this treatment he had the most vivid dream of his life, in which he was conscious that he was dreaming, but definitely asleep. He was faced by an authoritarian teacher in his dream, but he told the teacher, "I am a 68 year old man. I don't have to do what you say!" He interpreted this dream as "reclaiming his power". I saw him for a follow-up visit two and a half weeks later. At that time his pulses had stayed at their normal depths, but I confirmed a finding that I have mentioned as occurring in some of my other cancer patients: Although the pulses were at their proper depths, there was a slight widening at the Fire depth remaining on all the yáng Meridian pulses except for UB. I interpret this as a "prodromal sign" that the

body is battling to prevent the cancer from recurring. If that is true, then its absence might provide a useful indicator of when treatment has achieved as close to a cure as possible. Of course, all this is highly speculative at this point, however, I did repeat his treatment and this prodromal sign disappeared, along with a major strengthening of his O-Ring test. When San Francisco went into lockdown during the pandemic he decided to forgo acupuncture treatment. As of June of 2020 he was feeling quite well, when contacted by telephone, but when I spoke to him 6 months later, in January of 2021, he related that a slow increase in his PSA led him to repeat the PET scan. It showed that there was some recurrent tumor in the original site. He is still too worried about Covid-19 to come for acupuncture, but I hope he will change his mind. His clinical course reinforces my opinion that all cancer patients successfully treated with acupuncture should receive a preventative booster treatment once a month, at least until they have been cancer free for five years, as this time period is a common standard for comparison with other modalities of treatment. Those of my patients who have elected to get booster treatments once a month have so far remained cancer free. I will continue to follow his progress.

31) CA: My second case of prostate cancer was in a 70 year old man with a long history of intermittently elevated PSA tests. Ten years previously a prostate biopsy was read as benign, but in the interim his PSA had fluctuated to as high as 60, which is quite abnormal. He consulted me on 07/22/20 shortly after a second biopsy showed an adenocarcinoma with a 4 out of 5 Gleason scale reading, indicating a very aggressive type of tumor. He was otherwise in good health and the only notable finding was that his brother had been treated for a meningeal type of brain cancer. On examination, all of his yáng Meridian pulses except ST were at the Fire depth. On the right side RY>CK, and the zào quality was present bilaterally. The Subdoshas were of the Fire type bilaterally as well. He showed a positive tropism for banana. These findings suggested that his Constitution must be SI Deficiency (Fire Taeyangin), which is in the same Sasang family as that of the previous prostate cancer patient, but with a different Element organizing his Constitution. His touch treatment protocol was (LI 1, UB 67- and SI 5, UB 60+) plus (LI 1, GB 44- and SI 5, GB 38+) plus (TH 1- and TH 3+) plus (LI 1, UB 67- and SI 5, UB 60+) plus (LI 1, GB 44- and SI 5, GB 38+) plus (UB 66, SI 2- and GB 41, SI 3+) plus (UB 66, LI 2, and SI 5, LI 5+), all on the right side. Although his

pulse depths were back to normal right after this treatment, by the next week they had reverted to the metastatic pattern. This situation continued for 5 visits, which led me to suppose that it somehow correlated either with the aggressive cell type of his pathological diagnosis, or with his omnivorous diet, which I felt to be contraindicated for anyone with a Taeyangin Constitution. I encouraged him to avoid eating any land animal products. After his fifth treatment he changed his diet and also began receiving hormone injections to suppress testosterone. From then on he never again showed the metastatic signal in his pulses, however his sexual functioning and energy level were diminished, a common side effect of the hormonal treatment. His RYCK imbalance subsequently switched to the left side (normal) until he began radiation treatments in addition to the hormonal therapy. It may be coincidental, but the RYCK imbalance recurred on the right side soon after radiotherapy began. He received acupuncture twice a week after that finding, and eventually maintained a left sided RYCK imbalance until radiation was terminated. He had received 20 acupuncture treatments in total until then, at which time I switched the approach to a once per month booster treatment. His oncologist is following his PSA, which was reported as undetectable in March of 2021, and is still in the normal range as of September of 2021, and he continues feeling well. I can find no abnormalities in his pulses. I think there is a good chance that he is now free of cancer.

32) CA: A 68 year old accountant was referred on 4/6/21 by his podiatrist, who is also a patient of mine. He described his problem as pain from his right hamstring radiating to the groin, which had been present for 20 years, but in fact he had been diagnosed via biopsy with an aggressive prostate cancer 3 months prior, and was clearly upset about the options for treating it. The cancer was said to be at stage 3 with a Gleason score of 6 to 7, both of which indicate quite an aggressive tumor. His oncologist was undecided about treatment, as the patient absolutely refused the hormonal therapy that had been suggested. On examination he showed the metastatic signal of multiple yáng Meridian pulses at the Fire Element depth, but it is particularly difficult to perform a thorough pulse diagnosis on him due to a combination of continual tremors and extreme muscular tension. I believe his Constitution is GB deficiency, which I am treating on the right side with the following protocol: (ST 36, LI 11- and GB 41, LI 3+) plus (ST 36, UB

40- and GB 41, UB 65+) plus (ST 36, GB 34- and UB 66, GB 43+) plus (TH 1- and TH 3+) and (ST 36, LI 11- and GB 41, LI 3+) plus (ST 36, UB 40- and GB 41, UB 65+) plus (LI 1, GB 44- and UB 66, GB 43+) plus (ST 36, SI 8- and UB 66, SI 2+), all with touch treatment. This markedly reduced his muscle tension and changed his pulse depths to the prodromal levels initially and then to normal. He subsequently had surgery to remove the original lesion on 7/27/21, and the radiology report indicated that there were clear margins, so the entire tumor was successfully removed. A preliminary PET scan prior to surgery showed no spread beyond the prostate. His current pulses reflect the pattern of chronic pain (Z/D/N), but no signs of malignancy.

33) CA: I started treating this 64 year old woman in March of 2018. She wanted help with arthritis in her hands and knees, which had bothered her for more than 20 years (Fig. 49).

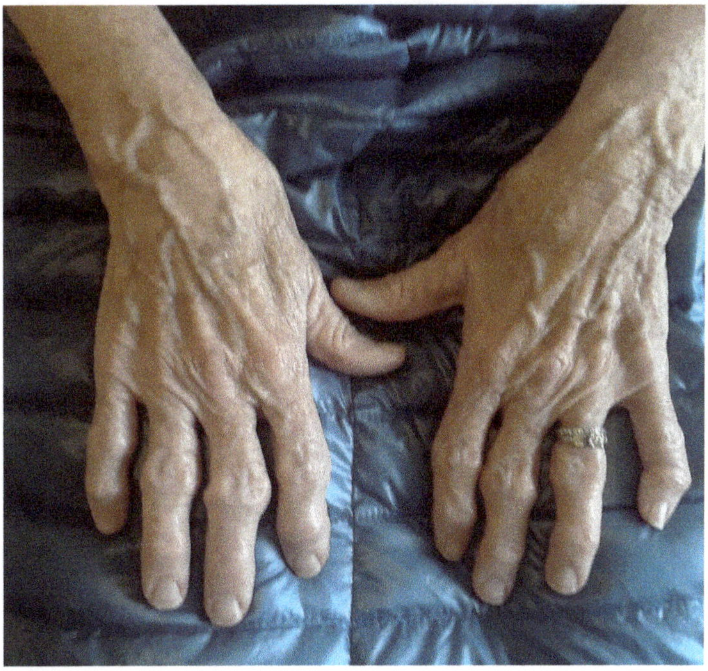

Figure 49: Arthritic Hands [261]

261 Source: Peter Eckman

She took anti-inflammatory drugs daily, although she felt they didn't help very much with the pain and stiffness. Her inter-phalangeal joints were obviously swollen, and x-rays confirmed the diagnosis of osteoarthritis. Lately she had been dropping things because her grip had been very weak. In taking her history, I noted that she had been diagnosed with melanoma 4 years previously, and that there had been a recurrence 2 years ago. Both episodes were treated surgically. Her mother also had a history of both arthritis and cancer. More than 30 years ago she had developed hyperthyroidism and was treated with radiation. Since then she has taken replacement thyroid medication. On examination her right side showed RY>CK with a ratio of 3/1 and no zào quality present. The *Màijīng* deviation was ulnar at the guān position, so her Constitution was SP Deficiency (Earth Taeumin). Both Root/Zàng inflammation and Degeneration/ Vitalization/Neuro treatments helped with the arthritis pain on different visits, and her grip strength improved. In January of 2019, ten months after her initial visit, I noticed that her yáng position pulses were all at the Fire Element depth, although she was generally feeling well, and had less arthritic complaints. I explained what I believed to be the significance of her pulse status as being a metastatic signal, and urged her to see her oncologist ASAP. A lump was found in her thoracic region and a biopsy revealed that she did have a recurrence of melanoma, now stage 4, with metastatic nodules in various places. I gave her 3 treatments of my original idea about the correct cancer protocol for Zàng Constitutions, on the left side, using regular needling with getting a dé qì response at each Point. The Point prescription was: (ST 36, GB 34+ and UB 66, GB 43-) plus (ST 36, SI 8+ and UB 66, SI 2-) plus (PE 7+ and PE 3-) plus (ST 36, GB 34+ and UB 66, GB 43-) plus (ST 36, SI 8+ and UB 66, SI 2-) plus (SI 5, ST 41+ and UB 66, ST 44-) plus (ST 36, UB 40+ and GB 41, UB 65-). Following these treatments her pulses returned to their normal depths. Meanwhile, she decided with her oncologist to begin an immune boosting type of chemotherapy (Keytruda), and 2 weeks later she started the intravenous treatment. There were no obvious side effects from the first 2 rounds of chemotherapy, but after the 3rd infusion her pulses once again showed the metastatic signal (after a lag of several weeks). I explained that I thought it possible that the chemotherapy was effective in killing her cancer cells, but possibly weakened her ability to prevent new cancer cells from developing. We decided to combine both acupuncture and chemotherapy in order to treat both issues. After

2 more anti-cancer acupuncture treatments, her pulse depths returned to normal again. She has not had a return of the metastatic pulse pattern since then (with a few brief exceptions to be described), but often shows the Zàng inflammation pattern typically found in arthritis, which I have treated as in her initial visits. On a subsequent appointment she was feeling very well, with only minimal arthritic discomfort, but I was able to detect the subtle prodromal pulse finding: Although her Fŭ Organ pulses were widest at their normal depths, there was a somewhat wider pulse at the Fire depth than at either the Metal or Earth depths on these pulses. I think this indicates that her body was struggling to prevent a new metastasis. According to the American Cancer Society, the 5-year survival rate for stage 4 melanoma is only 15–20 percent.[262] I discussed with her the option of using the anti-cancer protocol as ongoing preventive care. She was very happy with the idea and eager to try it. Treatment did diminish the width of the pulses at the Fire depth, but not eliminate it, so I felt encouraged. However, I have felt that patients with Zàng Constitutions have not had as lasting an improvement in their pulses as those with Fŭ Constitutions, so after careful attention to pulse depths and to the importance of mirror image symmetry between these two kinds of Constitutions, I revised the cancer protocol for patients with Zàng Constitutions, as was described in the chapter on cancer. The new prescription for her is (ST 36, GB 34+ and UB 66, GB 43-) plus (SI 5, ST 41+ and GB 41, ST 43-) plus (ST 36, LI 11+ and GB 41, LI 3-) plus (ST 36, UB 40+ and GB 41, UB 65-) plus (PE 7+ and PE 3-) plus (ST 36, LI 11+ and GB 41, LI 3-) plus (ST 36, UB 40+ and GB 41, UB 65-). The Points used in this version are only slightly different than the original ones, but the order is quite different. This protocol has been equally effective in restoring normal pulse depths in the short term, but further follow-up is needed to see if it can prevent or reverse any chemotherapy associated pulse changes. In June of 2019 she had a repeat PET scan that showed that her previous tumor nodules had shrunk in both size and number, and that they had also diminished in signs of malignancy (hypermetabolic signals). Additionally, there were no new metastases. At a later treatment I added dispersing UB 20, the Back Shū Point of the SP, (her Constitution), and this was finally successful in temporarily establishing equal RYCK pulses. It also immediately relieved a

262 https://www.medicalnewstoday.com/articles/322765.php accessed 5/17/19

persistent area of chest pain that her physician had not been able to diagnose. She has been very motivated to continue using acupuncture as frequently as possible (mostly twice a week), given her statistical prognosis. As I reported earlier, Kuon routinely used the same treatment repeatedly on his cancer patients almost daily, with good results, so I have followed his lead in that regard. As of March of 2020 she had continued the chemotherapy regimen for four months, and while her pulses continued to show prodromal signs consistent with residual inactive cancer lesions, there had been no recurrence of the earlier metastatic signal, so I expected that her next scan would show further tumor shrinkage. In fact, her PET scan on November 18, 2020 and a concurrent CT scan showed no active cancer indications, and her pulses did not show the metastatic signal, nor even prodromal indications. Her oncologist agreed that it was time to terminate the chemotherapy. Since then the RYCK pulse quality has not been zào on either side, which would be unusual if cancer was still present. I interpreted these findings as an indication that her cancer was at the very least in remission, and possibly cured. She has been off chemotherapy for almost a year now, and has been very motivated to continue using acupuncture as frequently as possible, given her statistical prognosis. I discussed her case with my personal dermatologist, and he was quite surprised that she was both alive and feeling well. There is however an important episode to include in her case history. During the initial wave of the Covid-19 pandemic I closed my clinic for a little over 2 months. She was OK for the first month, but then began to feel very weak and to lose weight, and a friend told me that she was essentially bed-bound. As soon as I reopened my clinic, she resumed treatment and rapidly returned to feeling normal. I received an email about her from her niece, who is an oncology nurse. Her niece was convinced that resuming acupuncture was the only explanation for the reversal in her aunt's declining health, having herself been involved in the care of many cancer patients. This patient's most recent scans, (PET on 5/18/21 and CT on 8/16/21), were again free of any sign of cancer, at which point I decided that she was apparently cured, and changed her treatment frequency to once a month, as a preventive measure. The importance of this preventive approach was demonstrated by the following observation: This woman has been on daily replacement medication with hydrocortisone, 15 mg per day, because of pituitary damage resulting from one of her chemotherapy drugs, Keytruda (a situation that was also true for my other melanoma patient who was treated with this drug). As an experiment to see if she could improve her residual mild fatigue, she tried increasing the hydrocortisone dose to 30 mg per day, but it did not change the fatigue, so she reduced the dose to 15 mg again. Unfortunately, on her next preventive visit her pulses surprisingly showed the metastatic signal again. It only took one treatment to return her pulses to normal, most likely because this change was caught right away. My best guess as to the reason for this incident was that corticosteroids suppress the immune system, and that increasing the dosage had interfered with the immune surveillance responsible for destroying new cancer cells. In any case, she has continued feeling well and shows no other signs of a return of cancer as of February 2022.

34) CA: This 68 year old woman is the sister of the previous woman with melanoma. There were 3 sisters all together. The third one had recently died from lung cancer. This new patient had in fact been feeling well, but was found to have high blood pressure. As part of the work-up for treating her blood pressure, she had a chest x-ray that showed a spot in the right lung field. A preliminary CT scan was read as normal, but a subsequent scan showed a suspicious nodule. At that point she consulted me. Like her sister who I am treating, she also had previously had 2 episodes of localized melanomas excised surgically in the past. Aside from high blood pressure and elevated cholesterol, for which she took medication, she otherwise had always been healthy. Here are her pulse findings: Left side RY>CK with a ratio of 3/1, but no zào quality present. All of her yáng position pulses were at the Fire Element depth. The *Màijīng* deviation was ulnar at the guān position, so her Constitution was SP Deficiency (Earth Taeumin), exactly the same as her sister. I told her that I believed she definitely had some type of cancer, and that an excisional biopsy was the best immediate course of action. This was carried out the following week, and the pathological diagnosis was non-small-cell lung cancer. I checked her pulses and found that the excision of the tumor had not changed her findings at all. Therefore I started treating her with acupuncture, using the same treatment as I am giving her sister. Immediately after this treatment her pulses reverted to their normal depths and RY became equal to CK on the left side. Kuon Dowon had repeatedly claimed that Constitutions were genetically inherited, as were susceptibility to certain illnesses, and this pair of cases seems

to illustrate these notions. This patient was out of town for a month during which she had no treatment of any kind, and on examining her on the next visit, she had perfectly normal pulse depths. A follow-up PET scan was also perfectly normal. She was unable to continue with acupuncture during the pandemic, but she felt well anyway. My impression is that the combination of early surgery and anti-cancer acupuncture has led to an apparent cure. Her oncologist does not think she needs any further treatment other than periodic monitoring, and I plan to take the same approach. In August of 2020 she returned because she had developed a case of sciatica. On examination she showed no signs of cancer in her pulses, and immediately felt better after a treatment for sciatica.

35) CA: This 65 year old woman was first seen by me in 2011, way before I had formulated the CCA methods of diagnosis and treatment. Two months prior to her first visit she had surgery to remove a malignant meningioma, but the surgeon was not able to excise all of the tissue involved. The pathology report classified her tumor as stage 2. Her presenting symptoms had been dizziness and weakness of the left arm and leg. These problems had improved after surgery, but she was experiencing increasingly severe headaches. In addition she had maturity onset diabetes that was not well controlled, elevated cholesterol and emotional stress, all of which were being treated pharmaceutically. I treated her intermittently for many years, but was not able to get her to follow a healthier diet. In spite of this, she always responded to acupuncture treatment with improvement in her symptoms. She didn't receive any acupuncture for 2 years, but periodic MRIs showed that the meningioma was slowly increasing in size, so in July of 2019 she underwent radiation treatment, and then returned to my office for evaluation. Having in the interim developed the CCA approach to treating cancer, I was curious to check her pulses. I diagnosed her as having a SP Deficiency Constitution, and all her yang position pulses were at the Fire depth. Her treatment was thus the same as the previous 2 cases, and her pulses normalized right after treatment. However, she was only able to come for treatment once a month, so that the metastatic signal in her pulse returned by the time of her next visit until 2/20/20 when the Fire depth pulses were only slightly widened, a hopeful sign. Then the pandemic caused her to suspend treatment until 7/20/20, and at that time the metastatic signal

was clearly present again, so the cancer protocol was repeated. An MRI 1 week later showed no tumor progression. She is still only able to come for treatment once a month. Her MRI evaluation in May of 2021 showed no progression of the meningioma, however the cancer protocol acupuncture has been effective in clearing her headache, dizziness and improving her mood and mental clarity. After several monthly protocol treatments, the metastatic signal has disappeared again.

36) CA: A 53 year old woman consulted me in March of 2015, two years after having been treated for Waldenstrom's myeloma, a type of lymphatic cancer. Her body hardly produced a certain type of gamma globulin (IgA), so she needed injections of gamma globulin as replacement therapy. She had undergone almost 9 months of chemotherapy, but still had abnormal blood tests and was experiencing neuropathy in her hands and feet, night sweats and fatigue. She was on an experimental type of chemotherapy when I first saw her, but eventually stopped it because of the side effects. Her medical history was complex, including treatment with steroids lasting 5 months for systemic lupus erythematosis (SLE) more than 10 years before developing cancer, and she had also had 3 surgeries for back pain. There was no family history of any kind of cancer, but both her parents had cardiovascular disease. I have been treating her now for 6 years (starting well before I had developed any protocols for treating cancer), and it has taken this long to get her to the point where she now feels basically normal again, and although her laboratory tests for myeloma are normal, she has other immune deficiencies by laboratory testing. In 2016 I did notice the presence of multiple yáng Meridian pulses at the Fire depth, which I later identified as the metastatic signal, but which at the time I simply treated by focusing on balancing the yáng Meridians. By the time I was able to accurately identify her Constitution, she no longer showed a metastatic signal, and has not had it return since. Her Constitution was not easy to diagnose, perhaps because of all the medications and surgeries. I now think her Constitution is UB Deficiency (Water Taeumin). At various times her oncologist has wanted to put her back on chemotherapy, but she refused, as it always made her feel worse without any observable benefit. The cancer has not returned during the six years that I have been working with her, and I plan to continue to monitor her pulses for any sign of a recurrence of the metastatic signal.

37) CA: A second woman with Waldenstrom's macroglobulinemia has recently started treatment with me. This 64 year old retired physician became ill two years ago, with the onset of peripheral neuropathy, shortness of breath, loss of appetite and hypersensitivity to all sensory stimulation. It took a year to discover the cause, finally revealed by a bone marrow biopsy. She has not gotten any Western medical treatment other than immunoglobulin replacement. Her pulses show left side RY>CK in a 3/1 ratio, ulnar deviation at the guān position, and GB, UB, LI and PE all at the Fire depth, so she is still showing the metastatic signal. This pulse pattern confirms her Constitution as LU Deficiency, and I have been treating her with touch technique repeated 4 times as follows: (GB 41, SI 3- and LI 1, SI 1+) plus (SI 5, LI 5- and ST 36, LI 11+) plus (SI 5, UB 60-, and LI 1, UB 67+) plus (SI 5, GB 38- and LI 1, GB 44+) plus (PE 7- and PE 3+) plus (SI 5, UB 60-, and LI 1, UB 67+) plus (SI 5, GB 38- and LI 1, GB 44+). She no longer is experiencing shortness of breath, she is sleeping better and the painful neuropathy is almost gone from her hands and somewhat diminished in her feet. She periodically has her immunoglobulin levels tested and had not experienced any improvement prior to starting acupuncture, nor have the levels changed in the few months since starting acupuncture. However, based on her symptomatic response I am hoping that this laboratory test will finally start to show recovery as a result of acupuncture treatment.

38) CA: I have been treating this 67 year old man for about 3 years (since June of 2018) as a result of a request from his wife, who is a long-term patient, insisting he see me. He had been diagnosed with stage 3 malignant melanoma. Initially he only had skin and lymph node involvement, which was treated surgically, but 6 months later he developed a metastasis in his neck. Otherwise he had been healthy except for arthritis in his hips. Concurrent with starting acupuncture, he also began an "immunotherapy" type of chemotherapy (Keytruda). On exam his left side showed RY>CK with a ratio of 3/1 and the zào quality was present. The deviation was ulnar at the guān position. Thus his Constitution is LU Deficiency (Metal Soyangin). His right side radial artery is aberrantly located, so I have had to rely only on his left side pulses for diagnosis. His left side Fǔ Organ pulses except UB were at the Fire depth, so I treated him, with simple insertion to get dé qì, using the original cancer protocol I proposed, on the left side as follows: (GB

41, SI 3- and LI 1, SI 1+) plus (GB 41, ST 43- and LI 1, ST 45+) plus (PE 7- and PE 3+) plus (GB 41, SI 3- and LI 1, SI 1+) plus (GB 41, ST 43- and LI 1, ST 45+) plus (SI 5, LI5- and ST 36, LI 11+) plus (SI 5, GB 38- and LI 1, GB 44+). It took 5 treatments before his pulses stopped showing the metastatic signal at the Fire depth, but since then they have mostly been at a normal depth. He felt well as time went on, and his concern shifted to the arthritis in his hips, which kept him from playing golf. I decided to treat him one time for the hip pain with the D/Z/N protocol on the right side, and he has been able to play golf again since then with almost complete relief of pain, but I decided to change to using the newer cancer protocol for Zàng Constitutions, since it appears to give more lasting results. As a preventive measure, I have been using the anti-cancer protocol about once a month, as the 5 year survival rate for stage 3 melanoma is only 23 percent.[263] His new prescription is: (GB 41, SI 3- and LI 1, SI 1) plus (SI 5, LI 5- and ST 36, LI 11+) plus (SI 5, UB 60-, and LI 1, UB 67+) plus (SI 5, GB 38- and LI 1, GB 44+) plus (PE 7- and PE 3+) plus (SI 5, UB 60-, and LI 1, UB 67+) plus (SI 5, GB 38- and LI 1, GB 44+). After many acupuncture treatments, he began to have weak, but perceptible pulses, on the right side. I partly decided to adopt this protocol change after identifying ST as the pulse that showed no tendency to widen at the Fire depth. Recently (8/31/21), more than 3 years after his initial visit, a repeat PET scan showed that there were no remaining tumors present. His oncologist agreed a while ago to terminate chemotherapy, believing that his cancer is cured. As a result of the Keytruda treatments, he has needed to be on replacement cortisone therapy indefinitely due to pituitary damage, a similar situation as in the other melanoma patient I have treated, but he is otherwise in good health. He has elected to continue preventive treatment once a month.

39) CA: This 67 year old woman has been a patient of mine for over 22 years. She has generally been in good health, with no serious illnesses, but a little over 2 years ago she had an abnormal mammogram, and needle biopsy diagnosed an early cancer. I definitely did not feel the "metastatic signal" of 5 Fǔ Organ pulses at the Fire depth. I discussed this finding with her and supported her decision to treat this with a simple lumpectomy procedure. The tumor was

263 https://www.cancer.org/cancer/melanoma-skin-cancer/detection-diagnosis-staging/survival-rates-for-melanoma-skin-cancer-by-stage.html accessed 5/17/19

found to be stage 0-1 (no invasion), and she has had no further issues in the following 2 years related to cancer. Based on her case, my interpretation has been that stage 0-1 tumors do not show what I decided to call the "metastatic signal". I did have one other patient with a similar situation, who elected to have surgery, chemotherapy and radiation, even after I shared my impression that her tumor was not acting in an invasive manner. This second patient had subsequently lost a significant amount of weight and all her hair by one month later, shortly after starting Western medical cancer therapy, and in fact she now did have the metastatic signal present! I wonder if her medical treatment had actually produced the opposite effect than was intended. In any case, she decided at that time to discontinue acupuncture, and I have not heard from her since. I hope she made the right choice for herself.

40) CA: This 61 year old man consulted me after a CT scan showed 2 nodules in separate areas of the lungs. His regular doctor was highly suspicious that these were malignant nodules and referred him to a pulmonologist for a PET scan. Before this could be arranged, the patient consulted me for evaluation. His pulse diagnosis was ST Excess Constitution (Earth Soyangin) with left sided C>R in a 3/1 ratio. The Conditional exam did not show Fire Element stagnation of the Fǔ Organ pulses, so I reassured him that the likelihood of cancer was extremely small, in my opinion, but that he should follow through with the PET scan and pulmonology consult. Two weeks later he wrote me that the pulmonologist thought the nodules were scarring from an old rib injury, and that there was no evidence of malignancy on the PET scan.

41) CA: This 62 year old man was diagnosed with squamous cell cancer of the tongue in August of 2020. He was treated surgically, needing three operations to remove all the malignant tissue, but was then told that he was cancer free. As I had been treating his wife for cancer for several years, he decided to consult me for a second opinion on his prognosis. His initial visit was 2/23/21, about three months after his final surgery. Unfortunately, his pulses did in fact show the metastatic signal, with all the Fǔ Organ pulses at the Fire depth except the UB, indicating that surgery had definitely not been curative. On the right side RY was wider than CK in a ratio of 3/1, and the deviation at CK was radial at the guān position. The left Subdosha was Fire and the right Subdosha was Pitta, so it was clear that his Constitution was LI Excess. I elected to treat him

with simple needle insertion (getting dé qì, but without needle retention) on the right side, using the protocol as follows: (LI 1, SI 1- and GB 41, SI 3+) plus (LI 1, ST45-, and GB 41, ST 43+) plus (TH 1- and TH 3+) plus (LI 1, SI 1- and GB 41, SI 3+) plus (LI 1, ST 45- and GB 41, ST 43+) plus (UB 66, LI 2- and SI 5, LI 5+) plus (LI 1, GB 44- and SI 5, GB 38+). This resulted in an immediate correction of all the abnormal pulse depths, and his kinesiology test showed improved neuromuscular coordination. Since he was not receiving any other cancer treatment, I was not surprised to find that on follow-up exam two weeks later his pulse depths had remained normal. They have continued to be normal for the ensuing seven months, with only once a month booster treatments, and he feels quite well in general. His case exemplifies why I prefer to prioritize surgery as the best Western approach for treating cancer, as long as it is accompanied by CCA acupuncture. Surgery does not appear to have many of the potential side effects of chemotherapy or radiation, including the possibility of triggering new malignancies. I am not opposed to the use of these other Western methods when surgery is not an option, but these other therapies seem to require longer and more frequent acupuncture in order to abolish the metastatic signal.

42) CA: This 69 year old woman was first seen by me 10/07/20 after battling lung cancer for more than 3 years. Multiple relatives including her mother had died from cancer. The initial diagnosis was non-small cell cancer detected on a CT scan for hip pain following a fall. There were already metastases to bones, making her cancer stage 4. She went through all the biomedical treatments including surgery, radiation and chemotherapy, which seemed to delay progression at first, but a CT scan in February 2020 showed that the lung tumor was growing larger. Biopsy revealed that the cell type had transformed into the more aggressive small cell variety. A month before consulting me a scan showed that there was tumor newly present in 2 lymph nodes near her pleura, and she was started on even more powerful chemotherapy. In spite of the seriousness of her illness and its treatment, her only complaints were mild nausea and fatigue. Her home was in the Virgin Islands and she was only temporarily in San Francisco to get medical treatment and visit her daughter. Like the other cancer patients I've treated, her pulses showed the metastatic signal, with all Fǔ Organ pulses at the Fire depth except for UB. The RYCK imbalance was on the right side and the left side Subdoshas reflected Vata/Kapha. These findings

were consistent with a Constitutional diagnosis of LI Deficiency. I treated her with touch treatment on the right side twice a week at the following protocol Points: (LI 1, SI 1+ and GB 41, SI 3-) plus (LI 1, ST 45+ and GB 41, ST 43-) plus (TH 1+ and TH 3-) plus (LI 1, SI 1+ and GB 41, SI 3-) plus (LI 1, ST 45+ and GB 41, ST 43-) plus (ST 36, LI 11+ and SI 5, LI 5-) plus (LI 1, GB 44+ and SI 5, GB 38-). After the initial treatment she reported that her energy was improved, as was her appetite. For the first time since beginning chemotherapy she was able to eat without needing THC supplements to stimulate her appetite. Amazingly, her pulses were at their normal depths at the time of her second visit, but then she had another round of chemotherapy, and the metastatic signal returned by her third visit. I was able to give her 13 treatments before she returned to the Virgin Islands, and by that time there had been no metastatic signal for several weeks, only the prodromal widening at the Fire depth. I encouraged her to try to return for booster treatments once a month, but that has not been possible with the pandemic still being very active. On January 2nd 2021 her daughter reported that her mom was maintaining her improved sense of well-being. She had both a CT scan and an MRI in April of 2021, and reportedly it showed no evidence of cancer, despite 4 months without further acupuncture. The statistics for stage 4 small cell lung cancer 5 year survival are only about 3% according to the American Cancer Society. According to this woman's oncologist, she had no patients survive for 5 years after experiencing a transformation from non-small cell to small cell lung cancer. I was therefore alarmed when this patient experienced a relapse in December of 2021 and was found to have 3 new tumors in her pleura. She had severe chest pain and cough, and appeared to be seriously ill. Her pulses showed the metastatic signal and improved somewhat from the same treatment protocol as previously given, and began chemotherapy the following day. Two days later she still had the metastatic signal in her pulse, and was so ill that she began investigating possible hospice facilities. I urged her to continue the acupuncture, and after the third treatment the metastatic signal was reduced to the prodromal level. By the 5th treatment she had much more vitality and then had a second round of chemotherapy, which was followed by reduced vitality and a recurrence of the metastatic signal. However further acupuncture again eliminated this cancer signal from her pulse, increased her vitality, and a follow-up CT scan showed "near total resolution of previously seen pleural masses, with no new or growing nodules." All her symptoms have been greatly reduced, but on the advice of her oncologist she decided to proceed with one more round of chemotherapy. I will evaluate her after that is completed. In any case, it is remarkable how someone so close to death could still turn their health around.

43) CA: This 35 year old woman had been diagnosed, a week before seeing me on 10/29/20, with a left sided aggressive and invasive type of ductal breast cancer[264]. It had been found on palpation, followed by ultrasound, mammography and finally biopsy. An MRI showed that the tumor was approximately 4 cm in every direction, and had already spread to internal mammary and subpectoral lymph nodes. She was still lactating having given birth to her second child 6 months previously, and she was as scared as I have ever experienced a patient to be in my clinical career. Apparently her doctors had given her a very poor prognosis. Fortunately I got to evaluate her before she started any biomedical treatment, so I could gauge her response to treatment without any complicating interventions. Her pulses revealed the following: On the right side RY>CK with a ratio of 2/1 and the zào quality was not present. The deviation at cùn was radial, and her KCA pulse pattern was Taeyangin - fǔ Organ type. The only possible Constitution with these findings is UB Excess. Conditionally, all her fǔ Organ pulses were at the Fire depth except for the GB, which indicated the typical metastatic pattern. The Subdoshas were Pitta/Kapha of the Water type on the right and Fire on the left, further confirming her Constitution. Based on her Water Element Constitution I knew that my major task would be reassuring her that her prognosis was not necessarily as bad as she had been led to believe. Her treatment was the standard protocol, given by touch treatment, 4 times for each component as follows on the right side: (UB 66, ST 44- and SI 5, ST 41+) plus (UB 66, LI 2- and SI 5, LI 5+) plus (TH 1- and TH 3+) plus (UB 66, ST 44- and SI 5, ST 41+) plus (UB 66, LI 2- and SI 5, LI 5+) plus (GB 41, UB 65- and ST 36, UB 40+) plus (UB 66, SI 2- and ST 36, SI 8+). As hoped for, after this initial treatment all her pulse depths returned to normal and the O-Ring kinesiology test confirmed an immediate improvement. These findings gave me the confidence to reassure her that her prognosis was actually a favorable one, based on my experience treating other cancer patients. When I saw her again 4 days later she was still just as panicked, but to my surprise her pulse

264 The pathology report classified the tumor as a poorly differentiated infiltrating ductal carcinoma with a very high "Tumor Cell Proliferation Index" of 80%.

depths had stayed normal after just that single treatment. I explained to her how favorable these findings were, and for the first time her eyes softened. I continued to give her the same treatment twice a week mostly, and after 5 treatments she began chemotherapy[265]. After 8 acupuncture treatments she had a biopsy of one of the abnormal lymph nodes. It turned out to show only necrotic tissue, indicating that the cancer cells there had already been killed. Since then she has had her RYCK imbalance shift to the left side (normal for her Constitution) and then become equal bilaterally, and there have been no abnormal pulse depths since her first treatment. After her pulses normalized in this fashion I surmised that she no longer had active cancer[266], and she felt well in spite of the chemotherapy except for mild fatigue, walking 3 miles for exercise without any problem. She was able to complete the full course of chemotherapy with minimal side effects. In fact, her oncologist didn't believe her when told that she had no symptoms of neuropathy. Apparently her oncologist had never seen a patient on these drugs avoid this problem before. On 5/26/21 she had surgery to remove the remnants of her original tumor plus 3 lymph nodes (per standard protocol). The pathology report showed a very small area of poorly differentiated cancer cells surrounded by a mass of scar tissue and necrotic cells at the original site, but no lymph node abnormalities. My impression at this point is that all the cancerous tissue has been removed, and that those remaining cells had simply been so poorly perfused by blood circulation that her body's immune system hadn't been able to reach them yet. Her oncologist told her that statistically, she now has a 25 percent chance of cancer recurrence within 5 years from the original diagnosis, but she is now optimistic (a dramatic reversal of her initial state of mind) that with continued acupuncture treatment and standard Western medical treatment she will not have a recurrence. I do not feel any abnormalities on her pulse, but after doing the anti-cancer acupuncture protocol, her kinesiology improves each time. I'm not sure how to interpret this finding, but it may be a useful guide to choosing treatment length and frequency in the future. Even though she was doing so well, her oncologist convinced her

265 The following 4 intravenous drugs were employed in her chemotherapy regimen: Adriamycin, Cyclophosphamide, Carboplatin and Taxol. They typically have side effects including severe bone marrow suppression, fatigue, peripheral neuropathy and an increased risk of developing other types of cancer.

266 By active cancer, I mean tumor cells that are continuing to grow and spread (metastasize). It may take a longer time for the body to destroy all the original cancer cells, but if they are not growing or spreading, then they presumably cannot cause a recurrence of cancer.

to add 6 weeks of radiation therapy. Midway through this course she developed a severe sinus infection, and her pulse at the same time showed a recurrence of the metastatic signal. Thankfully this signal disappeared after only one repetition of the anti-cancer acupuncture protocol. My sense of what happened is that the radiation weakened her immune system, killing some healthy immunological cells along with any possible latent cancer cells, but that was enough for new cancers to begin along with a bacterial infection. Both problems reflect a weak immune system, plus it is well established that radiation is carcinogenic by nature. Reversing this "side effect" has bolstered my belief that CCA acupuncture is a powerful way of undoing the neoplastic process, and I'm hopeful that this woman will now maintain a cancer free state. She decided to add a six month course of immunotherapy with Xeloda, and appears to be tolerating it well, although quite recently she has begun to experience a known side effect – sores on her feet and oral mucosae. These have coincided with more noticeable cancer prodromal signs in her pulse, although no definite metastatic signal. The prodromal signals disappear after each acupuncture treatment. Her case reinforces my conviction that chemotherapy or radiotherapy is a safer intervention when accompanied by CCA treatment on a regular basis. As of February 2022 there are no cancer related abnormalities in her pulse and she feels completely healthy.

44) CA: Just about a month after initiating treatment with the previous patient, I was consulted by a 34 year old woman who also had just been diagnosed with breast cancer. This lesion was in the right breast, and was diagnosed as invasive ductile carcinoma, that had already metastasized to one of her lymph nodes. Her initial visit was on 12/08/20, less than 3 weeks after her diagnosis. Like the previous patient, she also had 2 young children, but her affect was very different, even if both women faced similar situations. This woman appeared more stoic, but admitted to being prone to depression, including a prolonged bout of post-partum depression after her second child was born, and she complained of being more tired than normal. Her pulses reflected the difficulty of detecting cancer in patients with zàng Constitutions. The only fǔ Organ pulses that were abnormally at the Fire depth were UB, LI and ST. However, her PE pulse was also at the abnormal Fire depth. Knowing that she had already been diagnosed with cancer actually made it easier to determine her Constitution. Since her TH pulse was at the normal Water Element depth, she had to have

a zàng Organ Constitution. While it was normal for her to present with a SI pulse at the Fire depth, it was significant that her GB pulse had stayed at the Wood depth. The only explanation for this was if her Constitution was either HE Excess or HE Deficiency. She had a clear positive tropism for potato and her RYCK imbalance on the left side showed RY>CK in a 2/1 ratio, with bilateral presence of the zào quality. These findings confirmed her Constitution as HE Deficiency, which fit with her affect. She had not started any biomedical treatment yet, so I expected to see a significant improvement in her pulses following the initial touch treatment on the left side, as follows: (SI 5, UB 60+ and LI 1, UB 67-) plus (GB 41, SI 3+ and UB 66, SI 2-) plus (SI 5, ST 41+ and UB 66, ST 44-) plus (SI 5, LI 5+ and UB 66, LI 2-) plus (PE 7+ and PE 3-) plus (SI 5, ST 41+ and UB 66, ST 44-) plus (SI 5, LI 5+ and UB 66, LI 2-). Upon finishing treatment she smiled and said, "I feel lighter." All her pulses were at their normal depths and her O-Ring test was markedly stronger. When seen 2 days later she noted that her energy level was much higher and her pulses had stayed at their normal depths, however her RYCK imbalance was still on the left side. Two further treatments each resulted in an improvement in mood and energy, and by the sixth treatment there was no longer any RYCK imbalance, and she remarked, "I've never felt better." It was only then that she started biomedical treatment. This consisted of both intramuscular and oral drugs to alter her hormonal balance, and she immediately started to have night sweats. Her pulses reverted to showing the metastatic signal at first, but ongoing acupuncture quickly returned her pulses to normal. I was able to do 14 treatments, at which time she was forced to discontinue acupuncture due to loss of childcare assistance. I contacted her by telephone 2 months later (5/30/21), and she reported that her oncologist was so pleased with her progress in the shrinkage of the original tumor, that she probably would not need more aggressive chemotherapy, only excisional surgery after another 5 months of hormonal treatment. She was able to return for evaluation once more on 6/14/21, at which time all her pulses continued to remain in the normal range. A bone scan showed no abnormalities and she continues to feel well. When contacted subsequently by telephone, she reported that she also had a negative MRI, so hopefully she has been cured.

45): The two previous women were treated at Stanford Hospital, and the breast cancer patients there have been discussing their experiences with acupuncture. As a result, another young woman (40 years old) decided to start treatment with me. She originally was diagnosed with a stage 3, triple negative tumor measuring 8 centimeters in the left breast. She had since been through 6 rounds of chemotherapy, 30 sessions of radiation, and surgical resection of the tumor plus 16 lymph nodes, 3 of which were cancer positive, and was still on additional chemotherapy (Xeloda) when she first consulted me on 5/7/21. Both parents and a sister had developed cancers, attacking lungs and stomach. Although this was a more advanced case with lots of Western medical treatment, her pulses were practically identical to those of the other breast cancer patient I described with a UB excess Constitution. In spite of all the treatment she had received, there was a very clear metastatic depth pattern present on the initial pulse examination. Surprisingly however, it only took one treatment for the metastatic signal to disappear, and it has mostly stayed gone ever since, with treatments given approximately once a week. There initially had been a prodromal signal left, which disappeared after each acupuncture treatment, but reappeared before the next one. After several months of treatment, even the prodromal signal disappeared. A PET scan 7/21/21 showed no indications of cancer. Her oncologist has prescribed a course of Keytruda, which is an experimental protocol for breast cancer patients. The 2 patients in my practice on this drug both developed pituitary insufficiency, and have since needed to be maintained on hydrocortisone replacement therapy, so I will keep alert to the possibility of this side effect in this patient. When she started Keytruda, she developed itching all over her body, which I would assume was an allergic reaction. That was followed by a severe inflammation of her left knee with a pronounced joint effusion. At that point her pulses showed a return of the metastatic signal for the first time since achieving remission. I explained to her that I thought this situation was not a very good reason to pursue this type of experimental chemotherapy further, and she has for the moment delayed the second dose, usually given by injection every 6 weeks. With such a good report on her recent PET scan, I believe careful monitoring is better than more chemotherapy, especially as her pulses were slow to eliminate the metastatic signal initially, but on her visit 10/20/21 there was only a prodromal signal present. She is inclined to give the Keytruda one more try, and I never argue with patient's decisions if they have been informed about the risks and benefits. I do think she is a very high-risk patient considering her clinical course and family history, so

I plan to continue weekly acupuncture treatment in hopes of mitigating any adverse effects of the chemotherapy.

46) CA: This 69 year old woman was seen on 2/26/21 after a diagnosis, 4 months previously, of breast cancer. Her type was different than the other breast cancer patients reported here. It was treated with an excisional biopsy, and diagnosed as micro-invasive lobular carcinoma, but the surgical margins were not definitively clear of abnormal cells. She also had some musculoskeletal complaints that often bothered her more than the symptoms in her breast. In fact, she had a history of unresolved Lyme disease, and these musculoskeletal complaints appeared to me to have a neurological basis, consistent with chronic Lyme disease. Her daughter actually had a more severe case of cerebrospinal infection due to Borrelia, one of the organisms implicated in Lyme disease. This patient's pulse was initially typical of cancer, so that was the initial focus of treatment. After more than 3 months of acupuncture treatment, her oncologists had still not initiated any treatment, as the tumor board at her hospital couldn't agree on the next step. Meanwhile I have been treating her about twice a week, and she currently appears to be in remission at least, as there are no longer any cancer indications in her pulse. Both her parents died of cancer (father of lung cancer, mother of breast cancer), so I am planning to continue the anti-cancer protocol well into the future. Her Constitution is KI Excess, although it took several visits for me to confirm her pulse findings. The left side showed CK>RY with ulnar deviation at the chǐ position. The metastatic pattern showed as Fire Element depths on the pulses of SI, GB, UB, ST and PE. The fact that the GB pulse had shifted to the Fire depth, but the LI pulse had not, confirmed that her Constitution was KI Excess rather than HE Excess. The left Subdosha was Vata (Water type) and there was no zào quality, while the right Subdosha was Fire and the zào quality was present. Her treatment has been (SI 5, ST 41+ and UB 66, ST 44-) plus (ST 36, UB 40+ and GB 41, UB 65-) plus (ST 36, GB 34+ and UB 66, GB 43-) plus (ST 36, SI 8+ and UB 66, SI 2-) plus (PE 7+ and PE 3-) plus (ST 36, GB 34+ and UB 66, GB 43-) plus (ST 36, SI 8+and UB 66, SI 2-), with touch technique, all on the left side. Once I was confident that there were no cancer indications in her pulse any longer, I noticed that her left guān pulse in the yīn finger position was deviated radially from where it should have been. This indicated an Akabane imbalance on the LV Meridian. Before giving her the cancer protocol treatment that day, I

tonified LV 5 on the left side. The next visit she declared that her condition had "taken a 180 degree turn, and she now felt amazingly better". The neurological discomforts, including a hand tremor, were also immediately relieved and have stayed gone. Another impressive result from treatment followed my decision to treat the Cleft Point of her Constitutional Meridian when I perceived that her Lyme related complaints were no longer improving, taking into account that I had not been treating the Meridian of her Constitutional Organ. After stimulating KI 5, all her pulses became normal and stayed that way until her next visit. She said that she had never expected to feel normal again, but that now she does. Meanwhile, the tumor board has recommend either a wider surgical excision including the site of the original lesion or radiation. However, her preference is not to add any further biomedical treatment, given that she feels so well and her oncology consultants can't even agree with each other on a specific recommendation. As far as I can tell, she appears to be cancer-free, with neither a metastatic signal nor any prodromes of such a state.

47) CA: This 68 year old woman has been a patient of mine for over 20 years. During that time I have treated her for multiple issues including anxiety, irritable bowel syndrome, paresthesias, muscle twitches, and musculoskeletal pains, all resolving following acupuncture. In June of 2019 she had a check-up with her primary care doctor, who found some nodules in her thyroid gland on physical examination. An ultrasound exam confirmed the presence of nodules, and the radiologist led her to believe that there was a strong likelihood of cancer, so a biopsy was recommended. She scheduled an emergency appointment with me, as she was feeling extremely anxious about the diagnosis. Upon examination I was quite clear that she did not have the metastatic signal of 5 Fǔ organ pulses at the fire element depth, and told her that I thought the possibility of cancer was remote, but that a biopsy would be appropriate. An endocrinologist scheduled the biopsy right away, and the diagnosis turned out to be benign calcification. I've included this case to indicate the possibility of using the findings I've mentioned, not only in the diagnosis and treatment of cancer, but also as a valuable clinical guide to suggest the presence of benign lesions in patients who have been worried about the probability of cancer. This woman was very grateful for the reassurance and explanation I had given her prior to the biopsy. She has remained healthy for the 2 years since the thyroid nodules were first noticed.

48) CA: On September 18, 2019, a colleague trained in Korean Constitutional Acupuncture referred a 60 year old woman to me for diagnosis and treatment of a swollen lymph node in her groin. He had treated her for health maintenance after she had recovered from a previous bout of lymphoma. Three years before consulting me, the lymphoma had presented with a similar swollen inguinal lymph node. That first episode of lymphoma was successfully managed with a combination of surgery and chemotherapy, but the patient found the side effects of chemotherapy almost unbearable, which was why she had started acupuncture in the hope of strengthening her immune system. She was terrified of having to endure Western medical treatment again, as she had experienced severe side effects from the chemotherapy. My colleague's diagnosis of her Constitution had been KI deficiency. When I examined her, in addition to the metastatic signal of Fire depth pulses on all the fǔ Organs except ST, I also found that her Constitution was LU deficiency. Her left side showed RY>CK with a ratio of 3/1 and the zào quality was present. The deviation was ulnar at the guān position. In CCA, both KI deficiency and LU deficiency share the Young Yáng (Soyangin) classification developed as part of the Korean system of Sasang Constitutional Medicine (SCM), so I was not surprised that she felt well from treatment directed at KI deficiency, but at the same time I thought the subtle difference in Constitutional diagnosis might explain her susceptibility to recurrence of malignancy. She had recently lost weight (about 10 pounds), but aside from the enlarged lymph node and its psychological implications, she otherwise felt quite well. After examining her, my impression was that she did indeed have a recurrence of lymphoma, since all of her fǔ organ pulses except ST were at the fire depth. After explaining my diagnosis to her, I treated her on the left side with the same protocol as I used for the gentleman with melanoma who also had a LU Deficiency Constitution, but using touch treatment instead of simple stimulation to get dé qì. Although she was feeling well to start with, right after the first treatment she exclaimed that she was feeling "even perkier". Her pulses immediately returned to their normal depths. I gave her 4 treatments before she was able to consult with an oncologist, and he confirmed the diagnosis of lymphoma, with metastasis to the spleen on 10/16/19. He initiated a milder form of chemotherapy than she was given previously, starting after her 6th acupuncture treatment, and at that point her lymph node had begun to shrink and her pulses stopped reverting to the Fire depth. After

6 months of weekly acupuncture treatments plus chemotherapy, the lymph node was no longer enlarged and she was feeling well, except for a minor side effect of lip swelling, which she attributed to the chemotherapy. The plan was to continue this combination of acupuncture and chemotherapy for 2 years, given that her malignancy was a recurrence, with an advanced stage involving splenic metastasis. The Covid pandemic however motivated her to stay at home except for emergencies, so she stopped coming for acupuncture treatment. After a little more than 2 months she reported by telephone that she had continued to stay in good health. Unfortunately the lymphoma was only in remission and several more months later it recurred in her spleen. I explained the importance of regular acupuncture treatment while the tumor was still present, followed by monthly preventative treatment, but for now she feels that not missing work is more important to her, as she believes the chemotherapy she is receiving will provide sufficient treatment. I hope she changes her mind, given that she responded so well to acupuncture initially.

49) CA, 6L: This 76 year old man developed hematuria in September of 2020. His urinalysis showed that there was no infection to explain it, and a CT scan revealed an 8 mm tumor in his bladder. At that point he consulted me (on 11/16/20), as his wife was already a patient of mine. His history was notable for urinary tract infection and urethral stricture at the age of 19, treated by a series of dilatations, and he had mild symptoms of benign prostatic hypertrophy. There was no family history of tumors. At the time of his first visit he had not received any treatment for the tumor, and was feeling well. His pulses showed that his Constitution was HE Excess (ulnar deviation at chǐ, CK>RY with a ratio of 2/1 on the left side, and Fire Element Subdoshas plus the zào quality bilaterally). The Conditional pulses showed the metastatic signal, with Fire Element depths at UB, LI, ST, SI and PE. The GB pulse was at the normal depth, which confirmed that his Constitution was HT Excess rather than KI Excess, even though his symptoms pointed so strongly to the Water Element. I gave him touch treatment on the left side 4 times each at these Points: (SI 5, UB 66- and LI 1, UB 67+) plus (ST 36, SI 8- and UB 66, SI 2+) plus (SI 5, ST 41- and UB 66, ST 44+) plus (SI 5, LI 5- and UB 66, LI 2+) plus (PE 7- and PE 3+) plus (SI 5, ST 41- and UB 66, ST 44+) plus (SI 5, LI 5- and UB 66, LI 2+). Following this treatment all his pulse depths returned to normal. He subsequently had a biopsy that

showed a malignancy of high grade but without any spread beyond the initial lesion. I treated him twice a week with the same protocol, and by the third treatment his pulses had reverted to their normal depths and stayed there. By the 8th treatment his RYCK imbalance had shifted from the left to the right (normal) side, and later became equal on both sides. He still had not received any biomedical treatment as his oncologist wanted to do a second biopsy to confirm the staging of his tumor. This biopsy was done (prior to his 8th treatment) and showed only ductal carcinoma in situ (DCIS), a much lower grade of pathology than his initial biopsy had indicated. DCIS is considered stage zero and is classified as non-invasive. It has been my experience that stage zero lesions don't show the metastatic signal in the pulse, so I was relieved to hear his pathology report, and surmised that the acupuncture had already begun to reverse the neoplastic process. His oncologist recommended treating him with BCG bladder infusions to stimulate an immunological response to the tumor. He agreed, and started BCG infusions on 1/22/21, once a week for 6 sessions. He began to experience fatigue after the third infusion, and it increased over time. He was able to finish the initial series of BCG treatments on 2/26/21, and afterwards I continued the acupuncture twice a week until he had a follow-up cystoscopy on 3/25/21. The result was no detectable cancer. I changed his treatment frequency to once a month at that point, but his oncologist convinced him to continue with booster BCG treatments periodically. When he resumed BCG infusions, he immediately started having severe pain in the bladder and low back. Given his previous patholgy and cystoscopy reports, it is my impression that further BCG treatments are not clearly indicated, especially as his pulses show no metastatic signal. On 8/12/20 he had planned to stop the BCG treatment, but his oncologist told him that he had a 70 percent chance of recurrence without the BCG, and this frightened him into trying it again, even though he still was having Bladder irritation from the previous BCG treatments. On examination, his pulses showed the devitalization pattern for a HE excess Constitution: HE at the Water depth and UB at the Fire depth, with the left Subdosha Fire and the right Subdosha Vata (Water type). The RYCK qualities were zào on the left, but not on the right. One touch treatment on the right side finally eliminated the Bladder irritation using these Points: (SP 3, HE 7- and KI 10, HE 3+) 4 times plus (SI 5, UB 60- and LI 1, UB 67+, LI 1, UB 67+) 2 times. This is the vitality-c formula, which directs the effect to the lower body. My recommendation is for him to rely on twice a year cystoscopies to catch any possible recurrences early, but he has decided to give the BCG one more try first. He subsequently finished the third round of BCG infusions, and the Bladder irritation got even worse, but began to gradually improve with ongoing acupuncture. As of 9/23/21 there has been no return of the metastatic signal, and no reason to expect a recurrence in the future. Cystoscopy on 10/7/21 showed no cancer on both gross and microscopic examinations, results that were confirmed on repeat cystoscopy on 1/6/22. However, he was still experiencing severe bladder irritation symptoms. His pulses were zào on the left side with Fire Element Subdoshas and jìng on the right side with Vata (Water type) Subdoshas. These rindings in someone with a HE Excess Constitution could only reflect a problem in the Taiyang level, so I palpated for tender points on the UB and SI Meridians. SI 8 on the left and UB 65 on the right were tender and needled for 10 minutes. This treatment changed the pulse quality on both sides to zào and diminished his urinary tract symptoms. At his next visit he stated that he finally felt like himself again for the first time since starting the BCG treatments. He plans to continue the acupuncture cancer protocol even after any residual symptoms resolve, as he feels that acupuncture was the treatment that got rid of his original cancer. I think this is a very instructive case, illustrating the power of a doctor's words to dissuade a person from following their instinctive sense of what course of treatment is optimal for them.

50) CA: This 70 year old man was first seen on 5/18/21 after being referred by another patient of mine, for evaluation of his status following a prolonged course of chemotherapy for a "grapefruit sized" mediastinal large B cell lymphoma. It was first diagnosed 2 and a half years previously, with 1 lymph node also involved, and the treatment had left him with a painful case of neuropathy in both feet. His most recent CT and PET scans had been read as showing complete resolution of the tumor, and he felt well except for the residual neuropathy. According to his oncologist, he was in remission, and needed no further treatment. When I examined him however, he showed the typical "metastatic signal" of 5 fǔ Organ pulses at the Fire Element depth, all except for the GB. His left sided pulses showed RY equal to CK and the presence of the zào quality, while the right sided pulses showed RY>CK at a 2/1 ratio, but without the zào quality. The left

Subdosha was Fire and the right Subdosha was Pitta/Kapha of the Water type. These findings were alarming, but made me glad that he had decided to seek CCA evaluation and treatment. This pulse pattern is a typical presentation of cancer in someone with a UB excess Constitution, and while it does not indicate the presence of observable pathology, it seems to indicate that cure has not been achieved. My experience is that in cancer, both the Subdoshas and the RYCK quality reflect what is happening energetically on the opposite side of the body. The treatment I gave him on the right side was exactly the same as that described for the two women with breast cancer, who also had UB excess Constitutions. He experienced some relief of the neuropathy in his right foot as soon as the touch treatment protocol was finished, his kinesiology had also improved and all his pulse depths had returned to normal. When next seen 2 weeks later, the neuropathy had almost disappeared from his right foot, but his pulses once again showed the same metastatic signal. I took this finding as evidence that I had not made a mistake in his original pulse diagnosis. The same treatment was repeated, and he began to schedule weekly treatments. Since the second visit his pulses have consistently stayed at the normal depths, and additionally the neuropathy in his left foot has now also been decreasing. At the beginning of October 2021 a CT scan and blood tests showed no signs of cancer. His case is a powerful reminder to me that cancer can be an insidious disease, which may appear to have been resolved, only to reappear later with a vengence. I believe there is a good likelihood that with CCA treatment this tendency for relapse can be prevented.

The consistency with which these cancer patients have presented similar pulse patterns and have responded positively to treatment designed to address this pulse pattern leads me to suspect that there is a common pathogenic mechanism (bìng jī 病 機) in most, if not all, cases of cancer. This overarching finding is not one that I have encountered in either Western or Chinese medicine, and would seem to be a very important discovery, if confirmed. Again, I do not take credit for this discovery, as it is based on the work of Kuon Dowon, however I have altered, expanded and systematized his groundbreaking work. Since the targets of treatment seem to be widespread malfunctions of the Fǔ Organs, I imagine there must be something affecting the organism's yáng qì involved. I note that Ivan Cloud has written that "Tumors are yin accumulations, so from this perspective we need to focus on the yang qi," in a presentation of the various theories about cancer in Chinese medicine.[267] His comment certainly is consistent with my own impression. I have presented cases of cancer in patients with Earth, Metal, Water, Wood and Fire Element Fǔ Organ Constitutions and cases of patients with Fire, Earth, Metal and Water Element Zàng Organ Constitutions. My current impression is that no particular Constitution is more prone to cancer than any other, but perhaps the type of cancer and/or its location might correlate with a person's Constitution. A much bigger study is needed to investigate that question, but my results indicate that CCA acupuncture is extremely promising as a major advance in treating all types of cancer.

51) AKA: One of my patients is a fellow acupuncturist, originally trained in the Worsley style of Five Element Acupuncture. I first began treating her in 1992, when she was 48 years old, noting that she had already been treated by several FEA style practitioners, all of whom diagnosed her as having a Fire Element Constitution. According to CCA pulse diagnosis, she actually has a LI Deficiency Constitution. Her main symptoms relate to the digestive system, and are usually diagnosed as irritable bowel syndrome. She has continued to occasionally get FEA treatment of her Fire Element, and typically responds with panic attacks and palpitations. Tonifying Metal almost always helps. The reason I'm including her here is that when seen recently for a recurrence of her digestive problems, I felt that her pulses were quite good, except that her KI pulse was shifted to the left. I treated this as an Akabane imbalance, and tonified KI 4 on the left side. Only after explaing this treatment to her did she tell me that her doctor had found decreased kidney function on her lab tests. After the Akabane treatment her digestive symptoms cleared up and she wrote to tell me that she felt much calmer and grounded. That's an example of what one correctly placed needle can do.

52) AKA: I have been treating this 92 year old woman for the past 8 years. Her ongoing problems are labile hypertension and insomnia, both clearly

267 From "A multifaceted perspective on cancer" posted by Ivan Cloud, 5/16/19, on the Classical Chinese Medicine Facebook Group site, which he originated. Ivan discusses many of the theories about the etiology and treatment strategies for dealing with cancer proposed by both famous herbal practitioners and in the classical texts.

aggravated by different stressful circumstances in her life. A little more than a year ago she fell after a dog jumped on her, and she fractured her right hip in several places. Surgery was performed, and a number of pins were used to stabilize her bones. She recovered most of her mobility, but ever since the accident she had felt unstable and off balance. She has a KI deficiency Constitution, but none of the usual treatments seemed to make a differebce in this distressing condition. When I discovered how to identify Akabane imbalances via the pulse, I was finally able to eliminate this symptom. Her GB pulse was shifted to the right, almost in line with her UB pulse. After tonifying GB 37 on the right side, she was able to walk without any feeling of instability for the first time in over a year.

53) EPF: One of my patients is a 74 year old man whom I have treated off and on for more than 40 years. He has had multiple problems, mainly musculoskeletal issues, but he has also suffered from hemorrhoid bleeding for as long as I have known him. While it has not been difficult to help with the musculoskeletal pains, I have not figured out how to stop the bleeding until recently. When he was younger he had had surgery for hemorrhoids, which was one of the worst experiences of his life, so he never considered a repeat operation. The chronic bleeding left him often feeling weak, and his lab tests showed moderate iron deficiency anemia. His Constitution is SI Deficiency (Fire Taeyangin), and on this recent visit there was no RYCK imbalance, no Extraordinary Meridian pulse, none of the other Elements showed an abnormal relationship with Fire, and his Subdoshas were both Vata (Fire type). However, I did notice that the Subdoshas had spikes at positions 1 and 2 on the left index finger but positions 3 and 4 on the right index finger. Upon careful examination of the depths of all 12 Organs, the only abnormality was that the LI pulse was at the Fire depth instead of the Metal depth. This is an easy finding to miss because the two depths are not very different. I decided to treat this as a case of retained pathogenic energy (xié qì), and dispersed the Luò Point LI 6 on the right side (having tested laterality with a silver press pellet at the corresponding KHA Point). When I saw him a week later, he said, "I've had the best week in a really long time. I'm almost pain free, my spirits are much improved and I've had no bleeding at all!" I had treated many different Points on the LI Meridian in the past, in addition to symptomatic points of all kinds,

without affecting the bleeding, so I interpreted this result as attributable to clearing xié qì, specifically dryness (the Yangming climatic factor for which the LI Meridian is the Operator). In fact he had always mentioned that his bowel movements prior to bleeding episodes were hard and dry, but I finally understood the mechanism.

54) EPF: I have another patient with a SI Deficiency Constitution whom I have been treating for mostly musculoskeletal problems during the past 8 years, with very good results. Just as I was finishing this manuscript she scheduled an emergency appointment because she had a severe adverse reaction to a booster shot of the Moderna Covid-19 vaccine. Her first 2 shots were only mildly bothersome, but this time she had marked chills and fever the next day, and even after those symptoms resolved she developed a pressure type of headache, a worsening cough, an inability to take a deep breath, and lastly a swollen lesion about an inch in diameter under her left eye. Her pulses were mostly unremarkable, showing the zào quality and Vata (Fire type) Subdoshas bilaterally, and her SI pulse was at the normal Fire Element depth. However, I noticed that on the left side the Subdosha spikes were at positions 1 and 3 on my index finger, while on the right side they were at positions 2 and 4. This finding alerted me to the probability of an EPF, so I carefully checked to see if any other pulses had been displaced to the Fire Element depth. Indeed, only her SP pulse had moved to the Fire depth. Using silver press pellets at the KHA corresponding Points, I found that SP 4 on the left side corrected the SP pulse, but SP 4 on the right side did not. Her treatment therefore was to disperse SP 4 on the left with a retained needle for 10 minutes, after getting dé qì. When the treatment was finished she immediately said, “I can breathe normally again!” and her headache was gone, in addition to which her SP pulse was at the Earth Element depth. The SP is the Operator of Taiyin, which is responsible for transforming Dampness, one of the 6 perverse climatic factors. The swollen area on her face along with the other symptoms could be easily understood as resulting from a retained Damp pathogen due to the vaccine injection.

55) EPF: A 55 year old woman was referred by her primary care physician for bilateral Achilles tendinitis. She claimed her main problem was “adrenal burnout” lasting a number of years. She felt tense, anxious, tremulous

and hyperactive, plus she suffered from insomnia. Her emotional state caused her to overeat and she gained an unhealthy amount of weight. In taking her history I noted that she had always used athletics to cope with her feelings, but that with the tendinitis, following more than 100 miles of hiking in the preceding few months, she could not use this outlet anymore. On examination she had a positive tropism for cucumber. Her left side showed CK>RY without the zào quality in a ratio of 1.5/1. The deviation was radial at the chǐ position, so her Constitution was GB Deficiency (Wood Soyangin). She initially showed a GB pulse at the Earth depth and a ST pulse at the Wood depth, which represents a lack of vitalization. That made perfect sense in view of her chronically overusing her yang qì. I used touch treatment on the left side as follows: (LI 1, GB 44- and UB 66, GB 43+) 4 times plus (LI 1, ST 45- and GB 41, ST 43, GB 41, ST 43+) 2 times (Immune/Vitalization-c). She immediately felt more comfortable, with less pain and a much stronger O-Ring test. I advised her to change her diet to match her Soyangin Constitution. The next time I saw her was a month and a half later. She was feeling better emotionally and physically, but her tendinitis was still present and her legs felt hot, although she tended to feel cold in general. This time her pulses indicated the D/Z/N treatment on the left side: (KI 10, HE 3- and LV 1, HE 9+) plus (KI 10, SP 9- and LV 1, SP 1+) plus (TH 1- and TH 3+) with retained needles. On removing the needles she said, "I feel like a whole new person." 3 weeks later she had walked a lot with minimal soreness in her legs, and her Subdosha pulses no longer showed any Elemental imbalances, however the left Subdosha was Pitta/Vata (Wood type) and the right Subdosha was Kapha (Wood type), while her RYCK findings were the same as initially. I suspected a problem with retained pathogens, and checked all her pulses to see if any had shifted to the Wood depth. It was clear that the HE pulse was at the Wood depth, which is sufficiently different than the Fire depth to be notable. I asked her if she had ever been extremely overheated (the EPF of Shaoyin, for which the HE is the Operator). She answered that the only time she recalled was when she was competing in a horseback event 15 years previously when the hot temperature was almost unbearable. That time also coincided with the end of a relationship that left her "heartbroken." Such a situation represented a double attack on her HE, which is often enough to create a lasting imbalance, in my experience. I treated her with dispersion of HE 5 on the left side, which corrected the depth of the HE pulse and further diminished the Achilles pain. A month and a half later she reported that she felt during and after that treatment both relaxed and grounded, while increasing her energy, but she had not fully recovered from the tendinitis. At this visit there were no abnormal pulse depths and the Subdoshas on both sides were Pitta/Vata (Wood type), which matched her Constitution. The only remaining abnormality was the CK>RY of 1.5/1. Since her Constitution is GB Deficiency, I decided to use the Associated (背腧) and Alarm (募) Points system to address what was either yīn Excess or yáng Deficiency. With her response to dispersing HE in mind, and considering that I had already supplemented GB several times, I chose to disperse the Front Mù Point, GB 24, which would reduce the GB yīn, a pathology that had been transmitted from her HE. This simple treatment produced the biggest reduction in the tendinitis of any of my treatments, while also strengthening the O-Ring test and normalizing the RYCK balance.

56) 50 Cycles: I began treating a 41 year old woman in 2010 for recurrent upper respiratory infections, years before I developed CCA. Her Constitution is ST Excess, and she has stayed in excellent health for over a decade with only very intermittent acupuncture sessions. I'm including her case because it illustrates Simonov's contention that yáng excess of a yáng Meridian is naturally paired with yīn excess of its coupled yīn Meridian. Although she has been free of colds, she does get musculoskeletal discomfort in her upper back and neck. Since her pulses have stayed normal on examination, I have been treating her with contact needling per Simonov's method. When she comes for treatment in the morning (a time of yáng victory for her), if it is also during a yáng period according to his slide rule, I contact points on the wèi side of the ST Meridian. If she comes during a yīn period in the morning, I contact points on the SP Meridian on the yíng side. These measures decrease ST yáng by moving its yáng qì out of the Meridian and decrease SP yīn by moving yáng qì into its Meridian. Both types of treatment have been equally effective in relieving her aches and pains, while also strengthening her kinesiology O-Ring test.

57) 50 Cycles: This 59 year old woman has been in my practice for almost 20 years, with an original complaint of chronic migraine headaches. She is the only person I have ever encountered who has aberrant radial arteries on both left and right wrists, which makes diagnosis quite a challenge. By using an Ayurvedic Constitutional survey plus the Korean Hand Acupuncture abdominal exam, among other measures, I have been able to confirm that her Constitution is SI Excess (Fire Taeumin). Acupuncture based on that diagnosis has been very effective in keeping the migraines as a rare event. The choice of treatment is however always difficult, given that there are no radial pulses to guide me. I'm including her case because I have been able to use Simonov's Macrocosm/Microcosm balancing method, as it only depends on knowing her Constitution. When last seen 8/28/20 she was migraine free, although she had recently developed episodes of vertigo. Vertigo is a classic symptom of imbalanced Vata (Wind) Dosha, which is in turn characteristic of SI Excess. I saw her at 8 AM that day, and at that time, using Simonov's slide rule, I determined that she was experiencing a Yáng Period (from 7:55-8:24 AM). During the first cut (kè 刻) of this period (zhōu 周), the most receptive Meridians are respectively Shaoyang, Yangming and Taiyang for approximately 5 minutes each, in that sequence. As a female, she was in a Yáng Victory phase (the morning, from midnight to noon), so the strategy was to direct this cosmic Yáng qì into the Wèi territory of her SI trajectory. By increasing SI Wèi qì, the result is a decrease in SI Yíng (Róng) qì that is drawn out of the SI Meridian, and thus a dispersion of the SI tendency to be overactive. During this first cut, the right arm is in a Wèi state, so the treatment was carried out from 8:05-8:10 on the right arm pathway of the SI (Taiyang) Meridian, which is when it was most receptive to cosmic Yáng qì. An acupuncture needle was merely touched to the points along this Meridian for 5 minutes, holding the needle at each Point for about 7 seconds each, and touching as many Points on the Meridian as possible. Before the treatment her abdomen was tender at SP 15 on the right side, and Ren 4 and 5 in the midline. Also her O-Ring kinesiology test was markedly weak. After this simple contact treatment (no needles penetrated the skin), her abdominal tenderness had disappeared completely and her finger strength was increased by several orders of magnitude. The explanation is that contacting the skin with a steel needle acts as an antenna, allowing her body to absorb the cosmic Yáng qì. As this Yáng

qì was increasing SI Wèi qì, it pulled qì out of the Róng circulation, and so the SI Organ was effectively dispersed. She has subsequently felt well and not needed further treatment at this time.

58) Covid, 8E, E/E: I would like to finish this chapter of case histories by illustrating how CCA can successfully address a new illness that has challenged our whole world, the Covid-19 pandemic. A 63 year old Professor at the University of California (Berkeley) had come down with a textbook case of Covid-19, confirmed by PCR. She had a headache, fever, dizziness, loss of sense of smell, brain fog and severe anxiety. These symptoms mostly resolved during the first week, except that the cognitive disturbance and anxiety remained unchanged a month later, and she was unable to do even simple mathematical calculations. She felt desperate and incapacitated. This syndrome has been called Long-Covid, and its mechanism is poorly understood. Because her condition was not improving after a month, she was referred to me by one of my other patients. She had always been in good health except for hernias at age 3 and a tendency to grind her teeth at night. On exam she had a positive tropism for carrot. On the left CK>RY and on the right side CK>RY. This finding alerted me to check for an Extraordinary Meridian imbalance, and she displayed a Yīn Wéi Mài pulse on the left and a Chōng Mài pulse on the right. Using positions 4, 5 and 6 more proximally on the arm I could tell that the cùn position was deviated ulnarly. The left side had the jìng pulse quality while the right side had the zào pulse quality. These findings alone made the presumptive Constitutional diagnosis LV Excess and the Conditional diagnosis an Extraordinary Meridian disruption. I treated her on the right side with SP 4- connected to PE 6+ with ion cords, and on the left side with PE 6- connected to SP 4+, also with ion cords. After 15 minutes the needles were removed and she reported that the anxiety was completely gone, her head felt clear and she was much less fatigued. I next saw her 2 weeks later, and she felt more or less normal, with a return of a healthy appetite and minimal fatigue, plus she felt up to flying to Rome for a job interview! I didn't do any acupuncture, as nature seemed to be doing everything correctly. I then saw her again after another month. She felt quite well, however there was one new symptom: Her right leg had been feeling tight, almost in constant spasm, for about a week. On checking her pulses, there were no Extraordinary Meridian signals, but her LV pulse felt hollow

and her LU pulse felt solid. Left side was zào with Vata/Kapha Subdoshas and right side was jìng with a Kapha (Wood type) Subdosha. Had this simply been an imbalance between LV and LU Organs, these pulse findings should have been on opposite sides from where I felt them. This pattern then is the opposite of what one would expect with her Constitution, but I realized that I should interpret it as the "Reverse Entry/Exit" block that I described in the text. The treatment was to disperse LV 14 (Exit Point) and disperse LU 1 (Entry Point). When I removed the needles, her leg tension had completely disappeared. This final case emphasizes that there is always something new to discover by paying careful attention to pulse examination. I hope for many more discoveries by those using the methods I have outlined.

BIBLIOGRAPHY

Borsarello, J. et. al., *Traite d'acupuncture.* "Histoire de l'acupuncture contemporaine". pp. 11-17 by Peter Eckman, Masson: Paris, 2005.

Boym. M. (Daniels, S. trans.) *The Medical Key to the Doctrine of the Chinese on Pulses.* Vancouver: Soul Care Publishing, 2018.

Eckman, P. (1988), *The Book of Changes in Traditional Oriental Medicine.* Columbia, MD: Traditional Acupuncture Institute.

Eckman, P., (2007) *In the Footsteps of the Yellow Emperor; Tracing the History of Traditional Acupuncture* (Second edition). San Francisco: Long River Press.

Eckman, P., (2014) *The Compleat Acupuncturist.* London: Singing Dragon.

Eckman, P,. (2015) "Precision in finger placement for pulse diagnosis." *Journal of Chinese Medicine (UK)* 109, 49-55.

Eckman, P. (2017), *Grasping the Donkey's Tail; Unraveling Mysteries from the Classics of Oriental Medicine.* London: Singing Dragon.

Eckman, P., (2018) "Constitutional/Conditional Acupuncture: Basic Theory and Clinical Practice." *Journal of Chinese Medicine (UK)*, 117, pp. 45-53.

Eckman, P. (2021) "The Diagnosis and Treatment of Cancer with Constitutional Conditional Acupuncture." *Journal of Chinese Medicine (UK)*, 125, pp. 3-16.

Hammer, L., *Chinese Pulse Diagnosis; A Contemporary Approach.* Seattle: Eastland Press, 2001.

Hsu, E., *Pulse Diagnosis in Early Chinese Medicine.* New York: Cambridge University Press, 2010.

Huard, P. and Wong, M., *La Médicine Chinoise Au Cours des Siècles.* Paris: Les Editions Roger Da Costa, 1959.

Ki, S. *The Canon of Acupuncture Huang Di Nei Ching Ling Shu.* Los Angeles: Yuin University Press, 1985.

Kroll, P., *A Student's Dictionary of Classical and Medieval Chinese.* Boston: Brill, 2017.

Kuon D., "A Study of Constitution Acupuncture." *J. Internat. Cong. Acup. Mox. Japan Acup. Mox. Soc.* pp. 149-167, 1965.

Kuon D., "Studies on Constitution-Acupuncture Therapy." *Korean Central J. Med.* 25, No. 3, pp. 327-342, 1973.

Kuon D., "Eight Constitutions Medicine: An Overview." IMKS Occasional Paper No. 2, Institute for Modern Korean Studies, Yonsei University Press, 1999, pp. 601-623.

Kuon D.,"A Theoretical Basis for the Eight Constitution Acupuncture." *Advances in Medicine and Biology*, Nova Science Publishers, 2011, Volume 5, pp. 243-246.

Larre, C. and Rochat, E., *The Eight Extraordinary Meridians*. Cambridge (UK): Monkey Press, 1997.

Lavier, J. *Histoire, doctrine et pratique de l'acupuncture Chinoise.* Tchou. Paris, 1966.

Li, P. *Management of Cancer with Chinese Medicine*. USA: Donica Publishing, 2003.

Liu, L. *Classical Chinese Medicine*. Hong Kong: The Chinese University Press, 2019.

Lu, H. (trans.) *The Yellow Emperor's Classic of Internal Medicine and the Difficult Classic.* Vancouver: Academy of Oriental Heritage, 1978.

Mann, F. *The Treatment of Disease by Acupuncture*. London: William Heinemann Books, 1963.

Mann, F. *The Meridians of Acupuncture*. London: William Heinemann Books, 1964.

Matsumoto, K. and Birch, S. *Extraordinary Vessels*. Brookline, MA: Paradigm Publications, 1986.

Mathews, R. *Chinese English Dictionary.* Cambridge: Harvard University Press, 1979.

Niboyet, J., *Nouveau Traité d'Acupuncture*. Paris: Maisonneuve, 1979.

Otsuka, K., *30 Years of Kanpo*. Los Angeles: Oriental Healing Arts Institute, 1984.

Paik, M. et. al., "Altered Urinary Polyamine Patterns of Cancer Patients under Acupuncture Therapy." *Amino Acids*, 37 (2), 2008, pp. 407-413.

Paik, M. et. al., "Therapeutic Monitoring on Urinary Nucleoside and Polyamine Levels of Cancer Patients by CE and GC under Acupuncture Treatment." *Proc. of the Convention of the Pharm. Soc. of Korea*, 1, 2003.

Richardson, T., *Extraordinary Chinese Medicine*. London: Singing Dragon, 2018.

Schuessler, A., *ABC Etymological Dictionary of Old Chinese*. Honolulu: University of Hawaii Press, 2007.

Soulié de Morant, G., *Précis de la vrai acuponcture Chinoise*. Paris: Mercure de France, 1934.

Soulié de Morant, G., *Le Diagnostic par les pouls radiaux*. Paris: La Maisnie, 1983.

Soulié de Morant, G., *Chinese Acupuncture*. Brookline, MA: Paradigm, 1994.

Unschuld, P., (trans.) *Nan-Ching, The Classic of Difficult Issues*. Berkeley: University of California Press, 1986.

Unschuld, P., (trans.) *Huang Di nei jing su wen*. Berkeley: University of California Press, 2003.

Unschuld, P., (trans.) *Huang Di Nei Jing Su Wen*. Berkeley: University of California Press, 2011.

Unschuld, P., (trans.) *Huang Di Nei Jing Ling Shu*. Oakland: University of California Press, 2016.

Wang Shu-he (Yang Shou-zhong trans.) *The Pulse Classic (Mai Jing)*. Boulder: Blue Poppy Press, 1997.

Wieger, L., *Chinese Characters*. New York: Dover Publications, 1965.

Wu, J., *Ling Shu or The Spiritual Pivot*. Honolulu: University of Hawaii Press, 1993.

Yang Z. (Liu, L. ed.) *The Yellow Emperor's Inner Transmission of Acupuncture*. Hong Kong: The Chinese University of Hong Kong Press, 2020.

Yoo, T., *Koryo Hand Acupuncture* (Eckman, P., ed.) Eum Yang Mek Jin: Seoul, 1988.

Yuen, J., *The Eight Extraordinary Vessels*. New England School of Acupuncture, 2005.

ABOUT THE AUTHOR

Peter Eckman M.D., Ph.D., M.AC. (UK), located in the San Francisco Bay Area, has been practicing, teaching and writing about classical approaches to acupuncture for the past 50 years. His interest is in furthering the development of an Eastern science of traditionally based medicine by emphasizing its unique qualitative, inductive and synthetic modes of thought. These are complementary to the Western science modes typically characterized as quantitative, deductive and analytic. From this starting point he has developed Constitutional Conditional Acupuncture (CCA), which he has been teaching worldwide, including in China itself, birthplace of acupuncture. The main focus of CCA is in fostering a deep understanding and ongoing development of pulse diagnosis, a critical skill that is only possible to effectively transmit in a "hands on" manner. Together, both Eastern and Western branches of medicine can contribute to providing optimal health care for all people, but good professional training is essential for all practitioners of any medical discipline.

Dr. Eckman welcomes all inquiries regarding CCA, including invitations to teach wherever students or practitioners are interested in making suitable arrangements (once the Covid-19 pandemic is no longer a restriction to in person physical contact). His email address is healingmountain.eckman@gmail.com.

A synopsis of Dr. Eckman's professional career can be found on his website: www.petereckmanacupuncture.com.

ADVANCED PRAISE

*"Dr. Eckman has followed in the footsteps of this revered figure (*Sun Simiao, the Medicine King*) and integrated Ayurvedic lineage-based approaches and diagnostic concepts into his system of approach. Now what is even more important is keeping the identity of each model separate, so as not to conflate the meanings of the findings for each one. It is my great pleasure to see how Dr. Eckman analytically manages to avoid this conceptual trap where theories are mixed together, and then lose their individual meaning…Dr. Eckman has spent his time seeking clarity and making sure whatever he didn't understand continually becomes elucidated by the refinement of the diagnostic process. Examples range from the nuances of the finger positioning on the pulse, to the diagnostic truth of Ren Ying Cun Kou, to a dynamic Nanjing based 5 phase model, and most importantly to the theoretical and philosophical constructs underlying each one. This is paramount to understand, because without this foundational rationale, we will always be chasing a shadow and never understanding why we chase it in the first place. Peter ultimately invites the reader to do their own inquiry so they can attain the skill rooted in flexibility that he has. But his penetrating insight into matters that are simply not discussed much in the West should not be ignored, as they are of great utility to any aspiring clinician seeking clarity."*

—Ivan Cloud Zavala, Founder of Classical Chinese Medicine (Facebook group)

"Hi Peter, we met at a BAcC conference (UK) a few years back and exchanged a conversation over a glass of wine in the courtyard about diagnosis. Since then I have been inspired by your writings on pulse and can confirm that the methods you propose can lead strongly towards a CF diagnosis. As a result of this, pulse diagnosis for me has become my main focus of diagnosis, and it has launched a big interest for me in the Mai Jing*! Thank you for your writing!"*

—Jamie Hamilton, author of *Essential Pulse Diagnosis in Chinese Medicine*.

Design by Liliana Guia

ISBN 979-8-9859-0840-4 (paperback)
ISBN 979-8-9859-0841-1 (eBook)

www.ingramcontent.com/pod-product-compliance
Lightning Source LLC
LaVergne TN
LVHW070848210926
848482LV00016B/891
* 9 7 9 8 9 8 5 9 0 8 4 0 4 *